Fire Safe Use of Wood in Buildings

This book provides guidance on the design of timber buildings for fire safety, developed within the global network Fire Safe Use of Wood (FSUW) and with reference to Eurocode 5 and other international codes. It introduces the behaviour of fires in timber buildings and describes strategies for providing safety if unwanted fires occur. It provides guidance on building design to prevent any fires from spreading while maintaining the load-bearing capacity of structural timber elements, connections and compartmentation. Also included is information on the reaction-to-fire of wood products according to different classification systems, as well as active measures of fire protection, and quality of workmanship and inspection as means of fulfilling fire safety objectives.

This book:

- Presents global guidance on fire safety in timber buildings
- Provides a wide perspective, covering the whole field of fire safety design
- Uses the latest scientific knowledge, based on recent analytical and experimental research results
- Gives practical examples illustrating the importance of good detailing in building design

Fire Safe Use of Wood in Buildings is ideal for all involved in the fire safety of buildings, including architects, engineers, firefighters, educators, regulatory authorities, insurance companies and professionals in the building industry.

Dr Andy Buchanan is Principal at PTL Structural Consultants, and Emeritus Professor of Timber Design Christchurch at the University of Canterbury, New Zealand, and is the author of the *New Zealand Timber Design Guide* and *Structural Design for Fire Safety*.

Dr Birgit Östman is Emeritus in building technology at Linnaeus University, Sweden, and has recently won the Bernadotte Royal Award of the Swedish Forestry Association, and the L.J. Markwardt Award for Best paper in Wood and Fiber Science 2018, Forest Products Society, USA.

Fire Safe Use of Wood in Buildings
in Buildings
Global Design Guide

Edited by
Andrew Buchanan & Birgit Östman

CRC Press
Taylor & Francis Group
Boca Raton London New York

CRC Press is an imprint of the
Taylor & Francis Group, an **informa** business

First edition published 2022
by CRC Press
6000 Broken Sound Parkway NW, Suite 300, Boca Raton, FL 33487-2742

and by CRC Press
4 Park Square, Milton Park, Abingdon, Oxon, OX14 4RN

CRC Press is an imprint of Taylor & Francis Group, LLC

ISBN: 978-1-032-04039-4 (hbk)
ISBN: 978-1-032-04041-7 (pbk)
ISBN: 978-1-003-19031-8 (ebk)

DOI: 10.1201/9781003190318

Typeset in Sabon
by Deanta Global Publishing Services, Chennai, India

Contents

Acknowledgements

The authors wish to thank all their co-authors and a large number of colleagues and reviewers in many countries who provided much assistance in the preparation of this book. We wish to acknowledge the assistance of Melody Callahan who did all the line drawings and managed the delivery of text and images to the publisher.

We also acknowledge the large number of consulting firms, research institutions, universities and professional organisations around the world who supported this project through in-kind support of authors and reviewers. Generous financial contributions were made by FPInnovations, Softwood Lumber Board, BRANZ, Linnaeus University and SFPE NZ.

Contributors

Anthony Abu
University of Canterbury
Christchurch, New Zealand

David Barber
Arup
Melbourne, Australia

Daniel Brandon
RISE
Lund, Sweden

Andrew Buchanan
PTL Structural Consultants
Christchurch, New Zealand

Ed Claridge
Auckland Council
Auckland, New Zealand

Christian Dagenais
FPInnovations
Quebec, Canada

Gianluca De Sanctis
Basler and Hofmann
Zürich, Switzerland

Andrew Dunn
Timber Development Association
Sydney, Australia

Paul England
EFT Consulting
Melbourne, Australia

Andrea Frangi
ETH Zürich
Zürich, Switzerland

Kevin Frank
BRANZ
Wellington, New Zealand

Claudius Hammann
Technical University of Munich
Munich, Germany

Marc Janssens
Southwest Research Institute
San Antonio, Texas, USA

Robert Jockwer
Chalmers University
Gothenburg, Sweden

Alar Just
TalTech
Tallinn, Estonia

Koji Kagiya
Tohoku Institute of Technology
Sendai, Japan

Kamila Kempna
Majaczech
Bile Policany, Czech Republic

Michael Klippel
ETH Zürich
Zürich, Switzerland

Eugeniy Kruglov
State Academy for Fire Service in
  Russia
Moscow, Russia

Cristian Maluk
University of Queensland
Brisbane, Australia

Esko Mikkola
KK-Fireconsult
Espoo, Finland

Martin Milner
Milner Associates
Bristol, UK

Birgit Östman
Linnaeus University
Växjö, Sweden

Dennis Pau
University of Canterbury
Christchurch, New Zealand

Peifang Qiu
Tianjin Fire Science and
  Technology Research Institute
Tianjin, China

Lindsay Ranger
FPInnovations
Ottawa, Canada

Joachim Schmid
ETH Zürich
Zürich, Switzerland

Boris Serkov
The State Academy of Fire Safety
Moscow, Russia

Jan Smolka
Majaczech
Bile Policany, Czech Republic

Konstantinos Voulpiotis
ETH Zürich
Zürich, Switzerland

Colleen Wade
Fire Research Group
Hastings, New Zealand

Norman Werther
Technical University of Munich
Munich, Germany

Foreword

This book *Fire Safe Use of Wood in Buildings – Global Design Guide* is a welcome addition to the fire safety literature. It seeks to provide guidance and insights into the use of wood in construction, including fire science and international regulatory information. Because the volume focuses on areas where wood directly impacts fire performance and fire hazards, readers will also need to refer to more general guidance documents to complete the design of any project involving wood in the construction.

All forms of wood products from traditional use of dimensional lumber to modern wood products are covered in the volume. The role of wood included in the guide ranges from interior finishes to structural elements. The volume does not treat wood included in the mobile fire load, though fire loads in buildings are dominated by wood products in general. The volume includes excellent citations to the fire science and building literatures, so it is valuable as a gateway to the greater literature on the use of fire performance of wood in buildings. The guide makes excellent use of photographs to illustrate wood products and their use in construction.

The use of wood as a building material has a long history. The use of wood as a structural material in residential construction is traditional in many countries, as is the use of wood as exterior siding, roof coverings and interior finishes. The volume treats these traditional uses of wood. However, the real focus of the volume is on innovative use of wood and wood products as structural materials in larger buildings. The guide does a good job of addressing the contributions of wood construction to the fire load as well as the impact of fire on the wood structural assemblies.

The guide treats both the fire endurance of structural assemblies as well as means of preventing fire spread within the structure via fire-rated separation assemblies and penetration protection. It covers the effects of charring on structural members as well as encapsulation methods to delay or prevent the onset of charring of the members. It also addresses the design and fire performance of connections of wood assemblies.

Much of the fire science information is included in the chapter "Fire Dynamics," with some aspects of ignition, flame spread, burning rate and smoke/toxic gas production are included in the chapter "Reaction to Fire

Performance." While the chapter includes information about the generation of carbon monoxide by wood products, it lacks any information on hydrogen cyanide products which arise out of the use of adhesives in modern wood products.

This guide includes a range of calculational approaches to the determination of fire performance of wood products and assemblies, ranging from the simple to the computer based. While the treatment of advanced methods is necessarily limited, it provides an entry point for designers to further pursue mastering such methods. The guide is straightforward about what is and what is not well understood, providing summaries of areas that require additional research. This is valuable to readers at all levels of expertise.

The treatment of regulatory requirements for wood in construction is decidedly international in scope. The treatment includes the test methods used throughout the world and the required performance in these tests. While the guide is not a substitute for a good knowledge of the local regulatory system requirements, the treatment of regulatory requirements will assist those who are already familiar with local requirements for all buildings, but who lack specific experience with wood. It is also most useful to understand the diversity of requirements around the world and the differing approaches in achieving fire safety by fire regulations.

While active fire protection measures are addressed in the guide, the guidance is primarily on sprinklers. Topics like detection and alarm, and smoke management are merely touched upon. Sprinklers are treated as having an important role in timber buildings.

The chapter "Risk and Performance-based Design" gives a good general introduction to the topic. It provides specific information and approaches useful for timber buildings. As this remains an emerging area of fire safety design, the chapter serves as an entry point to the subject for those who may wish to pursue improving their skills in this area and provides good context for the use of performance-based design approaches in timber buildings that is useful to a general audience. The chapter "Robustness in Fire" continues to develop approaches to enhance the structural robustness of fire designs for timber buildings. It is valuable in instructing the reader on the vulnerabilities of all individual systems and how to use a multi-faceted approach to achieve the desired fire performance.

Finally, the guide treats two aspects of timber buildings that are both important and often overlooked. Fire safety during construction presents definite challenges when wood is exposed in ways it will not be in the completed structure and when active fire protection measures are not yet in place. In addition, the construction process introduces potential ignition sources that would be uncommon in completed buildings. The other aspect treated in the guide is firefighting in timber buildings. All building designers need to be cognizant of how firefighting will be conducted in their building so that such operations are properly supported by the design of the building and the construction process.

The authors and editors are to be congratulated on the production of this very important guide that should be read and understood by anyone engaged in the design, development and maintenance of timber buildings. It introduces the wide range of issues that need to be addressed in timber buildings and provide an excellent gateway into the technical literature on the various topics discussed.

Craig Beyler
April 2022

Introduction

OBJECTIVES

The past few decades have seen renewed worldwide interest in timber as a structural and architectural material for many types of buildings. There are many incentives for the increased global demand for timber buildings, including aesthetics, sustainability, prefabrication, construction speed, economy and seismic performance.

On the other hand, there are a number of issues being raised about the use of timber as a structural material. Because wood is relatively lightweight and a natural biomaterial, there are some concerns such as durability and acoustic performance of timber buildings, especially fire safety of timber buildings compared with buildings of non-combustible materials.

Modern engineered wood products can now be used to construct large and complex timber buildings. Contemporary engineering techniques are enabling construction of timber buildings that were once only possible using concrete and steel. This is pushing the boundaries of modern fire codes and the basis on which they were founded.

Concern about the fire safety of timber buildings is understandable because it is well known that exposed wood surfaces can contribute to the early stages of a fire and can add to the fuel load in the later stages of the fire. There are also questions about issues such as fire separations, flaming from windows and extinguishment of smouldering wood as the fire goes out. Despite these concerns, well-designed timber buildings can be just as safe as buildings of traditional materials.

BACKGROUND

This book uses the latest scientific knowledge to give guidance on the extended use of design codes and standards and principles of performance-based design to provide practical guidance with examples for fire safe design of timber buildings. Reference is made to recent international codes for fire safety, including Eurocode 5 and other similar codes.

The guide includes structural fire design by providing the latest detailed guidance on separating and load-bearing functions of timber structures. It also includes information on the reaction to fire performance of wood products according to different classification systems. The importance of proper detailing in building design is stressed by giving practical examples. Active measures of fire protection and quality of construction workmanship and inspection are presented as important means for fulfilling fire safety objectives.

The core audience is all those involved in the fire safety of timber buildings, including architects, engineers, firefighters, educators, regulatory authorities, insurance companies and others in the building industry.

The authors would be pleased to see this guide used in the future development of new fire safety regulations around the world.

HISTORY

This design guide has grown out of an earlier European guide, *Fire Safety in Timber Buildings – Technical Guideline for Europe*, edited by Birgit Östman (2010). Since that time, an informal international group known as the FSUW (Fire Safe Use of Wood) group has maintained regular communication to discuss problems and solutions of timber buildings in different countries, led by Birgit Östman, and promoted by Andrea Frangi. The deliberations of this group, under the chair of Michael Klippel, led to a proposal to write this Global Design Guide, which has been written over the last two years despite the difficulties of the COVID-19 pandemic.

ORGANISATION OF THIS DESIGN GUIDE

The chapters in this design guide are summarised briefly below. Some of the chapters or technical information can be skipped by readers looking for guidance on specific topics.

Chapter 1 gives an overview of wood-based materials and construction techniques.

Chapter 2 is a summary of design principles for providing fire safety in all buildings, with particular attention to timber construction.

Chapter 3 introduces the fire dynamics of burning wood, moving from basic physics to compartment fires, and calculation methods for assessing the contribution of exposed wood to the fuel load.

Chapter 4 gives a summary of international regulations for the fire safe use of structural timber elements and visible wood surfaces in interior and exterior applications, presented in tables and maps.

Chapter 5 describes the systems used for compliance with prescriptive regulations in different regions for internal and external wood surface finishes.

Chapter 6 gives design principles for timber used as fire-resistance-rated separating assemblies to provide compartmentation for life safety and property protection, including walls, floors and roof constructions.

Chapter 7 provides guidance for the structural design of load-bearing timber members exposed to a standard fire, with an overview of the principles needed to predict the effect of charring and heating. Simplified design models include design models from the proposed second generation of Eurocode 5.

Chapter 8 is an introduction to connection typologies, potential failure modes and structural design methods to provide fire resistance to connections in timber buildings.

Chapter 9 gives recommendations for design to prevent fire from spreading into, within and through timber structures, including detailing of construction joints and penetrations.

Chapter 10 covers the effects of active fire protection systems on design of timber buildings for fire safety.

Chapter 11 introduces performance-based design of timber buildings, with a summary of possible risk-based design methods.

Chapter 12 describes general approaches and design guidance to achieve structural robustness in the fire design of timber structures.

Chapter 13 provides guidance for design and construction processes to ensure that the fire safety of timber buildings is maintained during and after construction.

Chapter 14 describes firefighting practices that may differ in timber buildings compared with other structural building systems and addresses concerns of fire services specific to timber building construction.

AUTHORS

This guidebook of 14 chapters has been written by 13 lead authors, each responsible for one or more chapters. A number of co-authors have been invited to assist with each chapter. A list of authors, co-authors and their affiliations has also been provided. Birgit Östman and Andy Buchanan carried out the final editing.

FEEDBACK

Feedback on this design guide is welcomed. A website for comments is available at www.fsuw.com.

REFERENCES

Östman, B., Mikkola, E., Stein, R., Frangi, A., König, J., Dhima, D., Hakkarainen, T. and Bregulla, J. (2010). Fire Safety in Timber Buildings – Technical Guideline for Europe, SP Report 2010:19, SP Trätek, Stockholm, Sweden.

COST Action FP 1404, Documents available at: https://costfp1404.ethz.ch/ and www.fsuw.com

Chapter 1

Timber structures and wood products

Christian Dagenais, Alar Just and Birgit Östman

CONTENTS

DOI: 10.1201/9781003190318-1

SCOPE OF CHAPTER

This chapter presents an overview of the occupancy groups in buildings and the types of timber structures that can be used to design and construct these buildings. Obviously, the types of construction presented in this chapter may have different names in different countries, but the fundamentals and design principles remain essentially the same.

A description of the various timber and engineered wood products available in the market is also provided. It summarises the manufacturing processes, typical end uses and product certifications, when applicable. Given the large variety of timber products around the globe, some of the engineered wood products presented herein may not be available in all countries.

This chapter is not intended to provide an exhaustive historical review of timber constructions and wood products but rather aims to provide sufficient information for designers, builders, building officials/authorities and fire services to better understand and differentiate the various wood products and timber building systems available.

1.1 TYPES OF BUILDING OCCUPANCY

Building codes around the globe dictate the design and construction of buildings. In a prescriptive building code, the type of building occupancy, the building area (per floor basis, or total), the building height and the presence of an automatic sprinkler system will dictate whether a timber structure is permitted (see Chapter 4). For most buildings, designers will follow prescriptive code provisions to demonstrate code compliance. The prescriptive design allows for a straightforward design and reflects the academic training of most designers. However, some building codes allow the use of performance-based design to demonstrate code compliance. This design method is usually more complex but allows for greater flexibility in the selection of materials and systems. This subsection describes a number of building occupancies where timber structures can be used. Some building codes may allow the use of timber for other building occupancies. Further details on performance-based design can be found in Chapter 11.

1.1.1 Residential buildings

Residential buildings typically refer to buildings destined for sleeping purposes, whether the occupants are primarily transient or permanent in nature (ICC, 2021a). The National Building Code of Canada, NBCC (NRCC, 2015) defines a residential occupancy as "an occupancy or use of a building or part thereof by persons for whom sleeping accommodation is provided but who are not harboured for the purpose of receiving care or treatment and are not involuntary detained". Examples of such residential buildings are single-family dwellings, semi-detached houses, attached houses, hotels, motels and apartments. However, the term "residential buildings" may include other types of buildings based on the applicable building code. In some building codes, assisted living facilities may be classified as residential buildings. Residences offering care services to residents due to cognitive, physical or behavioural limitations would most likely not be included in this category.

Timber is dominant in residential construction in North America. According to a market analysis conducted by FPInnovations (Chamberland et al., 2020), timber structures represented 97% of the market share of one-to four-storey multi-family (residential) buildings constructed in 2018 in Canada and 94% in the United States. For multi-family five-and six-storey buildings in Canada, timber structures increased from 26% in 2014 to 65% in 2018. This sharp increase coincides with the changes in the NBCC to allow five- and six-storey light timber frame residential construction since 2015. In the U.S., similar buildings represent 63% of the market share. Similar market trends can be observed in many other countries. Figure 1.1 shows some examples of residential buildings using various types of timber structures.

Typical residential buildings will have a high degree of fire-rated compartments because the use of many separating elements, such as floors and walls, provides a certain fire resistance rating based on the applicable building code. A localised fire can nevertheless grow to a fully developed fire, and flashover conditions may be reached, while the fire is still contained to the room of fire origin. In a residential building, it is important to note that building codes usually do not prescribe or differentiate the occupants. Their capacity for self-movement, walking speed and need for a wheelchair are not regulated in a residential building. As such, a broad range of occupants may be found in a residential building and means of egress are to be appropriate. According to the International Code Council *Performance Code for Buildings and Facilities* (ICC, 2021b), occupants and visitors in a

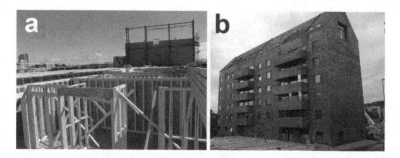

Figure 1.1 Residential buildings using a timber structure: (a) Light timber framed midrise building in Canada (photo Cecobois); (b) Residential building in Sweden (photo B. Östman).

residential building are assumed to be not awake, alert or capable of exiting without the assistance of others and are familiar with the building. If motels and hotels are classified in this type of occupancy, the same assumptions apply to occupants, visitors and employees, with the exception that employees are awake.

1.1.2 Office buildings

An office building can be defined as a *"building used principally for administrative or clerical work"* (ISO 6707-1). Examples of office buildings include administrative or professional businesses and commercial and low-level storage occupancies. Building codes may however classify such occupancy in another category.

While the aesthetic and biophilic advantages of timber are widely required by architects, timber has only a modest use in office buildings. With recent trends to construct taller and larger mass timber buildings around the globe, it is expected that the use of timber in office buildings will increase. Figure 1.2 shows examples of office timber buildings.

Office building design usually consists of large open spaces with moveable partitions, which result in long floor spans. In such an open-space concept, localised fires may be of primary concern as opposed to a fully developed fire. Travelling fires can also be an important risk to mitigate. In office buildings, it is assumed that occupants are awake, alert, predominantly capable of exiting without assistance from others and familiar with the building (ICC, 2021b). As such, evacuation can be initiated faster in an office building than in a residential building.

1.1.3 Educational buildings

Based on the applicable building code, educational buildings may be buildings where occupants are gathered for educational purposes, as well as day care services for children. In some other codes, they may be classified as

Figure 1.2 Office buildings using timber product: (a) First Tech Credit Union office in Canada (photo Structurlam Mass Timber Corp.); (b) Hybrid office building in New Zealand (photo A. Buchanan)

"assembly" buildings where occupants gather for civic, social, educational or recreational purposes.

Structural timber has a very low use in educational buildings, with some exceptions with low-rise buildings (one and two storeys) mainly due to prescriptive building codes in some countries imposing such limitations. Wood finish materials are however used in several locations in educational buildings for aesthetic reasons. There has, however, recently been a strong increase in structural timber for gymnasiums and sports complexes. Figure 1.3 shows some educational buildings where timber has been used both for structural elements and finish materials.

Construction of educational buildings is a combination of residential and office building types, where they may consist of large open-space concepts with moveable partitions and a high degree of fire-rated compartments between classrooms. Localised fires, fully developed fires and travelling fires are therefore potential risks that warrant mitigation. In educational buildings, it is assumed that occupants are awake, alert and familiar with the building (ICC, 2021b). Younger occupants (e.g. under the age of 10 years) are assumed to require assistance for safe egress, while older occupants will predominantly be capable of exiting by themselves.

1.1.4 Public buildings

A public building would essentially consist of an assembly occupancy where gatherings are taking place for recreational, commercial or mercantile purposes. Typically, building codes do not classify public buildings but will rather classify their type of assembly (e.g. performing arts, arena type or exterior gathering).

Similar to educational buildings, public buildings typically have low use of timber products mainly due to limitations imposed by prescriptive building codes in some countries. Nevertheless, wood finish materials are widely used for aesthetic reasons, with some low-rise buildings constructed with a

Figure 1.3 Educational buildings using timber products: (a) Université Laval in Canada (photo FPInnovations); (b) Atrium space in educational building in New Zealand (photo A. Buchanan)

Figure 1.4 Public buildings using timber products: (a) Formula 1 paddocks in Canada (Photo Nordic Structures); (b) Parking garage in Sweden (Photo AIX Architects).

timber structure. Figure 1.4 shows some examples of public buildings using various types of timber structures.

Localised fires would most likely be the main risk in public buildings with large open floor areas. Similar to office buildings, it is typically assumed that the majority of occupants are awake, alert and predominantly capable of self-evacuation with little to no assistance or prompting from others (ICC, 2021b).

1.1.5 Industrial buildings

As defined in the NBCC (NRCC, 2015), buildings intended for the assembling, fabricating, manufacturing, processing, repairing or storing of goods and materials would be classified as industrial buildings. Some building codes would further sub-divide industrial buildings based on the level of fire hazard represented by the flammable, combustible or explosive characteristics of materials that can be found within these buildings.

Industrial buildings may also be constructed with mixed occupancies, where industrial use would represent most of the building area, and other occupancies such as offices would be secondary occupancies. In such mixed-occupancy groups, most building codes will require fire-resistance-rated separations, or even sometimes firewalls, to separate one occupancy group from another.

Prescriptive building codes typically allow the use of structural timber for industrial buildings in relatively small areas. However, due to various reasons, such as misperception from insurance companies, timber has limited use in these buildings. Moreover, given that industrial buildings usually require high ceilings, light timber frame construction can be limited due to the available lengths of timber studs, unless engineered wood studs are used. Otherwise, post-and-beam construction could also be used, including for support of overhead cranes. Figure 1.5 shows some examples of industrial buildings using various types of timber structures. Special, active fire protection measures, such as deluge or foam sprinklers, can be used to

Figure 1.5 Industrial buildings using timber products: (a) Industrial building using LVL in New Zealand (photo A. Buchanan); (b) Glulam/CLT manufacturing plant in Canada (photo Nordic Structures ©Adrien Williams).

mitigate fire hazards associated with high-risk materials that can be found inside these buildings. Explosions or localised fires would most likely be the main risks, but fully developed fires can also be challenging. In industrial buildings, it is assumed that occupants are awake, alert and predominantly capable of exiting without assistance from others and familiar with the building (ICC, 2021b).

1.2 TYPES OF TIMBER STRUCTURE

Timber structures have historically been classified based on the type of system resisting gravity loads. In prescriptive building codes, the dimensions and configurations of the building systems would typically dictate the type of timber structure that can be used. While some building codes may classify all types of timber structures into a single category, some building codes allow a wider range of possibilities when using mass timber construction compared to light timber frame construction.

The following subsections provide a summary of the construction techniques of a number of timber structures, including the types of products typically used and comments on their fire performance.

1.2.1 Light timber frame construction

Light timber frame construction is the most dominant type of construction for residential buildings, at least in North America for buildings up to six storeys. It essentially consists of repetitive small-size structural elements made of sawn timber, engineered wood products and structural sheathing.

Balloon-framing was mainly used in the early days of the 20th century. This type of light timber frame construction allowed for rapid housing construction. The wall studs were continuous over the storeys, and the floor joists were supported on horizontal ledgers placed inside notches in the studs. This type of construction could have inherent concealed spaces, forming potential paths for a fire to spread from one storey to another, unless construction details were made to provide adequate fire stopping (Figure 1.6a). The lateral loads were typically resisted by either structural panels or diagonal bracing.

Then came platform-framing. In this type of light timber frame construction, the gravity loads remain supported by wood studs, but each wall is assembled one storey at a time and floor framing is installed at every storey (Figure 1.6b). Each wall is enclosed with top and bottom sill plates, creating inherent fire stopping between storeys. Floor framing is also enclosed by rim boards made of sawn timber or engineered wood products, which also create an inherent fire stopping within the floor. Typically, building codes will require that any openings in assemblies be required to provide fire resistance to be properly sealed by fire-stop materials. Guidance on preventing fire spread is given in Chapter 9. The lateral loads are taken by structural panels for both the floor diaphragms and shear walls. Blocks of sawn timber can be used at mid-height between every wall stud to provide additional nailing to the structural panels. When these blocks are used, an additional inherent fire stopping is created within the wall cavities to limit vertical fire spread.

In platform-framing, the wall studs are generally of sawn timber and may be of structural composite lumber (SCL) to increase the axial compression

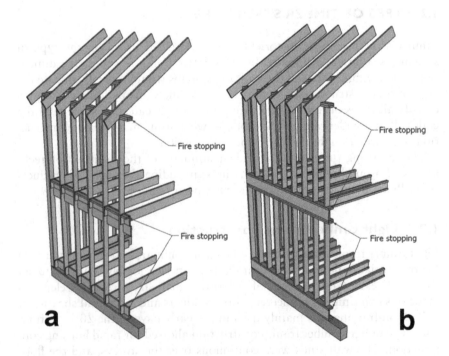

Figure 1.6 Light timber frame construction: (a) Balloon-framing, (b) Platform-framing.

strength of the studs and/or limit shrinkage of the sill plates due to varying moisture content. Using SCL for wall studs can also allow taller walls where sawn timber would otherwise be limited in length, as previously mentioned for industrial buildings. Nowadays, engineered wood products such as pre-fabricated wood I-joists and metal-plated trusses have replaced many of the traditional sawn timber floor and roof joists. These products allow for increased load-bearing capacities, longer spans for open-space concepts and better dimensional stability.

Given the small dimension of the structural elements, the fire perfor-mance of light timber frame construction is typically provided by protec-tive membranes, such as fire-resistance-rated gypsum plasterboards. Service penetrations made through these protective membranes need to be properly protected using fire-stopping devices tested according to the applicable test method in each country. Fire resistance rating of up to 2 hours, and more, can be achieved when tested in accordance with standard test methods such as ASTM E119, CAN/ULC S101 and EN 1363-1, among others.

Light timber frame assemblies can provide excellent fire performance, provided that they are detailed and constructed appropriately. Further guidance on proper detailing is provided in Chapter 9.

1.2.2 Post-and-beam construction

The modern post-and-beam timber construction is the logical evolution of the traditional system called "timber frame". Traditionally, post-and-beam construction, or "heavy timber" construction, used timber structural ele-ments of large dimensions and cast-iron caps to transfer the loads from one storey to the other, as well as connections between main to secondary beams using timber embedment, wood pegs and dovetails, as examples. It then evolved by using metallic fasteners such as bolts, dowels and hangers for side connections, as would be done in steel framing.

Nowadays, post-and-beam timber construction is taking full advantage of timber embedment strength for connections as well as the use of inno-vative fasteners, such as long and slender self-tapping screws. Engineered wood products such as glued laminated timber and structural composite lumber are now widely used in lieu of sawn timber. Floors and roofs typi-cally consist of timber decking or panels made of glued laminated timber, structural composite lumber or mass timber. With the advances in com-puter numerical control (CNC), machining of timber elements for drilling holes for fasteners and embedding metallic plates for concealed connec-tions, it is much easier to design and install this type of construction with a high level of precision.

Similar to steel framing, the lateral loads are typically resisted by braced frames or moment-resisting connections (although less popular), while the floors and roof elements act as diaphragms. Figure 1.7 shows examples of post-and-beam timber construction.

Figure 1.7 Post-and-beam timber constructions: (a) Old timber frame in Canada (Photo FPInnovations); (b) Modern post-and-beam construction using braced frames in Canada (Photo A. Buchanan).

Traditional post-and-beam timber construction has a long history demonstrating its inherent fire performance. In some building codes, "heavy timber" construction can be used in many applications where a non-combustible construction would otherwise be required. The large dimensions of the structural elements allow for maintaining their structural strength for long fire exposure. The load-bearing performance of timber elements can easily be calculated using their charring rates and other design assumptions, as detailed in Chapter 7. Information on the fire performance of connections can be found in Chapter 8.

1.2.3 Mass timber construction

Mass timber construction is a new type of timber construction that originated with the strong market acceptance and penetration by European cross-laminated timber (CLT) and was then rapidly adopted by other countries. While the term "mass timber" is relatively new, it is not necessarily a new type of construction as it was traditionally used in old buildings made of post-and-beam construction. The floor construction called "mill floor" consisted of sawn timber elements placed on edge, side-by-side, and nailed together, creating a massive thick timber slab (also called nail-laminated timber (NLT)).

Mass timber construction is the logical continuation of the post-and-beam timber construction detailed above, but with larger and longer plates used as wall and floor panels similarly to precast concrete construction. Cross-laminated timber is among the first modern timber products used in mass timber construction, where it is used for load-bearing walls, partitions, as well as floors and roofs. With the desire to increase the diversity of mass timber panels, mechanically laminated timber, such as nail-laminated timber and dowel-laminated timber, is now slowly gaining popularity. Figure 1.8 shows examples of mass timber panel construction.

Figure 1.8 Mass timber panels construction: (a) Old mill floor in Canada (Photo FPInnovations); (b) Sara Cultural Centre and hotel, 19 storeys, Skellefteå, Sweden (Photo Jonas Westling).

Nowadays, mass timber panels are used in conjunction with post-and-beam construction to reduce the amount of timber, limit the cost and offer greater design flexibility, such as open-space concepts. Engineered wood products such as glued laminated timber and structural composite lumber are used for gravity loads (columns and beams), while mass timber panels are used for floors and roofs, as well as lateral load-resisting systems.

An inherently high level of fire resistance is provided in a building made of mass timber panels, especially when it is fully built with mass timber walls, roof and floor panels. As with post-and-beam construction, the large dimensions of the structural elements allow for maintaining their structural strength for long fire exposure. Panelised elements provide the separating function to limit heat transmission and passage of flames, in addition to the load-bearing performance. Additional information on the separating function and load-bearing performance can be found in Chapters 6 and 7, respectively. Information about detailing of mass timber panels for fire safety can be found in Chapter 9.

1.2.4 Long-span structures

Timber structures also have a long history for long-span structures, in sports complexes and in industrial buildings. These applications however require a high level of expertise and knowledge in timber design and structural engineering so that loads are transferred adequately and long-term serviceability performance, including creep, dimensional changes and durability due to moisture content, is ensured.

From their structural design efficiency, curved arches made of glued laminated timber have widely been used for long-span applications and of various geometric shapes (simple or multiple curvatures). They allow for high roof clearance, as required for ice rinks, soccer stadiums and indoor water parks. Depending on the span, they can be of single, double or multiple elements.

A great engineered timber building is the Moffett Field hangar II near San Francisco, U.S. Completed in 1943, it consists of trusses made of large timbers constructed during World War II to serve the US Navy blimp surveillance programme. The timber structure follows a parabolic shape that is 328 m long by 90.5 m wide and 52 m in height to accommodate the profile of the airships contained in it.

Another structural system used in timber is the grid shell. This system allows for long spans and open-space concepts. An example of such system is the Odate dome built in Japan in 1997. The entire structure has a height of 52 m and an impressive span of 178 m along the major direction and 157 m in the minor direction. Grid shell systems have been used in some projects in Europe and recently for the three domes at the Taiyuan Botanical Garden in China. The domes range from 43 to 88 m in diameter and from 12 to 30 m in height. The dome design team claims that the largest of the three domes is the longest clear span timber grid shell in the world. Figure 1.9 shows examples of long-span timber structures.

As with post-and-beam construction, the large dimension of the structural elements used for long-span applications allows for maintaining their structural capacity for long fire exposure. The load-bearing performance of timber elements exposed to fire can easily be calculated using their charring rates and other design assumptions, as detailed in Chapter 7. In most buildings where arches are used, it is unlikely that a localised, travelling or fully developed fire can generate sufficient hazard to challenge the members and their connections at the top of the building, so more attention should be made to the lower ends. Moreover, it is likely that these buildings would require protection by automatic sprinkler systems.

Figure 1.9 Long-span timber systems: (a) Soccer stadium in Canada (67.6 m span) (Photo Nordic Structures © Stéphane Groleau), (b) Timber grid shells in China (43 to 88 m span) (Photo StructureCraft).

As such, some building codes may not require all the structural elements to be fire resistance rated.

1.2.5 Hybrid structures

All structures are essentially hybrid, as they consist of various materials used together to form a distinct system or structure. Hybrid structures can consist of any mix of materials at various locations within a building. A hybrid structure can consist, for example, of a gravity system made of timber and a lateral load-resisting system made of reinforced concrete core walls or steel braces. Using lateral load-resisting systems made of concrete or steel braces typically allows for using greater ductility and strength capacities. However, some mass timber panels can also provide the same lateral performance as that of concrete and steel.

Hybrid structures can also be horizontal elements made of timber, concrete and steel, such as a timber slab or beam connected to a concrete slab, steel joists or beams connected to a timber slab, etc. The use of hybrid horizontal systems typically allows for longer spans by positioning each material at its best location to take full advantage of its mechanical resistance. They also enhance serviceability performance such as acoustics, floor vibrations and deflections.

Long-span structures, as detailed in the previous subsection, can also be hybrid where timber would be positioned where the elements are solicited mainly in axial compression and steel tendons would be solicited in axial tension. This allows for pre-stressed systems to enhance serviceability performance.

When designing for fire resistance, each material needs to be considered, along with the potential impact from one to the other. As an example, the timber component of a timber–concrete composite slab exposed to fire will char and the residual timber will reduce in size with time and change the stress distribution between the two materials as well as the shear connectors used to fasten them together. Heat transfer between materials may also be a challenge where, as an example, a steel beam connected to a timber slab will accelerate localised charring in the vicinity of the steel beam due to heat conduction. Figure 1.10 shows examples of hybrid structures using timber, concrete and steel.

1.2.6 Prefabricated elements and modules

Industrialised building systems for multi-storey timber construction are being used increasingly in northern Europe during the first decades of the 2000s. They emerged from a long tradition of prefabricated single-family houses starting in the early 1900s. Still about 90% of all single-family houses in Sweden are built in timber. A whole house, or two-dimensional building elements, mainly walls, are built in a factory and brought to the

Figure 1.10 Hybrid structures: (a) Brock Commons in Canada – mass timber construction and reinforced concrete vertical shafts (photo Naturally Wood), (b) Meadows Recreational Center in Canada – Glued laminated timber and steel roof structure (photo Western Archrib).

Figure 1.11 Prefabricated elements and modules: a) Prefabricated mass timber floor in Northern Sweden, late 1990s (Photo Martin Gustafsson), (b) Modular houses in Norway (Photo Kodumaja).

building site. This technique has a lot of advantages, including close control of the building process, dry conditions, and a fast building process. Figure 1.11 shows examples of prefabricated elements and modules.

When taller timber buildings became allowed in Sweden in the late 1990s, it was natural to adopt the prefabricated system for multi-storey design. Different techniques have been applied and two-dimensional elements are now often made with CLT panels, while three-dimensional (3-D) volumetric modules are mainly timber frame structures. The 3-D modules may be load-bearing themselves or integrated into a separate load-bearing structure e.g. with post and beam. The latter is the case for the 14-storey high Treet building in Norway.

Prefabricated volumetric modules were initially used for small apartments e.g. accommodation for students, but they are now used for larger apartments consisting of several volumetric components, where the kitchen and/or bathrooms are built as separate modules and put together at the building site. One limiting factor is the size of elements or modules to be road transported.

1.3 STRUCTURAL TIMBER PRODUCTS

There is a wide variety of structural timber products available in the market. In the past few decades, many engineered wood products (EWPs) have been developed and commercialised as a substitute for traditional wood products. These EWPs are designed and manufactured for better use of the raw material, eliminating natural characteristics of timber that may have a negative impact (i.e. knots, wane, etc.), reducing waste from timber sawmills and reducing the amount of timber required for manufacturing a homogenous and stronger product.

Provided that the wood feed-stock is obtained from renewable forestry operations, all of these structural timber products provide great benefits for low carbon construction. The sequestered carbon stored in structural timber far exceeds the small amount of fossil fuel energy required to manufacture the wood products, and this can be used to offset the carbon released in manufacturing the other components of a building. Timber buildings hence have a much lower carbon footprint than similar buildings made from traditional materials such as steel and concrete.

The following sections describe some of the various structural timber products available in the market. The products presented below are largely based on current technologies and products available in North America and Europe.

1.3.1 Sawn timber

Sawn timber is among the oldest construction material. Sawn timber, called lumber (or dimension lumber) in some countries, is defined by ASTM D9 as a product of the sawmill and planing mill, usually not further manufactured other than by sawing, resawing, passing lengthwise through a standard planing machine, crosscutting to length, and matching. In some countries, the term timber can also refer to a wood element of minimum dimensions, differentiating them from smaller elements called lumber. In North America, the structural elements are named based on their nominal dimensions rather than actual sawn dimensions. As examples, a nominal 2″ × 10″ lumber joist is actually 38 × 235 mm (1½″ × 9¼″), and a nominal 6″ × 6″ timber beam is 140 ×140 mm (5½″ × 5½″). Other countries typically specify the actual (net) dimensions rather than the nominal dimension. For structural applications, building codes typically require that sawn timber has a moisture content no greater than 15% to 19% at the time of installation. As such, it is usually dried to a suitable moisture content prior to installation. In light timber-framed buildings of five and six storeys, dimensional changes due to drying during the service life can be significant and considerations should be given to limiting such shrinkage.

There are various types of sawn timber (lumber) used in construction and available on the market. Typically, softwood species are used for structural applications, while hardwoods are used for finishing materials. In some jurisdictions, hardwoods may however be used in structural applications, including the manufacturing of engineered wood products such as glued laminated timber. Structural products are required to be evaluated by their respective standards, such as those of the National Lumber Grades Authority (NLGA) in Canada, the American Lumber Standard Committee (ALSC) in the U.S., the European standard EN 15497 and the Australian/New Zealand standards AS 2858 and AS/NZ 1748.

The most common type of sawn timber is visually graded, which is sometimes also categorised within specific wood species groups. Based on visual observations by a trained inspector, the boards are visually graded into various classes, which are assigned mechanical properties based on regular quality control monitoring by the grading agency. This ensures that the grading is being made properly and that the mechanical properties published in wood design standards are maintained. Some of the visual characteristics used for classifying timber are the slope of grain, moisture content, knots and wane. Distortion of timber boards due to bow, crook, cup and twist also affect their grading. Standard test methods usually specify how to address the potential strength and stiffness reduction factors.

Some sawmills use mechanical grading of sawn timber, such as mechanically stress rated (MSR) and mechanically evaluated lumber (MEL). Both MSR and MEL refer to structural timber that has been graded for stiffness by means of a non-destructive test and subjected to similar visual grading as the visually graded timber. These testing techniques allow for a better evaluation of the raw material by non-destructively testing mechanical properties, mainly the modulus of elasticity. They also allow mills to sort timber exhibiting higher mechanical properties, thus providing a higher structural grade for stronger timber. Non-destructive testing is also widely used in the manufacturing of EWPs so that manufacturers can ensure that the timber used in the manufacturing process meets or exceeds the quality control criteria.

Lastly, sawn timber (lumber) can also be remanufactured into various products, such as finger-jointed lumber, face-glued lumber or edge-glued lumber. These types of EWPs allow for eliminating natural defects that may be present in visually graded lumber by remanufacturing smaller and/or shorter pieces together to form long and dimensionally stable products. The resulting products are widely used in the manufacturing of EWPs, such as those detailed in the following sections. When finger-jointing, face-gluing or edge-gluing is used, the fire performance of the adhesives should be properly evaluated so that the adhesives do not become the weak link in the fire resistance of the resulting product (see Chapter 7).

Due to their small cross-sections, the fire performance of typical sawn timber relies on the use of claddings or membranes (e.g. fire-resistance-rated gypsum plasterboard), unless the applicable building code allows them to remain exposed (unprotected). Otherwise, the load-bearing performance of larger timber elements can be calculated using their charring rates and other design assumptions, as detailed in Chapter 7.

1.3.2 Wood I-joists

Since the creation of prefabricated wood I-joists, the market has rapidly grown as an alternative to solid sawn timber joists and roof rafters, especially in light timber frame construction. A prefabricated wood I-joist is defined as *"a structural member manufactured using sawn or structural-composite lumber flanges and structural panel webs, bonded together with exterior grade adhesives, forming an "I" cross-sectional shape"* (ASTM D9).

Wood I-joists were first commercialised by the American company Trus Joist Corporation in the 1960s (Williamson, 2002). The main advantages of prefabricated wood I-joists are their light weight, longer allowable spans and low cost when compared to traditional sawn timber joists. They are typically used as floor joists and in some applications as roof joists. With an increasing demand for energy-efficient building envelopes, we are now seeing prefabricated wood I-joists used as wall studs. Their depths allow for a greater insulated cavity.

The I-shape cross-section allows for more efficient use of the timber resource, with flanges subjected to axial stress and web panel subjected to shear stress. Flanges are typically made of finger-jointed sawn timber or structural composite lumber (see Figure 1.12). They have various dimensions, resulting in varying bending resistance and stiffness. Web panels used to be made of plywood or hardboard but have changed to oriented strand boards (OSB) over the years. Some producers commercialise wood I-joists

a   **b**

Figure 1.12 Prefabricated wood I-joists: (a) Sawn timber flanges (photo FPInnovations); (b) LVL flanges (photo APA Wood).

with web materials from other types of panels such as high-density fibre-board (HDF).

Wood I-joist manufacturers usually offer their products in standardised dimensions. The available depths typically vary from 235 to 406 mm (9¼" to 16"), with some special deeper joists. Prefabricated wood I-joists are required to be manufactured and evaluated according to specific standards, such as ASTM D5055 for North America. As an example, ASTM D5055 provides the minimum requirements with respect to procedures for estab-lishing, monitoring, and re-evaluating structural capacities such as shear, reaction (bearing support), bending moment, and stiffness. Requirements for adhesives performance used for flange finger joints, web-to-web joints and web-to-flange joints are typically also provided. While there is currently no standard in Europe, I-joists may conform to the European Assessment Document (EAD 130367-00-0304) for CE-marking.

Due to their inherently small cross-section, the fire performance of pre-fabricated wood I-joists typically relies on either the use of claddings or membranes (e.g. fire-resistance-rated gypsum plasterboard) or web protec-tion materials, unless they are specifically allowed to remain unprotected by the applicable building code. Manufacturers can provide floor and roof assemblies made with prefabricated wood I-joists that can achieve up to 2 hours of fire resistance. Given the proprietary nature of these products, it is recommended to consult with the manufacturers for proper detailing. Some general guidance is given in Chapter 7.

1.3.3 Metal plate wood trusses

Similar to prefabricated wood I-joists, metal plate timber trusses are used as an alternative to solid sawn timber joists and roof rafters in light timber frame construction. Their main advantages are light weight, longer allow-able spans and low cost when compared to traditional sawn timber joists. They are typically used as floor trusses and widely used as roof trusses in North America.

A typical truss consists of top and bottom chords (flanges) and diagonal webs forming a triangular shape using sawn timber or structural composite lumber. Junctions between chords and webs are fastened together using proprietary galvanised steel plates, also called truss plates. Usually, a floor truss would have parallel chords positioned flatwise (i.e. wide dimension of the timber parallel to the floor plan), while roof trusses will have the chords positioned edgewise (narrow dimension parallel to the roof plan) either par-allel or sloped. Figure 1.13 illustrates metal plate trusses and some truss plates available on the market.

Some countries have enforced quality control standards for the man-ufacturing of metal plate timber trusses. Trusses can be designed and manufactured in almost infinite shapes and spans. Given the long roof spans that can be achieved by metal plate timber trusses, proper lateral

Figure 1.13 Images of metal plate trusses: (a) Floor trusses (photo FPInnovations); (b) Roof trusses (photo Naturally Wood).

bracing is crucial to ensure the stability of the compression chords and webs against buckling. The structural design is typically in accordance with the applicable timber design standard and proprietary metal plate design information. As an example, the Truss Plate Institute of Canada (TPIC) and Standards Australia publish standards that establish minimum requirements for the design and construction of metal plate timber trusses, including the materials used in a truss (both lumber and steel), the design procedures for truss members and joints as well as manufacturing and material variances and erection tolerances (TPIC; AS 1720.5). Guidance for lateral bracing is also typically provided in truss design standards. In Europe the metal plate web trusses are produced according to EN 14250 and designed according to Eurocode 5 (EN 1995-1-1 and EN 1995-1-2).

Similar to prefabricated wood I-joists, the fire performance of metal plate timber trusses typically relies on the use of cladding or membranes (e.g. fire-resistance-rated gypsum plasterboard), unless specifically allowed to remain unprotected by the applicable building code. Some manufacturers have floor and roof assemblies made with metal plate timber trusses that can achieve up to 2 hours of fire resistance. Given the proprietary nature of these products, it is recommended to consult with the manufacturers for proper detailing.

1.3.4 Structural composite lumber

Structural composite lumber (SCL) is a generic category of structural engineered wood products that includes laminated veneer lumber (LVL), parallel strand lumber (PSL), laminated strand lumber (LSL) and oriented strand lumber (OSL), as illustrated in Figure 1.14. Structural composite lumber (SCL) is defined as *"a composite of wood elements (for example, wood strands, strips, veneer sheets, or a combination thereof), bonded with an exterior grade adhesive and intended for structural use in dry service conditions"* (ASTM D9). See also Section 1.3.6 on mass timber panels.

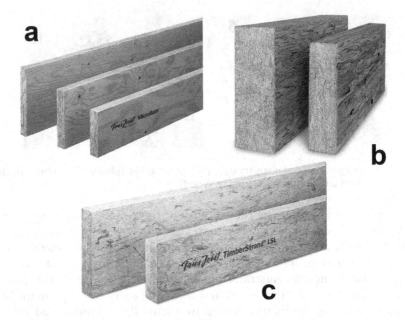

Figure 1.14 Structural composite lumber (photos courtesy of Weyerhaeuser): (a) LVL – Laminated Veneer Lumber; (b) PSL – Parallel Strand Lumber; (c) LSL – Laminated Strand Lumber.

LVL and PSL were first introduced into the market in the 1970s and 1980s, respectively (Williamson, 2002). LSL and OSL were introduced shortly after PSL. Their main advantages are the efficient use of the timber resource, higher strength and stiffness and longer spans. They are typically used as beams, columns, lintels and joists, with some applications as chords in metal plate trusses. SCL is also used as studs in mid-rise light timber frame construction (five and six storeys) where greater axial capacity is required at the lower levels, as well as sill plates for limiting building vertical displacement due to moisture shrinkage. Being manufactured at an initial low moisture content, SCL products tend to be more dimensionally stable than traditional sawn timber when subjected to varying degrees of moisture content during service.

Laminated Veneer Lumber (LVL) is defined as "a composite of wood veneer sheet elements with wood fibres primarily oriented along the longitudinal axis of the member, where the veneer element thicknesses are 0.25 in. (6.4 mm) or less (ASTM D9)."

LVL is manufactured in a similar manner as plywood, with the exception that the wood grain of the veneers is mostly oriented longitudinally to the main strength direction (i.e. towards the LVL length). LVL is often manufactured in a continuous process so the resulting products can be longer and stronger than traditional sawn timber.

Parallel Strand Lumber (PSL) is defined as

> a composite of wood veneer strand elements with wood fibres primarily oriented along the longitudinal axis of the member, where the least dimension of wood veneer strand elements is 0.25 in. (6.4 mm) or less and their average lengths are a minimum of 300 times the least dimension of the wood veneer strand elements.

(ASTM D9)

PSL is manufactured by gluing wood strands to form a condensed thick piece of timber in such a way that the wood grain of the strands is oriented longitudinally to the main strength direction (i.e. towards the PSL length). Wood strands may be cut from the residue of plywood or LVL manufacturing plants.

Laminated Strand Lumber (LSL) is defined as

> a composite of wood strand elements with wood fibres primarily oriented along the longitudinal axis of the member, where the least dimension of the wood strand elements is 0.10 in. (2.54 mm) or less and their average lengths are a minimum of 150 times the least dimension of the wood strand elements.

(ASTM D9)

Manufacturing process of LSL is somewhat like that of OSB. It requires, however, a higher degree of strand orientation and greater pressure to form the thick piece of timber. As with PSL, wood grain of the strands is oriented longitudinally to the main strength direction (i.e. towards the LSL length). Wood strands may be cut from the residue of plywood, LVL or PSL manufacturing plants. LSL usually has lower strength and stiffness than LVL and PSL.

Oriented Strand Lumber (OSL) is defined as

> a composite of wood strand elements with wood fibres primarily oriented along the longitudinal axis of the member, where the least dimension of the wood strand elements is 0.10 in. (2.54 mm) or less and their average lengths are a minimum of 75 times the least dimension of the wood strand elements.

(ASTM D9)

The manufacturing process of OSL is similar to that of LSL, with the exception that shorter strands are used. As with LSL, the wood grain of the strands is oriented longitudinally to the main strength direction (i.e. towards the OSL length). Wood strands may be cut from the residue of plywood, LVL, PSL or LSL manufacturing plants. OSL usually has lower strength and stiffness than LSL.

As with prefabricated wood I-joists, SCL manufacturers offer products in standardised dimensions ranging from 89 to 508 mm (3½″ to 20″) in depth, 38 to 178 mm (1½″ to 7″) in width and up to 18 m (60′) in length. SCL products are required to be manufactured and evaluated according to specific standards, such as ASTM D5456 in North America, which provides the minimum requirements with respect to initial qualification sampling, mechanical and physical tests, analysis, and design value assignments. Requirements for adhesive performance at elevated temperatures and/or fire conditions are typically also provided. While there are currently no other standards equivalent to ASTM D5456 applicable to all SCL products, LVL products are to be evaluated per European standard EN 14374 and Australia/New Zealand standard AS/NZS 4357.0. A European LVL Handbook is also available to provide design information for code compliance (LVL Handbook, 2020).

SCL products can be used as a single element or as built-up elements using nails, screws or bolts. When used as a single element, from either a single and large piece of SCL or an SCL obtained from a secondary face-gluing process, their fire performance and charring behaviour are similar to traditional sawn timber (Dagenais, 2014; White, 2000; O'Neill et al., 2001), provided the adhesive used for secondary gluing is a structural adhesive meeting the requirements to resist elevated temperatures and/or fire conditions. Structural fire resistance of SCL can therefore be determined based on the same design principles as those detailed in Chapter 7.

However, built-up elements made with metallic fasteners may not have the same fire performance as a single element of the same dimensions. Connections used to secure SCL elements together may not prevent the individual elements from separating when exposed to fire, which can lead to increased localised charring between the SCL elements (O'Neill et al., 2001). Proper caution should be taken when built-up SCL elements are required to provide some level of fire resistance.

1.3.5 Glued laminated timber

Glued laminated timber, also called glulam, can be defined as *"a product made from suitable selected and prepared pieces of wood bonded together with an adhesive whether in a straight or curved form with the grain of all pieces essentially parallel to the longitudinal axis of the member"* (ASTM D9). Its manufacturing allows for small or large structural elements, either straight or curved.

Glulam is one of the oldest engineered wood products and still much used in the timber construction market. According to Williamson (2002), glued laminated timber was first patented in Switzerland in the 1890s. It was then first used in the United States in the construction of the USDA Forest Products Laboratory in Madison (WI).

As with other EWPs, the main advantages of glulam are the efficient use of the timber resource, higher strength and stiffness and longer spans. Glulam is typically used as beams, columns, lintels and joists, with some applications as planks and decking in post-and-beam and mass timber constructions (see Figure 1.15). Given their flexibility to meet various shapes, they are also the most widely used EWP for the design and manufacturing of long-span arches. Moreover, being manufactured at an initial low moisture content, glulam tends to be more dimensionally stable than traditional sawn timber when subjected to varying degrees of moisture content during service.

The layup (or configuration) of glued laminated timber is based on the theory of composite materials, where each lamination has its own strength and stiffness characteristics and is positioned to result in effective strength and stiffness of the finished cross-section. Typically, laminations with the greatest mechanical properties are positioned towards the outer surface (also called tension laminations), where the axial stresses are at their maximum in a flexural element. Lower quality timber is used within the core (also called core laminations), with some intermediate requirements in between.

The manufacturing of glued laminated timber is usually regulated by the applicable building codes and standards. In Canada, glued laminated timber is manufactured in accordance with CSA O122 standard and manufacturing plants are to conform with CSA O177. In Europe the product standard is EN 14080, and in Australia/New Zealand AS/NZS 1328.1. These standards provide the minimum requirements for the materials to be used such as the timber and adhesives, as well as the minimum requirements

Figure 1.15 Structural glued laminated timber: (a) Post-and-beam construction using glulam and prefabricated wood I-joists (photo FPInnovations); (b) Curved beams and decking at ATCO commercial centre in Canada (photo Western Archrib).

for qualification testing and quality control. With respect to the fire performance of adhesives used in glued laminated timber, an international survey highlighted significant differences exist between countries (Wiesner et al., 2018). As such, it is strongly recommended to consult the appropriate standards accreditation bodies for assessing whether imported glued laminated timber is suitable and conforms to the applicable building codes and standards in the importing country. Effects of glueline fire performance can be found in Chapters 2, 3, 6 and 7.

Glued laminated timber has excellent inherent fire performance. The large cross-section allows for structural elements to char slowly and at a predictable rate, allowing them to sustain the applied loads for a long duration. Structural fire resistance of glued laminated timber can therefore be determined based on the same design principles as those detailed in Chapter 7.

Moreover, some countries such as Canada and the United States provide special provisions for fire-resistance-rated glued laminated timber beams, without specific calculations being necessary. For example, a beam requiring a one-hour fire resistance rating when exposed to fire from three sides (top is protected) shall be manufactured to the layups specified in the manufacturing standards, except that one core lamination shall be removed and one 38 mm thick outer tension lamination added on the bottom (see Figure 1.16). When such special manufacturing is made, the glued

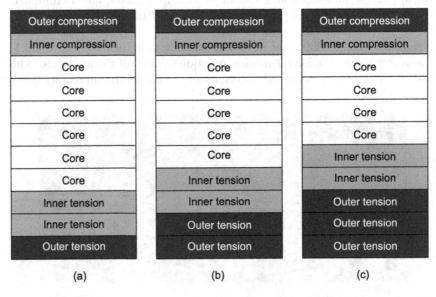

Figure 1.16 Manufacturing provisions for fire-resistance-rated glulam beams, as presented in CSA O86: (a) No fire resistance rating; (b) 1-hr fire resistance rating; (c) 2-hr fire resistance rating.

laminated timber beams should have a mark (stamp) specifying their fire resistance rating.

1.3.6 Mass timber panels

Mass timber panels, or plates, are essentially large timber panels used as floors, roofs and wall panels. They were traditionally used in old timber buildings made of "mill floors", but, with advances in timber engineering and manufacturing processes, new mass timber panels have recently emerged such as cross-laminated timber (CLT), mechanically laminated timber (MLT), mass plywood panels and mass OSB panels.

CLT is defined as *"a prefabricated engineered wood product made of at least three orthogonal layers of graded sawn lumber or structural composite lumber (SCL) that are laminated by gluing with structural adhesives"* (ANSI/APA 2019). In Europe the definition is similar, but the need for fire-resistant adhesives is not mentioned (EN 16351). Laminating orthogonally allows for enhanced dimensional stability and for bi-directional structural elements. However, CLT panels are mainly used as uni-directional structural elements where the laminations oriented along the strength axis carry most of the applied stress. Typically, the strength axis, or major direction, is oriented towards the longitudinal dimension of the CLT (e.g. span of a floor panel or height of a wall panel). Figure 1.17a illustrates a typical CLT panel.

As with glued laminated timber, manufacturing standards such as ANSI/APA PRG 320 for use in North America CLT specifies the minimum requirements for the materials to be used such as the timber and adhesives, as well as the minimum requirements for qualification testing and quality control. In Europe, CLT should comply with EN 16351, although it is not yet adopted by the European Commission as a formal European standard. Canadian, American and Swedish handbooks are also available to provide design information until the product becomes recognised in building codes (Karacabeyli & Douglas, 2013; Karacabeyli & Gagnon, 2019; Swedish Wood, 2019).

CLT is a relatively new EWP, and most building codes and standards do not fully address the use of this product. The international survey referenced

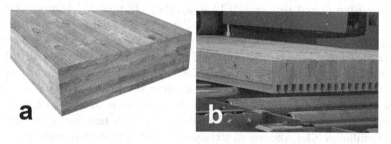

Figure 1.17 Mass timber panels: (a) Cross-laminated timber (Photo APA Wood); (b) Dowel-laminated timber panels (Photo StructureCraft).

in Section 1.3.5 with respect to fire performance of adhesives in engineered wood products highlighted significant differences between countries (Wiesner et al., 2018). It is therefore strongly recommended to consult the appropriate standards accreditation bodies for assessing whether imported CLT is suitable and conforms to the applicable building codes and standards in the importing country. Effects of glueline fire performance can be found in Chapters 2, 3, 6 and 7.

CLT has excellent inherent fire resistance. The large cross-section allows for the elements to char slowly and at a predictable rate, allowing them to sustain the applied loads for a long duration. Structural fire resistance of CLT can therefore be determined based on the same design principles applicable to timber as those detailed in Chapter 7, where the thermal performance of adhesives is explained. Additional information on the separating function performance can be found in Chapter 6, while information about the detailing of CLT structures can be found in Chapter 9.

Mechanically laminated timber (MLT) is an engineered wood product made by connecting graded timber laminations on edge with mechanical connectors that are inserted through the wide face of the laminations. MLT panels are typically used as one-directional structural elements and can be manufactured with various profiles for aesthetic purposes or to improve acoustic performance. While technical guides about best practices are available (BSLC, 2017a and 2017b), there are currently no manufacturing standards for MLT, with the exception of a Canadian standard under development covering the manufacturing, testing and quality control of MLT (CSA O125), planned for publication in 2022.

The oldest form of MLT panel is most likely nail-laminated timber (NLT), which was used as "mill floor" in historic timber buildings. NLT is a solid wood structural element consisting of lumber planks oriented on edge and fastened together with nails. NLT is usually tightly manufactured with lumber of a moisture content no greater than 19 %, which can result in some gaps appearing between boards once the product is conditioned during its service life. Some manufacturers have stringent manufacturing requirements for a lower moisture content and might not exhibit similar dimensional changes.

Dowel-laminated timber (DLT) is a relatively new MLT product that has recently emerged in Canada (see Figure 1.17b). DLT is a solid wood structural panel created by placing lumber planks oriented on edge and friction-fastening the laminations together with hardwood dowels. It does not require any adhesives or metallic fasteners. DLT is usually tightly manufactured with lumber of a moisture content no greater than 19% at the time of inserting the wood dowels, which can result in some gaps appearing between lumber boards once the product is conditioned during its service life. Similar to NLT, the use of stringent manufacturing requirements with a lower moisture content might result in smaller dimensional changes.

NLT and DLT may have a slightly less reliable fire performance than glued wood panels. NLT and DLT which are manufactured with tightly clamped laminations char at a slow, predictable rate, and their structural fire resistance can therefore be determined based on the same design principles as those detailed in Chapter 7. However, due to potential dimensional changes and gaps forming between boards, additional precautions might be needed to fulfil the separating function. See Chapter 6 for additional information on the separating function performance. It is recommended to consult with the NLT and DLT manufacturers for guidance on gap tolerances and dimensional changes for fire design.

The final categories of mass timber panels are those made of plywood, LVL or OSB layers bonded with a structural adhesive and pressed to form a solid panel. The resulting product is similar to CLT, with the exception that they are usually parallel laminated (not orthogonal). Mass plywood panels can also be made of LVL layers so that they are all oriented in the longitudinal (strength) direction. These products are typically manufactured as built-up elements obtained with a secondary face-gluing process. Their fire performance and charring behaviour can be assumed to be similar to traditional sawn timber, provided that the adhesive used for secondary gluing is a structural adhesive meeting the requirements for elevated temperatures and fire conditions. Structural fire resistance of mass plywood and OSB panels can therefore be determined based on the same design principles as those detailed in Chapter 7. There are currently no manufacturing standards for glued mass timber panels made from plywood, LVL or OSB.

1.3.7 Wood-based panels

The last category of wood products refers to the thin wood-based panels typically used in light timber frame construction. These are panels made from veneers, strands and wood fibres, or a combination of these materials. Wood-based panels can be used as floor and roof sheathing, floor and roof diaphragms, wall sheathing and shear walls, as well as a manufacturing component such as the web panel in prefabricated wood I-joists. Figure 1.18 shows some wood-based panels commonly used in timber construction.

Plywood was the first glued wood-based panel ever used, with apparently a background in ancient Egypt. Plywood is manufactured using layers of veneers bonded orthogonally with a structural and moisture-resistant adhesive. It is usually made of an odd number of layers where the outer layers and all odd-numbered layers are oriented in the direction of the panel length, i.e. the strength direction (ASTM D5456). Its orthogonal configuration allows for minimising dimensional changes while maximising strength and stiffness. CLT has essentially been designed based on the principles of plywood but using much thicker layers of timber as opposed to thin veneers. Structural plywood panels are manufactured in accordance with

Figure 1.18 Wood-based panels: (a) plywood (photo APA wood), (b) OSB (photo APA Wood).

regional standards such as CSA O151, PS 1, EN 13986 and AS/NZS 2269.0. Several types of structural plywood can be found depending on the species group and grade of the veneers and its bond classification (interior, exterior, marine, etc.). Plywood can also be a decorative wood panel intended for interior use only. When used as an interior finish material, the surface veneer typically consists of hardwood and is bonded to an assembly of softwood veneers, timber, particleboard or medium-density fibreboard (MDF).

The second type of wood-based panel is the Oriented Strand Board (OSB). This product is comprised primarily of wood strands bonded with a moisture-resistant adhesive under heat and pressure (ASTM D1038). Following a similar manufacturing principle as plywood, OSB is fabricated of compressed strands arranged in orthogonal layers, where the strands in the face layers are generally aligned in the direction of the panel length, i.e. the strength direction. OSB panels typically have a non-skid surface on one side for safety on the construction site for roof applications. In addition to floor, roof and wall applications, OSB is also widely used as rim boards in light timber frame construction. When combined with engineered wood joists (I-joists or trusses), OSB rim boards are cut to the exact depth and exhibit a better dimensional stability than a traditional sawn timber rim board. Structural OSB panels are manufactured in accordance with regional standards, such as CSA O325, PS 2 and EN 13986.

The last category of wood-based panels is medium-density fibreboards (MDF), high-density fibreboards (HDF), and particleboards. MDF and

HDF are composite panel products composed primarily of wood fibres bonded with adhesives and cured under heat and pressure. At the time of manufacturing, MDF density is usually between 500 and 1000 kg/m³ (ASTM D1554). HDF has a higher density than MDF, with no specific targets. Particleboards are similar in manufacturing to MDF but use wood particles rather than fibres. MDF, HDF and particleboards are usually used as decorative panels. When used as structural panels, such as webs in I-joists, they need to be tested accordingly so that their mechanical properties are evaluated and determined correctly. The European product standard for wood-based panels is EN 13986.

Wood-based panels are usually manufactured thinner than panels of timber, SCL, glued laminated timber and mass timber and tend to exhibit faster charring rates than the other wood products detailed in this chapter. As an example, EN 1995-1-2 specifies a one-dimensional charring rate of 0.90 mm/min for a wood-based panel of 450 kg/m³ and at least 20 mm in thickness, while timber with a characteristic density of 290 kg/m³ or greater would have a rate of 0.65 mm/min. Chapter 7 provides the charring rate adjustment factor when a wood-based panel is less than 20 mm in thickness. Their performance against flame-through is also of utmost importance so that the separating function of a floor, roof or wall assembly is maintained adequately. In Europe a test method (EN 14135) is specified to determine the fire protection ability of coverings, with more information in Chapter 6. The flammability/reaction to fire characteristics is explained in Chapter 5.

1.4 CONCLUSION

This chapter introduces timber structures and wood products. Some building codes may limit the use of timber and wood products, either for structural elements or interior finish materials, but these materials are being used throughout the world in many types of buildings and occupancies. With the increasing demand for sustainable buildings and performance-based design, it is expected that timber will gain even more popularity in the near future. Fire performance of timber structures and wood products can be evaluated by the guidance and design methods detailed in the following chapters.

One of the main advantages of timber structures is the variety of systems that can be designed and constructed to suit almost any need and to provide the level of fire performance required in building codes. Traditional light timber frame construction remains the most economical system, widely used in low-rise and mid-rise buildings. Innovative systems such as modern post-and-beam construction, mass timber construction, long-span and hybrid structures allow for expanding the use of timber in impressive and innovative structures, such as taller buildings. Prefabrication of timber elements and modules is also gaining popularity, due to the speed of construction, increased building control and waste reduction at the job site.

Another factor facilitating the use of timber in buildings is the variety of products available to designers. A broad range of structural engineered wood products has been developed over recent years to provide high-valued timber products through more efficient use of the raw material. For most countries, timber and engineered wood products are required to be manufactured, tested and evaluated by applicable standards. Quality control procedures are usually required to ensure high-quality end products and buildings with acceptable fire safety.

REFERENCES

ANSI/APA PRG 320 *Standard for Performance-Rated Cross-Laminated Timber.* ANSI/APA APA – The Engineered Wood Association, Tacoma, WA.

AS 1720.5 *Timber Structures – Part 5 Nail-Plated Timber Roof Trusses.* Standards Australia, Sydney, NSW.

AS 2858 *Timber – Softwood – Visually Stress-Graded for Structural Purposes.* Standards Australia, Sydney, NSW.

AS/NZS 1328.1 *Glued Laminated Structural Timber – Performance Requirements and Minimum Production Requirements.* Standards Australia, Sydney, NSW.

AS/NZS 1748 *Timber – Mechanically Stress-Graded for Structural Purposes.* Standards Australia, Sydney, NSW.

AS/NZS 2269.0 *Plywood – Structural – Part 0: Specifications.* Standards Australia, Sydney, NSW.

AS/NZS 4357.0 *Structural Laminated Veneer Lumber – Part 0 Specifications.* Standards Australia, Sydney, NSW.

ASTM D9 *Standard Terminology Relating to Wood and Wood-Based Products.* ASTM International, West Conshohocken, PA.

ASTM D1038 *Standard Terminology Relating to Veneer, Plywood, and Wood Structural Panels.* ASTM International, West Conshohocken, PA.

ASTM D1554 *Standard Terminology Relating to Wood-Base Fiber and Particle Panel Materials.* ASTM International, West Conshohocken, PA.

ASTM D5055 *Standard Specification for Establishing and Monitoring Structural Capacities of Prefabricated Wood I-Joists.* ASTM International, West Conshohocken, PA.

ASTM D5456 *Standard Specification for Evaluation of Structural Composite Lumber Products.* ASTM International, West Conshohocken, PA.

ASTM E119 *Standard Test Methods for Fire Tests of Building Construction and Materials.* ASTM International, West Conshohocken, PA.

BSLC (2017a) *Nail-Laminated Timber – Canadian Design & Construction Guide v1.1.* Binational Softwood Lumber Council, Surrey.

BSLC (2017b) *Nail-Laminated Timber – US Design & Construction Guide v1.0.* Surrey Binational Softwood Lumber Council, Surrey.

CAN/ULC-S101 *Fire Endurance Tests of Building Construction and Materials.* Underwriters Laboratories of Canada, Toronto, ON.

Chamberland, V., Kinuani, N., Ngo, A. & Li, J. (2020) *Expanding Wood-Use towards 2025 Trends in Non-Residential and Multi-Family Construction (Project No. 301013618)*. FPInnovations, Pointe-Claire, QC.

CSA O86 *Engineering Design in Wood*. CSA Group (Product Certification & Standards Development), Mississauga, ON.

CSA O122 *Structural Glued-Laminated Timber*. CSA Group, Mississauga, ON.

CSA O125 *Mechanically Laminated Timber – Production and Qualification Specifications (Draft)*. CSA Group, Mississauga, ON.

CSA O151 *Canadian Softwood Plywood*. CSA Group, Mississauga, ON.

CSA O177 *Qualification Code for Manufacturers of Structural Glued-Laminated Timber*. Canadian Standards Association, Mississauga, ON.

CSA O325 *Construction Sheathing*. CSA Group, Mississauga, ON.

Dagenais, C. (2014) *Analysis of Full-Scale Fire-Resistance Tests of Structural Composite Lumber Beams (Project No. 301009338)*. FPInnovations, Pointe-Claire, QC.

EAD 130367-00-0304 *Composite Wood-Based Beams and Columns*. EOTA European Organization for Technical Assessment, Brussels.

EN 300 *Oriented Strand Boards (OSB) – Definitions, Classification and Specifications*. CEN European Committee for Standardization, Brussels.

EN 1363–1 *Fire Resistance Tests – Elements of Building Construction – Part 1: General Requirements*. CEN European Committee for Standardization, Brussels.

EN 13986 *Wood-Based Panels for Use in Construction – Characteristics, Evaluation of Conformity and Marking. European Standard*. CEN European Committee for Standardization, Brussels.

EN 14081 *Timber Structures – Strength Graded Structural Timber with Rectangular Cross Section – Part 1: General Requirements*. CEN European Committee for Standardization, Brussels.

EN 14135 *Coverings – Determination of Fire Protection Ability*. CEN European Committee for Standardization, Brussels.

EN 14250 *Timber Structures – Product Requirements for Prefabricated Structural Members Assembled with Punch Metal Plate Fasteners*. CEN European Committee for Standardization, Brussels.

EN 14374 *Timber Structures – Laminated Veneer Lumber – Requirements*. CEN European Committee for Standardization, Brussels.

EN 16351 *Timber Structures – Cross Laminated Timber – Requirements*. CEN European Committee for Standardization, Brussels.

EN 1995-1-2 (2004) *Eurocode 5 – Design of Timber Structures, Part 1–2: General – Structural Fire Design*. CEN, European Committee for Standardization, Brussels.

ICC (2021a) *International Building Code*. International Code Council, Washington, DC.

ICC (2021b) *Performance Code for Buildings and Facilities*. International Code Council, Washington, DC.

ISO 6707-1 *Buildings and Civil Engineering Works – Vocabulary – Part 1: General Terms*. International Organization for Standardization, Geneva.

Karacabeyli, E. & Douglas, B. (2013) *CLT Handbook – US Edition*. FPInnovations, Pointe-Claire, QC.

Karacabeyli, E. & Gagnon, S. (2019) *Canadian CLT Handbook – 2019 Edition.* FPInnovations, Pointe-Claire, QC.

LVL Handbook (2020) *Europe*, 2nd Edition. FFWI Federation of the Finnish Woodworking Industries, Helsinki.

NBCC (2015) *National Building Code of Canada.* National Research Council Canada, Ottawa, ON.

NLGA *Standard Grading Rules for Canadian Lumber.* National Lumber Grades Authority, Vancouver, BC.

O'Neill, J., Carradine, D., Moss, P. J., Fragiacomo, M., Dhakal, R. & Buchanan, A. H. (2001) Design of Timber-Concrete Composite Floors for Fire Resistance. *Journal of Structural Fire Engineering*, vol. 2 (3), pp. 231–242.

PS 1 *Structural Plywood, Voluntary Standard.* NIST National Institute of Standards and Technology, Gaithersburg, MD.

PS 2 *Performance Standard for Wood Structural Panels.* NIST National Institute of Standards and Technology, Gaithersburg, MD.

Swedish Wood (2019) *The CLT Handbook.* Stockholm, Sweden.

TPIC *Truss Design Procedures and Specifications for Light Metal Plate Connected Wood Trusses.* Truss Plate Institute of Canada.

White, R. H. (2000) *Charring Rate of Composite Timber Products.* Forest Products Laboratory, Madison, WI.

Wiesner, F., et al. (2018) Requirements for Engineered Wood Products and Their Influence on the Structural Fire Performance. *Proceedings WCTE 2018 World Conference on Timber Engineering*, Seoul.

Williamson, T. G. (2002) *APA Engineered Wood Handbook.* McGraw-Hill, New York.

Chapter 2

Fire safety in timber buildings

Andrew Buchanan, Andrew Dunn,
Alar Just, Michael Klippel, Cristian Maluk,
Birgit Östman and Colleen Wade

CONTENTS

DOI: 10.1201/9781003190318-2

SCOPE OF CHAPTER

This chapter provides an overall description of the strategy for delivering fire safety in timber buildings. As in the design of all buildings, the goals are to provide life safety for occupants, safe access for firefighters and protection of affected property. It is essential to control the severity and duration of any accidental fire and prevent it from spreading elsewhere in the building. An important design objective for timber buildings is to control the burning or charring of exposed timber or protected timber, because this can add to the fuel load, and it will reduce the load capacity of structural timber members due to loss of cross section. Many of the topics introduced here are expanded on in the following chapters.

2.1 FIRE SAFETY GOALS

The primary goal of building design for fire safety is to manage the consequences of an accidental fire by reducing the probability of death or injury for occupants and enabling appropriate firefighting intervention (Buchanan and Abu, 2017). Secondary fire safety goals may relate to business interruption, controlling property loss, protection of heritage values or environmental protection.

The balance between life safety, property protection and other goals may vary in different countries, depending on the type of building and its occupants, and the objectives of the local building code and other stakeholders.

A summary of fire codes and regulations in different regions of the world is given in Chapter 4. Chapter 11 describes relevant concepts of performance-based design and risk assessment for fire safety in buildings. Fire dynamics in enclosures and reaction to fire performance for materials and assemblies are covered in Chapters 3 and 5, respectively. Structural performance of timber is covered in Chapters 7 and 8, and prevention of fire spread within buildings is covered in Chapters 6 and 9.

2.1.1 Life safety

The main life safety objectives are to ensure safe escape paths for occupants and the safety of firefighters. For safe escape, it is necessary to alert occupants to the fire, provide suitable escape paths, and ensure that they are not adversely affected by fire or smoke while escaping through those paths to a place of safety. Safe conditions in escape paths can be enhanced by compartmentation (see Chapter 6) and by limiting the use of combustible wood surfaces in escape routes, as explained in Chapter 5.

Many important aspects of life safety in human design for fire safety are beyond the scope of this text. Readers should look elsewhere for guidance on topics such as detection and alarm systems, design of egress routes, smoke control and tenability. An excellent reference is the *SFPE Handbook of Fire Protection Engineering* (SFPE, 2016).

In some buildings, it is necessary to provide safety for occupants with reduced mobility or other disabilities – for example, in hospitals, age-care or child-care centres, or in refuge areas during a phased evacuation. This requires more stringent fire safety precautions to make sure that the spread of fire and smoke is adequately controlled.

In the fire design of all buildings, it is essential to consider the safety of firefighters who may need to enter the building to carry out rescue and/or firefighting activities.

2.1.2 Property protection

The main objective of property protection is to protect the building structure itself (load-bearing and non-load-bearing structure) and the contents inside the building from fire damage. This need may also extend to neighbouring buildings or other adjacent infrastructure. Additional measures of protection may be necessary to minimise disruption of the building's operation after a fire – for example, in a hospital building or a fire station.

Other objectives may be to prevent fire damage to heritage buildings. Wider impacts to the environment and to the community may also need to be considered. Most building codes do not provide guidance when it comes to post-fire reinstatement of buildings.

2.1.3 Insurance views

Insurance companies have reacted in different ways on modern timber buildings. Some companies have insured in the same way as for similar buildings with traditional materials, while others have been reluctant to insure larger timber buildings. There are very few guidelines being issued, probably due to competition between insurance companies. However, a recent UK summary of insurance views was published by RISC Authority (2022). Canadian wood industry views have also been published (McLain et al., 2021).

One issue is that insurance companies do not use the same categories as building codes, as they have their own classification systems with more interest in "property protection" than "life safety." In addition to fire safety, a major insurance concern is the risk of water damage from plumbing leaks or poor weathertightness, but little data is available. One example is the Canadian experience in the rehabilitation of mass timber following fire and sprinkler activation (Ranger, 2019).

The two main strategies for preventing severe fire damage to property are (1) ensuring the fire is contained to the compartment where it originates and (2) preventing loss of structural capacity resulting in the collapse of any part of the structure. These are achieved by providing adequate fire resistance to key parts of the building, as described below.

2.2 SPECIAL CONSIDERATIONS FOR TIMBER BUILDINGS

The following sections describe major differences between timber buildings and buildings of non-combustible materials, which need to be addressed in the design for fire safety.

2.2.1 Influence of exposed timber surfaces

The rate of growth, intensity, duration, and extent of a fire in a timber building is influenced by the amount of timber surface exposed inside the building. Non-structural timber surfaces may include flat or decorative linings on walls and ceilings. Exposed timber floors also need to be considered. Structural timber elements can have large or small surfaces exposed in mass timber buildings. Figures 2.1 and 2.2 show typical rooms in timber buildings with different amounts of exposed structural timber.

Figure 2.1 Buildings with exposed structural timber panels: a) wall and ceiling timber panels exposed; b) timber ceiling exposed (photos A. Buchanan).

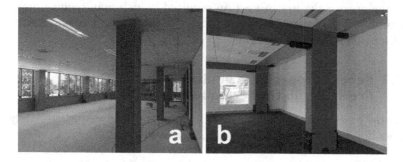

Figure 2.2 Buildings with exposed structural timber members: a) timber columns only exposed; b) timber beams and columns exposed (Photos A. Buchanan).

Building codes in different countries restrict the area of visible timber in different ways, with three main objectives:

1. During the incipient phase or early growth phase of a fire, it is important to control flame spread over timber surfaces. This could require the timber to be protected with limited-combustible or non-combustible lining materials (which do not need to be tested for fire resistance) or treated with a fire retardant or intumescent coating as described in Chapter 5
2. During the fully developed stage of the fire, it is important to prevent or reduce the charring of timber, which will result in additional fuel load and therefore increase the intensity and duration of the fire, as described in Chapter 3. This may require timber surfaces to be protected with encapsulation material, as discussed in Section 2.9 and in Chapter 6
3. Towards the end of the predicted design fire, it is essential to ensure that charring is not sufficient to cause the collapse of critical structural timber elements or connections, as discussed in Chapters 7 and 8

2.2.2 Exposed timber

There is often an aesthetic dilemma in the design of timber buildings. Many building owners, occupants and their architects want to see as much wood as possible, structural or non-structural, whereas the fire engineers may need to protect some or all of the timber using some form of encapsulation to meet the fire safety strategy. This need for encapsulation is to prevent or reduce burning or charring wood from becoming an excessive additional fuel load. Law and Hadden (2020) suggest it is essential that designers recognise the feedback loop between the timber structure and the fire dynamics, and either provide full encapsulation or demonstrate that burnout will occur. As buildings become taller, the associated risks, both in terms of the likelihood of fire and consequences of failure, also increase. This leads to the conclusion that the taller the building, the greater the fire protection required, which may require more of the timber to be covered with fire protective materials or other strategies, as discussed later in this chapter.

2.2.3 Recent reports and guidance on fire safety in timber buildings

Over the past decade, several international reports addressing fire safety in timber buildings have been published, including those listed below:

- Technical Guidelines for Europe (Östman et al., 2010)
- Fire Safety Challenges of Tall Wood Buildings (Gerard et al., 2013)
- Tall Wood Buildings in Canada (FPI, 2013a)
- Use of Timber in Tall Multi-Storey Buildings (Smith and Frangi, 2014)
- Fire Resistance of Timber Structures NIST White Paper (Buchanan et al., 2014)
- Fire Safety Challenges of Tall Wood Buildings (NFPA, 2018)
- Fire Safety Challenges of Green Buildings and Attributes (Meacham and McNamee, 2020)

There are several more specialised guidance documents for fire safety in timber buildings constructed from cross-laminated timber (CLT), including the following:

- US CLT Handbook (FPI, 2013b)
- Cross-laminated timber construction – an introduction (STA, 2015)
- Canadian CLT Handbook (FPI, 2019a)
- Swedish CLT Guide (SW, 2019)
- Engineered Wood Construction Guide (APA, 2019)
- Structural Timber Buildings Fire Safety in Use Guidance. Volume 6 – Mass Timber (STA, 2020)

All these reports confirm that well-designed timber buildings can be designed to an equivalent level of safety to that usually obtained for non-timber buildings. Careful design is needed to ensure that safety is achieved in all credible fire scenarios. All of the reports recommend automatic fire-sprinkler systems for tall timber buildings.

These reports generally recommend some of the structural timber be protected by full or partial encapsulation, but there is no consensus on the amount of encapsulation required. Most of the reports do not consider design to withstand burnout after the decay phase of an uncontrolled fire. Some of them refer to design for self-extinguishment, but they do not consider the practical difficulties of achieving this or its lack of definition, see Section 2.10.3.

2.3 FIRE DEVELOPMENT

The process of fire development in a typical fire can be illustrated by a time–temperature curve. This section provides a brief description of fire behaviour within a compartment, which is explained in more detail in Chapter 3.

2.3.1 Time–temperature curve

A time–temperature curve can illustrate the process of fire development in a typical fire. Figure 2.3 shows a typical time–temperature curve for the complete process of fire development inside a small compartment, e.g. a flat, assuming the fire is not suppressed or extinguished in any way. Not all fires follow this development because some fires go out naturally and others do not reach flashover, especially if the fuel item is small and isolated or if there is insufficient air to support continued combustion. Table 2.1 shows a summary of the main stages of fire behaviour relative to the active or passive design features that can be put in place.

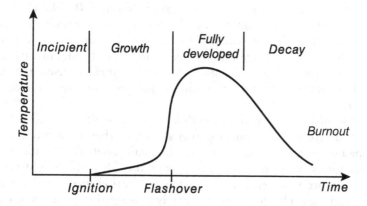

Figure 2.3 Indicative time–temperature curve for full process of fire development.

Table 2.1 Summary of stages of typical fire development in a small compartment

Stage Design feature	Incipient	Growth	Fully developed	Decay
Fire behaviour	Heating of fuel	Fuel controlled burning	Ventilation-controlled burning	Fuel controlled burning
Human behaviour	Prevent ignition	Extinguish by hand, alert others, escape	Untenable	
Fire detection	Smoke detectors	Smoke detectors, heat detectors, etc.	Smoke and flame visible externally	
Active protection	Prevent ignition	Extinguish by sprinklers or firefighters. Control smoke	Control by firefighters	
Passive protection	Control of materials	Select materials with resistance to flame spread	Provide fire resistance to contain the fire, prevent collapse and add robustness	

In the *incipient* stage of fire development, heating of potential fuel takes place with the area of heating remaining small, confined and typically undetected. *Ignition* is the start of flaming combustion, marking the transition to the *growth* stage. During the growth stage, a typical fire will spread at a rate that depends on the type of fuel and its distribution across the floor plan. The growth of the fire will essentially be driven by the ignition of unburned fuel, which is heated by radiation from the flaming combustion of burning fuel items.

Hot gases will rise by convection and spread across the ceiling, forming a hot upper layer that radiates heat to fuel items lower in the compartment. If the upper-layer temperatures exceed about 500–600°C, the fuel at the ground level will ignite rapidly, resulting in a rapidly spreading fire, leading to *flashover,* which is the transition from the *growth* stage to the *fully developed* stage (often referred to as "full room involvement" or a "post-flashover" fire). Combustible timber surfaces used as wall or ceiling linings can contribute to rapid fire growth and contribute to early flashover.

The rate of burning in the growth stage is generally controlled by the nature and layout of the burning fuel surfaces, whereas during the fully developed stage, the intensity of the fire is usually controlled by the ventilation conditions, as a "ventilation-controlled fire." It is the fully developed stage of the fire that generally impacts structural elements and compartment boundaries. The duration of the fully developed stage depends on the ventilation and the amount of fuel available, including any contribution of the burning timber structure to the fuel load, as described in Chapter 3. If the fire is left to burn, eventually the available fuel will be consumed and temperatures will drop in the *decay* stage, when the rate of burning again becomes a function of the fuel itself rather than of the available ventilation. The reduction of gas temperatures during the decay phase will depend on the amount of ventilation available. Structural failure can occur during the decay stage if charring continues to reduce the adequacy of structural members during this period.

The term *burnout* has been used to describe the end of an uncontrolled fire in a compartment when all the available fuel has been consumed and the compartment temperatures continue dropping to near ambient. For burnout to be a part of a successful fire safety design, the fire must be contained in the fire compartment with no structural collapse and with no spread of fire through the compartment boundaries (i.e., walls, ceiling and floor). However, in timber structures, charring wood may continue to smoulder slowly after all other fuel is consumed at the end of the decay stage, so final extinguishment will need intervention and application of water by firefighters.

The time to the end of the decay stage cannot be determined accurately because there are so many variables, but it can be estimated with several calculation methods some of which were developed for small compartments constructed with non-combustible materials. In a building with exposed structural timber, any burning or charring of wood must be added to the fuel-load calculations.

2.4 DESIGNING FOR FIRE SAFETY

The following sections briefly describe the overall means of designing for fire safety in buildings, with reference to Figure 2.3 and Table 2.1. For fire engineering design, buildings are often divided into "fire compartments" or "fire cells" of fire-resisting construction. A fire compartment may be a single room, a whole apartment or tenancy with several rooms, or the entire floor of a building.

2.4.1 Human behaviour

Occupants in the compartment where the fire starts may take action if they see or smell unusual signs of potential fire during the incipient stage when exposed fuel is being heated by some heat source. Many fires can be averted by occupants who prevent ignition by removing the fuel or eliminating the ignition source in the incipient stage. After ignition, the fire will grow, giving occupants in the vicinity of the fire the opportunity to extinguish it while it is small if they are awake and mobile. Once the fire grows and begins to involve one or more items of furniture, it becomes more difficult to extinguish by hand. Occupants who are mobile will have time to escape, if smoke or fire has not blocked the escape routes. Sprinklers activating in the room of fire origin will control the fire size or extinguish it completely. Even if sprinklers do not extinguish the fire, they will reduce the likelihood of the fire spreading rapidly, increasing the time available for occupants elsewhere in the building to escape safely.

Conditions in the compartment of fire origin become life-threatening during the growth stage. Survival after flashover is extremely unlikely because of the extreme conditions of heat, temperature and toxic gases. Hazardous conditions may occur for occupants elsewhere in the building if they are not alerted to the fire and instructed to evacuate. It is important to note that compartmentation (i.e., the fire being contained to the compartment of origin) is sometimes not possible due to design characteristics (e.g., atriums) or due to unexpected failure of a fire-rated compartment (such as a wall, door or services penetration).

To ensure life safety in a building, it is essential that the fire is detected and the occupants are alerted with sufficient time to reach a safe place before conditions become untenable.

2.4.2 Access and equipment for firefighters

As well as designing for the safe evacuation of occupants, it is critical that firefighters have safe access to enter the building and undertake search and rescue and firefighting activities. This is especially important for timber buildings where the load-bearing timber elements may continue to burn even after all the building contents have burned away.

Using appropriate equipment, firefighters are able to operate in environments that could be life-threatening to normal building occupants, but their safety remains paramount. Before entering a building, firefighters will ensure that there are safe paths for retreat, so the securing of safe entry and exit points from the building is of extreme importance. It is also important to provide nearby and safe street access for external firefighting and rescue by ladder trucks and other fire appliances which can reach up to about eight storeys. See Chapter 14 for a detailed description of firefighting considerations.

2.4.3 Fire detection

In the incipient stage of a fire, human detection may be possible by sight, smell or sound. A smoke-detector activating during this stage will alert occupants in the building that are not intimate with the fire. After ignition, a growing fire can be detected by the occupants (if present) or by a heat detector. For typical burning fuel in a building, smoke detectors are more sensitive than heat detectors, especially for smouldering fires where there may be life-threatening smoke but little heat produced. Automatic fire sprinkler systems are generally activated by heat. After flashover, neighbours may detect smoke and flames coming out of windows or other openings.

2.4.4 Active fire protection

Active protection refers to some fire control action taken by a person or an automatic device in the event of a fire. The most effective form of active fire protection is an automatic fire sprinkler system, which discharges water over a local area under one sprinkler head when it is activated by high temperatures in that locality. More than one sprinkler head will be activated if temperatures increase over a wider area of the ceiling. Well-designed sprinkler systems will prevent small fires from growing larger and will extinguish most fires completely. A sprinkler system must operate early in a fire to be useful because the water supply system is designed to tackle only a small or moderate fire, well before flashover occurs. Active fire protection is covered in more detail in Chapter 10.

Active control of smoke movement requires the operation of fans or other devices to remove smoke from certain areas or to pressurise stairwells. This may require sophisticated control systems to ensure that smoke and toxic products are removed from the building and not circulated to otherwise safe areas.

Occupants can prevent ignition if they become aware of hazardous situations or if they extinguish relatively small fires before they spread further and grow in size. Firefighters can control or extinguish a fire, but only if they arrive before it gets too large for the capacity of their equipment. Time is critical because it takes time for detection, time for notification of the firefighters, and then time to travel to the fire, to locate the fire in the building and set up water supplies. Firefighters usually have insufficient water to extinguish a large post-flashover fire, in which case they can only prevent the fire from spreading and wait to extinguish it during the decay stage.

2.4.5 Passive fire protection

Passive fire protection refers to the systems that are built into the structure or fabric of the building, not requiring external operation by people or automatic controls. For pre-flashover fires, passive control includes the selection of suitable materials for building construction and interior linings that do not support rapid flame spread or smoke production in the growth stage (see Chapter 5). In fully developed fires, passive fire protection is provided by load-bearing or non-load-bearing structures and assemblies which will perform appropriately in the event of a fire – preventing the spread of the fire beyond the room of origin (i.e., compartmentation; see Chapter 6) and preventing the partial or complete collapse of the structure (see Chapters 7 and 8).

2.5 CONTROLLING SPREAD OF FIRE

Several facets of fire protection are aimed at preventing small fires from growing in size or spreading into rooms outside the room of origin. The control of fire spread throughout the building is discussed below and split into four categories: (1) within the room of origin; (2) to other rooms on the same level; (3) to other storeys of the same building; and (4) to other buildings.

2.5.1 Fire spread within room of origin

The spread of fire within the room of origin depends largely on the heat release rate of the initially burning objects as well as the proximity and properties of any nearby combustible objects. Initial fire spread can result from flame impingement or radiant heat transfer from one burning item to another. As the fire grows, the movement of buoyant hot gases under the ceiling can cause the fire to spread to other parts of the room. The rate of internal fire spread will be increased if the room is lined with combustible materials susceptible to rapid flame spread on the walls and especially on the ceilings. Most countries have prescriptive codes that place limits on the combustibility or flame spread characteristics of linings in particular buildings or parts of buildings, especially in areas used for fire evacuation by the building occupants (see Chapter 4).

Unprotected wood-based materials are traditionally safer than most common plastic or synthetic materials used in furniture inside buildings, because they have a higher critical heat flux for ignition and a lower rate of flame spread. The early fire hazard properties of timber structures can be improved using fire retardant paints or chemical treatment, but these are not usually considered to improve the fire resistance of timber structures during fully developed fires. These topics are explored in more detail in Chapter 5.

2.5.2 Fire spread to adjacent rooms on the same level

The spread of fire and smoke to adjacent rooms has historically been a major factor resulting in deaths in building fires. The movement of fire and smoke depends very much on the layout and construction of the building. Open doors can provide a path for smoke and toxic combustion products to travel from the room of fire origin into the adjacent rooms or corridors. These hot gases can pre-heat the next area leading to the subsequent rapid spread of fire. People often die from smoke in an area remote from the room of fire origin.

Consequently, most national building codes restrict the area in which a fire can develop so that it can be contained in one fire compartment (or fire cell). In a residential building this fire compartment is often the whole apartment or residential unit consisting of several rooms. To contain the fire to one fire compartment, it must be surrounded by fire-resisting construction (also known as fire barriers or fire separations). Openings through fire barriers must have fire-rated closures such as fire doors to maintain the containment function of the barrier, both for smoke control and fire resistance. Self-closing doors must have reliable operating mechanisms.

Concealed spaces at the interface between compartments are one of the most common paths for spread of fire and smoke. A hazardous situation may occur if there are concealed spaces that allow the spread of fire and smoke to adjacent fire compartments or to other rooms some distance from the fire. This is covered in Chapter 9.

Fire can spread to adjacent rooms by penetrating the surrounding walls. Walls can be designed with sufficient fire resistance to prevent the spread of fully developed fires, but they must be constructed and maintained with attention to details if fire performance is to be ensured. Fire-resisting walls must extend to meet the horizontal fire separation or roof above. Walls at roof level should be extended above the roof line to form a parapet, or the roof can be fire-rated for some distance on either side of the top of the wall to inhibit the fire spread to the adjacent compartment.

2.5.3 Fire spread to other storeys

The vertical spread of fire from storey to storey is a hazard in all multi-storey buildings, with the potential consequences becoming more severe as the height of the building increases. Fire can spread to other storeys by a variety of paths, inside and outside the building. Internal routes for fire spread include failure of the floor/ceiling assembly, and fire spread through service penetrations, vertical ducts, shafts or stairways. Vertical services must either be enclosed in a protected duct, have fire-resistant closers or other approved fire-stopping measures at each floor level. The potential fire spread through internal void spaces and connections can be a particular problem for new types of modular construction (see Chapter 9).

Vertical fire spread can also occur outside the building envelope, via combustible materials within or on the exterior walls or via windows and cavities. Continuous combustible cladding susceptible to rapid flame spread should not be used on the exterior of any tall buildings unless further fire safety measures are applied. The fire performance of wood-cladding materials and facades is covered in Chapter 5.

The vertical spread of fire from window to window is a major hazard. Some building codes mitigate this risk by using sprinklers to reduce the likelihood of post-flashover fires. Vertical fire spread can be partly controlled by keeping windows small, well separated, and by using horizontal projecting "aprons," which project out horizontally above window openings helping to deflect the flame away from the wall. Flames from small narrow windows tend to project further away from the wall of the building than flames from long wide windows, leading to a lower probability of storey-to-storey fire spread (Drysdale, 2011).

In buildings with exposed interior timber surfaces, the performance of the façade must be critically assessed because exterior flaming will typically be greater with higher radiation levels in comparison to buildings with only non-combustible interior surfaces.

Many building codes prescribe specific test standards that must be used to demonstrate the acceptability of façade systems for taller buildings, especially when the façade system includes combustible components. Typical full-scale façade testing methods are based on specific and consistent radiation exposure to the façade system, with different levels applied in different countries. Some of these testing standards may not adequately address the greater heat fluxes which have been measured for fires in timber-lined compartments. See Chapter 9 for more information on preventing the vertical spread of fire.

2.5.4 Fire spread to other buildings

Fire can spread from a burning building to adjacent buildings by flame contact, radiation from openings such as windows, or flaming brands. Large areas of exposed timber on internal walls and ceilings are known to increase the severity of external flaming outside the openings. Fire spread can be prevented by providing a fire-resisting barrier or by providing sufficient separation distances between the buildings. If there are openings in the external wall, the probability of fire spread depends greatly on the distances between the buildings and the size of the openings. Exterior fire-resisting walls must have sufficient structural fire resistance to remain in place for the duration of the fire. This becomes a problem if the structure that normally provides lateral support to the walls is damaged or destroyed in the fire. Outwards collapse of exterior walls can be a major hazard for firefighters and bystanders and can lead to further spread of fire to adjacent buildings.

2.6 FIRE SAFETY DESIGN METHODS

2.6.1 Prescriptive codes and performance-based codes

In the past, design for fire safety in most countries was based on *prescriptive* building codes, with little or no opportunity for designers to take a rational knowledge-based engineering approach to design. Many countries have optional *performance-based* building codes that allow designers to use specific fire engineering to demonstrate that the performance requirements of the building code can be achieved. In general terms, a prescriptive code states how a building is to be constructed whereas a performance-based code states how a building is to perform under a wide range of conditions. Performance-based codes are described in more detail in Chapter 11 and prescriptive design in Chapter 4.

Some prescriptive building codes give the opportunity for performance-based selection of structural assemblies. For example, if a code specifies a floor with a fire resistance rating of two hours, the designer has the freedom to select from a wide range of approved floor systems that have sufficient fire resistance. This guide provides tools for assessing the fire performance of structural timber elements which have been tested and gives calculation methods for elements with different sizes, loads or fire exposure from those which have been tested.

In the development of new codes, many countries have adopted a multi-level code format such as that shown in Figure 2.4. At the higher levels, there is legislation specifying the overall goals, the objectives and the required performance. At the lower level, there are three alternative options for achieving those goals and objectives.

The three most common options are shown as:

1. A prescriptive design (often called an Acceptable Solution, a Deemed-to-Satisfy Solution or an Approved Document)
2. An approved standard calculation method to verify a design (sometimes called a Verification Method)
3. A performance-based design (sometimes called an Alternative Design or Performance Solution) which is a more comprehensive fire engineering design from first principles

Standard calculation methods are still being developed for widespread use, so compliance with performance-based codes in most countries is usually achieved by simply meeting the requirements of prescriptive design rules, with options 2 and 3 being used for special cases or very important

Figure 2.4 Typical hierarchical relationship for fire safety design.

buildings. Performance solutions can sometimes be used to justify variations from the prescriptive design in order to provide improved safety, to
achieve cost savings or to meet other design objectives.

Where a performance-based approach is used, most fire safety designs
use a mixture of a prescriptive design and a performance-based design. For
example, consider the situation where the design of the fire-rated barriers
follows prescriptive design rules, but the distance of travel for the escape of
occupants is increased using performance-based design.

The building code environment is similar in the UK, Australia, New
Zealand and some Scandinavian countries where performance-based
design is permitted. Even then, most designs are based on prescriptive rules.
Moves towards performance-based codes are being taken in the United
States, Canada and Switzerland. Codes are different around the world, but
the objectives are similar: to protect life and property, and to provide safety
for firefighters, as described in Chapter 4.

Performance-based fire codes are not simple to produce, or to use, because
fire safety is only part of a complex system of many interacting variables.
There are so many possible strategies that it is often not simple to assess
performance in quantitative terms, and there is a lack of information on the
behaviour of fires and the performance of people and buildings exposed to
fires. This is especially the case for timber buildings where many current performance-based design methods are not sufficiently validated for the building
geometries of most practical interest. See Chapter 11 for more information.

2.6.2 Trade-offs/alternative fire design

A major difficulty in design for fire safety is "trading off" some fire protection measures against others. For example, some prescriptive codes
allow fire resistance ratings to be reduced, or fire compartment areas to
be increased, if an automatic sprinkler system is installed. Travel distances may be increased when smoke or heat detectors or sprinklers are
installed. Trade-offs do not apply in a totally performance-based environment, because the designer will produce a total package of fire protection
features contributing to a required level of safety or a target failure probability. However, in practice, most designs are based on prescriptive codes,
so trade-offs are often useful. In some countries this process is called an
"alternative fire design."

The use of trade-offs for reducing fuel load, as a benefit of installing
sprinklers, is described in more detail in Chapter 10.

2.7 FIRE SEVERITY

Fire severity is a measure of the destructive potential of a post-flashover fire
applied to load-bearing structures and other construction elements used for

compartmentation during fires. Fire severity is usually assessed in a period of exposure to the standard test fire, but this may not be appropriate for real fires which have quite different characteristics, often similar to the indicative time–temperature curve shown in Figure 2.3. The fire severity used for a particular design will depend on the requirements of the local building code or sometimes on the design fire scenario selected by the fire engineer.

2.7.1 Code environment

In a *prescriptive* code environment, the design fire severity is usually prescribed as 30, 60, 90, 120 or 180 minutes of standard fire exposure, with little room for discussion.

In a performance-based code environment, the design fire severity will need to be assessed considering the size, use, configuration and construction of the fire compartment. The designer may consider a range of different fire scenarios. The design fire may be a parametric fire that predicts the full process of a realistic fire until burnout or an equivalent time of standard fire exposure. The most important measure of fire severity is the duration of the fully developed stage and the decay stage, but the fire temperatures are also important, as discussed by Buchanan and Abu (2017).

2.7.2 Fire design time

The term *fire design time* is the time of fire exposure for which the building is designed. The definition of *fire design time* depends on the type of design being undertaken:

For a prescriptive design, the *fire design time* is the duration of the fire resistance rating specified in the applicable building code, expressed as a specified time of exposure to the standard fire; often 30, 60, 90 or 120 minutes.

For a performance-based design, depending on the use of the building, the requirements of the owner, and the consequences of a structural collapse or spread of fire, the *fire design time* will be selected by the designer as one or more of the following predicted times:

1. The time required for occupants to escape from the building
2. The time for firefighters to carry out rescue activities
3. The time for firefighters to surround and contain the fire
4. The time at which the fire severity exceeds the fire resistance, after which the fire may spread and/or the structure will collapse
5. The duration of burnout in the fire compartment

Building codes in various countries use these times in different ways for different occupancies. Many small single-storey buildings may be designed to protect the escape routes and to remain standing only long enough for the occupants to escape (Time 1), after which the fire could destroy the building. Alternatively, very tall buildings, or buildings where people cannot easily escape, should be designed to prevent the major spread of fire and structural collapse after burnout in one or more fire compartments (Time 5). Times 2, 3 and 4 are intermediate times that may be applied to medium-sized buildings to provide appropriate levels of life safety or property protection.

2.7.3 Calculation methods

Chapter 3 describes calculation methods for quantifying the fire severity, including the standard fire, parametric fires and natural fires continuing to burnout. Calculations are based on the available fuel load and the ventilation provided when the windows break. The available fuel load is the estimated fuel from movable items plus the additional fuel from the burning or charring of timber or other combustible surfaces in the fire compartment. Fire severity calculations are very sensitive to the fuel load and the number of windows that break at flashover, both of which add uncertainty to the calculations. Designers should allow for this uncertainty in their calculations by considering applicable statistical distributions of fire load and ventilation when selecting appropriate design values. Chapter 3 also gives current state-of-the-art methods for assessing the contribution of timber structural materials to the fire load.

2.8 FIRE RESISTANCE

Fire resistance is the main tool used to provide fire safety to occupants, firefighters and property in a fully developed fire after flashover occurs. Providing appropriate building elements with sufficient fire resistance is essential to meeting the objectives of containing a post-flashover fire and preventing structural collapse. *Fire resistance* is determined by exposure to the standard fire test or by an equivalent calculation.

2.8.1 Objectives of fire resistance

The objectives for providing fire resistance need to be established before making any design, recognising that fire resistance is only one component of the overall fire safety strategy. Construction elements can be provided with fire resistance for controlling the spread of fire or preventing structural collapse, or both, depending on their function:

- To prevent internal spread of fire, most buildings are divided into "fire compartments" or "fire cells" with barriers which prevent fire spread for the *fire design time*

- To reduce the probability of fire spread to other buildings, boundary walls must have sufficient fire resistance to remain standing and to contain a fire for the *fire design time*
- To prevent structural collapse, structural elements must be provided with sufficient fire resistance to carry the applied loads for the *fire design time*. In all tall buildings, prevention of collapse is essential for all load-bearing structural members and for load-bearing barriers which provide containment
- Prevention of collapse is also essential if there are people or property to be protected elsewhere in the building, remote from the fire
- Specific consideration may need to be given for repair and reinstatement, rather than demolition, after a possible large or small fire

2.8.2 Components of fire resistance

The three components of fire resistance are the three failure criteria used in fire resistance testing:

- Structural adequacy
- Integrity
- Insulation

Structural adequacy

To meet the structural adequacy criterion, a structural element and its connections must perform their load-bearing function and carry the applied loads for the duration of the test without structural collapse. Calculation of structural adequacy of timber elements is described in Chapters 7 and 8.

Integrity

The integrity and insulation criteria are intended to test the ability of a barrier to contain a fire, to prevent fire spreading from the room of origin. To meet the integrity criterion, the test specimen must not develop any cracks or fissures which allow smoke or hot gases to pass through the assembly. Depending on the applicable fire test method or national standard, different criteria for integrity may apply such as no passage of flame, no development of gaps exceeding the specified size or no passage of hot gases sufficient to ignite a cotton pad.

Insulation

To meet the insulation criterion, the temperature of the cold side of the test specimen must not exceed a specified limit, usually an average increase of 140°C and a maximum increase of 180°C at a single point. These

Table 2.2 Typical fire resistance criteria for construction elements

	Structural adequacy (R)	Integrity (E)	Insulation (I)
Partition		X	X
Door		X	X
Load-bearing wall	X	X	X
Floor-ceiling assembly	X	X	X
Beam	X		
Column	X		
Fire-resistant glazing		X	

temperatures represent a conservative indication of the conditions under which fire might be initiated on the cool side of the barrier.

In fire resistance tests, all fire-rated construction elements must meet one or more of the three criteria, as shown in Table 2.2, depending on their function. Most fire-resistant glazing needs only to meet the integrity criterion because it is not load-bearing, and it cannot meet the insulation criterion. However, some special types of insulated glass can resist radiant heat transfer.

Most international fire codes specify the required fire resistance separately for structural adequacy (R)/integrity (E)/insulation (I), in that order. For example, a typical load-bearing wall may have a specified fire resistance rating of 60/60/60 (REI 60), which means that a one-hour rating is required for structural adequacy, integrity and insulation. If the same wall was non-load bearing, the specified fire resistance rating would be –/60/60 (EI 60). A fire door with a glazed panel may have a specified rating of –/30/–(E 30), which means that this assembly has an integrity rating of 30 minutes, with no fire resistance for structural adequacy or insulation.

2.8.3 Structural fire resistance

The provision of structural fire resistance, or structural adequacy in fire, may be essential, or unimportant, or somewhere between these two extremes. On the one hand, there may be a major role for the structure so that collapse is unacceptable even in the largest foreseeable fire. This may occur where evacuation is likely to be slow or impossible, where great value is placed on the building or its contents, or where the collapse of the building would represent an unacceptable safety risk to neighbouring buildings or communities. See Section 2.11 on tall timber buildings. On the other hand, there may be virtually no role for the structure so that structural collapse is acceptable after some time of fire exposure, where a small building can be readily evacuated, or there is little value placed on the building and there is no fire threat to adjoining properties.

Design for structural fire resistance is generally a matter of establishing that the fire resistance is greater than the fire severity, or more precisely, ensuring that the structural capacity exceeds the expected loads for a certain time of fire exposure, usually the *fire design time*, with a suitable safety factor. For buildings of traditional non-combustible materials, fire severity and fire resistance are uncoupled, so each can be assessed independently.

For timber buildings, assessment of structural fire resistance is more difficult because of the coupling between the fire load and structural fire resistance. Charring of structural timber in a severe fire not only adds to the fuel load but also reduces the structural performance of the residual cross section. A more detailed discussion of fire development in timber buildings is given in Chapter 3. Calculation of structural fire resistance is described in Chapter 7 and structural connections in Chapter 8.

2.9 TIMBER PROTECTION

Timber protection is a critical part of designing a large or complex timber building for fire safety, as described in the following sections. Timber protection refers to fire-resistive materials covering the timber structure to delay or reduce the charring of the underlying timber. There are several possible levels of timber protection, with definitions discussed by Schmid et al. (2021). A useful set of Canadian guidelines on encapsulation are given by FPI (2019b).

The required time for timber protection depends on the design strategy and the local building code requirements for the particular building. To be effective, the protective material must be designed to stay in place without significant deterioration for the *fire design time*.

2.9.1 Encapsulation

Encapsulation of a timber element describes protection with enough layers of protective material to prevent any ignition or charring during the *fire design time* or until burnout of the fire compartment. Encapsulation will ensure that the structural performance of the timber element is not compromised in the design fire, and there will be no significant addition to the available fuel load.

Encapsulation of all the timber surfaces in a fire compartment will enable fire severity and fire resistance to be calculated in a similar fashion as for any non-combustible material. As an example of full encapsulation, the eighteen-storey Brock Commons Building in Canada (see Figure 2.5) has all structural timber surfaces protected with three layers of 16 mm Type X gypsum plasterboard.

Protective materials providing encapsulation need to have their performance proven through fire testing, given that local failure can occur at the material joints and interfaces. Standard test methods for encapsulation have been developed in several countries. Tests for encapsulation are available

Figure 2.5. Typical examples of recent tall timber buildings: a) Light timber frame apartment building, Canada; b) Ascent Building, Milwaukee; c) Brock Commons, Canada; d) Mjøstårnet Building, Norway.

in Canada (CAN/ULC-S146) and in Europe (EN 14135). Encapsulation is covered in more detail in Chapter 6.

2.9.2 Partial encapsulation

Partial encapsulation of a timber element is protection not sufficient to provide full encapsulation. Partial encapsulation will prevent rapid flame spread on wood surfaces, but it may not prevent charring or ignition of the underlying timber later in the fire. Wood surfaces with partial encapsulation are expected to start charring when the wood temperature under the protective layer reaches about 300°C and to char at an increased rate after the protective material falls off. The design time for partial encapsulation will depend on the fire design strategy used for the building. Calculation methods for loss of cross section due to charring are given in Chapter 7.

With partial encapsulation, any charring under the protective layers, or after the protective layers fall off, will add to the available fuel load, increasing the severity of the expected fire. In the fire design of a compartment to withstand burnout, it is essential to calculate the impact of charring under the protective layers and accelerated charring after the protective layers fall off, because both add to the available fuel load. Design methods to include exposed timber in the fire load calculations are given in Chapter 3.

Some rooms of buildings will have encapsulation on only some of the timber surfaces, with the remaining surfaces exposed to view or partially encapsulated. The fire engineer must ensure that the fire load calculations for a given fire compartment are consistent with the surface areas of wood completely exposed, areas fully encapsulated and those areas which are only partially encapsulated. The wood surface includes linings as well as structural elements.

2.9.3 Time to start charring and encapsulation falloff times

Information on the time to start charring and the expected time to fall off of gypsum plasterboards under exposure to the standard test fire is given in Eurocode 5 (EN 1995-1-2, 2004) and by LaMalva and Hopkin (2021). Similar information is available from many manufacturers of gypsum plasterboard. This information can be included in iterative calculations of fire severity, considering progressive charring of partially encapsulated timber, with the rates of charring described by Eurocode 5.

For real fire exposure, rather than standard fire exposure, it is necessary to know the time–temperature or heat flux curve for the duration of the fire. The time to onset of char can then be determined by heat transfer calculations of the time for the temperature of the wood surface under the protective layer to reach 300°C. The time to gypsum plasterboard falloff can also be calculated, provided that the critical falloff temperature for the board is known (see Chapter 7).

2.10 DESIGN FOR THE FULL DURATION OF THE FIRE

2.10.1 Burnout

In this guide, the term *burnout* is used to describe the end of an uncontrolled fire in a compartment after all the available fuel has been consumed, and the room temperatures drop to allow firefighters safe access to carry out fire suppression activities. The expression "design to withstand burnout" can be misused unless it is clearly defined. Alternative expressions such as "design to withstand consumption of the available fuel" or "design for the full duration of the fire" can also be used, but they also need clear definitions.

Some modern codes require that certain buildings be designed to withstand burnout, especially tall buildings or other buildings where occupants or firefighters may be at risk. Such a design means that the fire scenarios considered in the fire engineering design must consider a fire duration that depends on the availability of fuel and ventilation in the compartment. Design to withstand burnout is a key element of a performance-based design strategy that should be performed by a competent fire safety engineer to ensure that a building can survive a post-flashover fire including the decay stage, with no contribution from automatic sprinklers and no firefighting intervention until late in the decay stage.

Designers of non-combustible steel and concrete buildings often assume a burnout scenario, whereby compartmentation prevents any spread of fire, and the structure is designed for adequate fire resistance. In such cases, a fully developed fire will be confined to the initial fire compartment, the fire-affected structure will continue to carry all expected loads, and the fire will go out after all the fuel is consumed. The structure will then cool to ambient temperatures over a few hours or days as the remaining heat is dissipated.

A burnout scenario is less certain for timber buildings because there will always be some fuel present in the timber structure, leading to the possibility of timber continuing to smoulder or char locally, long after flaming combustion has ceased. Designers and regulators should assume that any localised smouldering and glowing combustion towards the end of the burning stage will need to be extinguished manually by firefighters. This may require the removal of large areas of protective layers of gypsum plasterboard, see also Chapter 14.

2.10.2 Design to withstand burnout

For buildings with exposed wood on internal surfaces, design to withstand burnout requires calculations to demonstrate that temperatures will drop to low levels during the decay stage of the fire. There is considerable on-going research investigating the conditions under which this will happen.

Design to withstand burnout can be demonstrated by using the results of compartment fire tests, including the decay stage, or by calculations, to show that the radiant heat exposure or the calculated mass loss rate is consistent with low compartment temperatures at the end of the decay stage. The overall heat losses from the compartment must be greater than the energy generated from the burning of any remaining fuel towards the end of the fire, as described in Chapter 3. The contribution of charring of any exposed timber structure must be added to the design fire load.

Law and Hadden (2020) point out that a rapid reduction in fire temperatures in the decay stage is most effectively achieved when the energy losses from a compartment are maximised, i.e., with large ventilation openings. However, it should be recognised that large windows to maximise these energy losses will also mean that more energy is available to promote vertical fire spread via external flaming during the fully developed stage of the fire.

Recent research suggests that tenable conditions for fire service access are more likely if the areas of exposed timber surfaces are limited. A current proposal is that a wood ceiling can be fully exposed, together with wall areas no more than 40% of the floor area, provided that no two wood walls butt up to each other in corners of rooms (with a distance between two adjacent walls more than 4 m) to prevent radiant exposure across the corner (Brandon and Smart, 2021). This has to be verified by further research.

2.10.3 Self-extinguishment

The term "self-extinguishment" (or auto-extinction) of wood is often used in the literature on fire safety in timber buildings (Schmid et al., 2021). This misleading term is not well-defined, so its use is strongly discouraged in this Design Guide. Self-extinguishment is a deprecated term in the international standard ISO 13943 Fire safety – Vocabulary.

For example, referring to a compartment fire that is allowed to burn itself out, the term self-extinguishment could be used to describe the end of flaming combustion, the end of the decay stage, or the end of smouldering combustion. There is a big difference between these definitions, as flaming combustion may stop early in the decay stage, whereas full extinguishment of smouldering combustion may need the application of water by firefighters very late in the fire.

2.10.4 Structural design to withstand burnout

Structural design of timber members to withstand burnout requires that the total depth of charring is calculated for the full process of fire growth, burning and decay. Iterative fire severity calculations described in Chapter 3 give an estimate of the final depth of charred wood at the end of the fire. Structural designers must subtract a zero-strength layer from the residual cross section before calculating the residual strength. The thickness of the zero-strength layer has been traditionally taken as 7 mm, but this may need to be increased to account for the thermal wave which continues to travel into the timber after the fire has effectively gone out (Wiesner et al., 2019; prEN 1995-1-2, 2021). As with steel and concrete structures, timber structures are vulnerable during the decay stage of a fire because the structural capacity of heated members can continue to decrease during this time.

Any structural steelwork inside a fire compartment with exposed timber structural members (e.g. a steel skeleton system supporting CLT floor panels) must be designed for the same time of fire exposure as the timber structure. Protection of structural steelwork to a critical temperature of

600°C or 700°C is not acceptable if the steel is in contact with wood which will begin to char at 300 °C.

Whatever the fire design strategy, it is important to provide full details of the fire scenarios, the fire design methods and underlying assumptions. Structural design for standard fire exposure is covered in Chapter 7.

2.10.5 Glueline failure

For all engineered wood products (EWPs), which consist of small pieces of wood glued into larger components, the fire performance of the adhesives in gluelines, may be essential for the fire safety in the building. An increasing range of adhesives is becoming available for a wide range of EWPs, as described in Chapter 1.

The fire performance of the adhesive is more critical for cross-laminated timber (CLT) than for other EWPs, because the gluelines in CLT are parallel to the fire-exposed surface. A number of large-scale compartment fire tests with CLT floor-ceiling assemblies have shown that glueline failure of some thermo-plastic adhesives can result in fire-exposed boards falling off, adding additional fuel to a decaying fire, causing a second flashover and eliminating the possibility of the fire decaying (Brandon & Dagenais, 2018). A fire-resistant adhesive can be used to maintain glueline integrity and prevent char layer fall off. Designers of tall or complex timber buildings may insist on the use of fire-resistant adhesives. The North American manufacturing standard for CLT (ANSI, 2018) requires the use of such fire-resistant adhesives to prevent the falloff of charred or partially charred lamellae during fires. The current draft of Eurocode 5 (prEN 1995-1-2, 2021) includes a method to assess the glueline integrity of engineered wood products.

2.11 SPECIAL PROVISIONS FOR TALL TIMBER BUILDINGS

Regardless of the building materials used, fire safety becomes much more important in tall or very tall buildings (Buchanan, 2015). Table 2.3 gives approximate definitions of building height, with numbers of storeys and an indication of typical levels of fire resistance prescribed in many countries.

Table 2.3 Approximate definitions of building height and typical fire resistance

	Height range	Number of storeys	Typical fire resistance
Low-rise	H < 12 m	Less than 4 storeys	30 to 60 minutes
Medium rise	12 m < H ≤ 25 m	4 to 8 storeys	60 to 120 minutes
High-rise	25 m < H ≤ 60 m	9 to 20 storeys	90 to 180 minutes
Very high-rise	60 m < H	More than 20 storeys	120 minutes or more

Building height is usually measured to the top floor level. See Chapter 4 for more detailed requirements in different countries.

Figure 2.5 shows a few recent tall timber buildings. The taller the building, the greater need for special provisions for fire safety because of the large number of possible occupants and the significant time required for occupant escape and firefighter access.

There is no simple way of assigning different levels of fire protection to timber buildings based solely on the height of the building. The use of the building and the mobility of the occupants are also important. Fire precautions may be more stringent for open-plan office buildings with no fire separations on each floor. International building codes are only beginning to address these issues.

High-rise and very high-rise buildings may have full encapsulation so that no timber is exposed. If any significant area of timber is exposed, the design should ensure that the fire compartment can withstand burnout, even in the unlikely event of sprinkler failure and unavailability of firefighters. For medium-rise timber buildings, codes may allow more timber to be exposed, with no requirement to withstand burnout.

Some countries allow relaxation in fire precautions if the sprinkler system has a secondary water supply to ensure that the sprinklers have water, even if the street mains are rendered inoperative. See Chapter 10 for more on sprinklers.

For every tall timber building, the fire designer must identify the structural fire safety objectives, the routes to code compliance, and the design solutions to be used for each building. This requires consulting with all the relevant stakeholders, including the local fire services.

As tall timber buildings become more popular around the world, it will be necessary for code writers in different countries to adopt requirements that reflect these ideas in a rational way. Any changes to the regulatory environment should be based on clear design objectives, following recent research on structural fire design to withstand burnout. More research will help to further define the options, including quantitative risk assessment. Guidance on performance-based design is given in Chapter 11.

2.12 FIRE SAFETY DURING CONSTRUCTION

Fire safety during construction is a hazard for all timber buildings. Light timber frame buildings under construction are especially vulnerable before protective linings, and other fire safety design features have been installed. Severe fires during construction have caused large financial losses in several countries. The construction fire hazard may be less severe in mass timber structures than in light timber frame buildings, but comprehensive fire precautions are essential. Management to control fires during construction is covered in Chapter 13.

2.13 RESEARCH NEEDS

Future research in the following areas will help design engineers and code writers make good decisions about some of the unresolved issues raised in this chapter:

1. Fire severity in compartments with exposed timber surfaces, including travelling fires
2. Charring rate of timber as a function of fire exposure, and its contribution to the fire load
3. Conditions needed to ensure access by firefighters after burnout
4. Extinguishment of charring or smouldering timber
5. Fire performance of encapsulated or partially encapsulated timber
6. Dangers of wood used in façade systems
7. Effect of different combinations of passive and active fire protection
8. Quantitative risk assessment of fire safety in tall timber buildings
9. Risk assessment for property protection to meet the needs of the insurance industry.

REFERENCES

ANSI (2018) *ANSI/APA PRG 320 Standard for Performance-Rated Cross-Laminated Timber (American National Standard).* APA – The Engineered Wood Association, Tacoma, WA.

APA (2019) *Engineered Wood Construction Guide.* APA – The Engineered Wood Association, Tacoma, WA.

Brandon, D. and Dagenais, C. (2018) *Fire Safety Challenges of Tall Wood Buildings – Phase 2: Task 5 – Experimental Study of Delamination of Cross Laminated Timber (CLT) in Fire.* Report FRPF-2018-05, Fire Protection Research Foundation, Quincy, MA.

Brandon, D. and Smart, J. (2021) Expanding Mass Timber Opportunities through Testing and Code Development. American Wood Council. https://cdn.ymaws .com/members.awc.org/resource/collection/7C07FAE7-6A6B-4CAE-8687 -5D905C066CA0/DES612COLOR.pdf.

Buchanan, A.H. (2015) Fire Resistance of Multi-Storey Timber Buildings. Keynote presentation. *The 10th Asia-Oceania Symposium on Fire Science and Technology*, Tsukuba, October 2015.

Buchanan, A.H. and Abu, A.K. (2017) *Structural Design for Fire Safety.* 2nd Edition. John Wiley and Sons, Chichester, UK.

Buchanan, A.H., Östman, B. and Frangi, A. (2014) *Fire Resistance of Timber Structures.* NIST White Paper. National Institute of Standards and Technology, Washington, DC.

CAN/ULC-S146 *Standard Method of Test for the Evaluation of Encapsulation Materials and Assemblies of Materials for the Protection of Structural Timber Elements.* Underwriters Laboratories of Canada, Toronto, ONT.

Drysdale, D. (2011) *An Introduction to Fire Dynamics.* 3rd Edition. John Wiley & Sons Ltd., Chichester, UK.

EN 1995-1-2 (2004) *Eurocode 5 – Design of Timber Structures, Part 1–2: General – Structural Fire Design. European Standard.* CEN European Committee for Standardization, Brussels.

EN 14135 *Coverings Determination of Fire Protection Ability. European Standard.* CEN, European Committee for Standardization, Brussels.

FPI (2013a) *Technical Guide for the Design and Construction of Tall Wood Buildings in Canada. Chapter 5 – Fire Safety and Protection* (pp. 223–282). FPInnovations, Vancouver, BC.

FPI (2013b) *CLT Handbook – US Edition.* FPInnovations and Binational Softwood Lumber Council, Pointe-Claire, QC, Canada.

FPI (2019a) *CLT Handbook – Canadian Edition.* FPInnovations, Pointe-Claire, QC. https://web.fpinnovations.ca/clt/.

FPI (2019b) *Encapsulated Mass Timber Construction – Guidelines for Encapsulation Details and Techniques.* FPInnovations and Morrison Hershfield Ltd., Pointe-Claire, QC, Canada.

Gerard, R., Barber, D. and Wolski, A. (2013) *Fire Safety Challenges of Tall Wood Buildings.* Arup North America Ltd., San Francisco, CA, and Fire Protection Research Foundation, Quincy, MA.

ISO 13943 *Fire Safety – Vocabulary.* ISO International Organization for Standardization, Geneva.

LaMalva, K. and Hopkin, D. (Eds) (2021) *International Handbook of Structural Fire Engineering.* The Society of Fire Protection Engineers, and Springer Nature Switzerland, Gaithersburg, MD.

Law, A. and Hadden, R. (2020) We Need to Talk about Timber: Fire Safety Design in Tall Buildings. *The Structural Engineer.* March 2020. I. Struct. E. UK.

McLain, R. and Brodahl, S.G. (2021) *Insurance for Mass Timber Construction. Assessing Risk and Providing Answers.* Wood Works, Wood Products Council, Ottawa, ON, Canada.

Meacham, B. and McNamee, M. (2020) *Fire Safety Challenges of Green Buildings and Attributes.* NFPA Research Foundation, Quincy, MA.

NFPA (2018) *Fire Safety Challenges of Tall Wood Buildings-Phase-2 (Five Separate Reports).* https://www.nfpa.org/News-and-Research/Data-research -and-tools/Building-and-Life-Safety/Fire-Safety-Challenges-of-Tall-Wood -Buildings-Phase-2.

Östman, B., Mikkola, E., Stein, R., Frangi, A., König, J., Dhima, D., Hakkarainen, T. and Bregulla, J. (2010) *Fire Safety in Timber Buildings – Technical Guideline for Europe, SP Report 2010:19.* SP Trätek, Stockholm, Sweden.

prEN 1995-1-2 (2021) *Final Draft Eurocode 5 – Design of Timber Structures, Part 1–2: General – Structural Fire Design. European Draft Standard.* CEN European Committee for Standardization, Brussels.

Ranger, L. (2019) *Rehabilitation of Mass Timber Following Fire and Sprinkler Activation.* FPInnovations, Pointe-Claire, QC. Canada.

RISC Authority (2022) *Insurance Challenges of Massive Timber Construction and a Possible Way Forward.* RISC Authority, Fire Protection Association, Moreton-in-Marsh, Gloucestershire, UK.

Schmid, J., Barber, D., Brandon, D. and Werther, N. (2021) The Need for Common Terminology for the Fire Safe Design of Timber Structures. *AOSFST 2021 – 12th Asia-Oceania Symposium on Fire Science and Technology,* Brisbane.

SFPE (2016) *SFPE Handbook of Fire Protection Engineering*. 5th Edition. Society of Fire Protection Engineers, Gaithersburg, MD.

Smith, I. and Frangi, A. (2014) *Use of Timber in Tall Multi-Storey Buildings, Structural Engineering Document SED 13*. International Association for Bridge and Structural Engineering IABSE, Zurich, Switzerland.

STA (2015) Cross-Laminated Timber Construction – An Introduction. *Structural Timber Engineering Bulletin* 11, Structural Timber Association. http://www.structuraltimber.co.uk/assets/InformationCentre/eb11.pdf.

STA (2020) Structural Timber Buildings Fire Safety in Use Guidance. *Volume 6 – Mass Timber Structures; Building Regulation Compliance* B3(1). https://www.structuraltimber.co.uk/assets/STA_Vol_6_Fire_Safety___Mass_Timber_Compliance_Oct_2020_v1.1.pdf.

SW (2019) *CLT Handbook*. Swedish Wood (Svenskt Trä), Stockholm.

Wiesner, F., Bisby, L.A., Bartlett, A.I., et al. (2019) Structural Capacity in Fire of Laminated Timber Elements in Compartments with Exposed Timber Surfaces. *Engineering Structures*, 179, pp. 284–295.

Chapter 3

Fire dynamics

Colleen Wade, Christian Dagenais, Michael Klippel,
Esko Mikkola and Norman Werther

CONTENTS

DOI: 10.1201/9781003190318-3

SCOPE OF CHAPTER

This chapter provides information on fire dynamics in timber buildings. It summarises the fire behaviour in compartments with a focus on buildings with exposed timber structures and wood linings. It includes basic information on the pyrolysis and charring of wood, along with fire dynamics in compartments and the impact of having exposed timber surfaces. A description of common approaches to characterising post-flashover fires with parametric time–temperature curves is provided with guidance on a simplified design method to account for exposed timber surfaces based on parametric fire curves. Limitations in current knowledge are highlighted.

3.1 INTRODUCTION

Traditional prescriptive approaches for fire design of buildings do not generally require the fire dynamics of building fires to be considered. The commonly applied standard fire resistance test does not describe the actual expected conditions in a compartment during a fire, yet has served a useful role over many years to ensure building elements possess an ability to withstand relatively severe fire conditions for a defined period of time (Law and Bisby, 2020). However, designers should be aware that, while common approaches that rely on standard fire resistance tests have limitations for all buildings, there are additional considerations needed for timber buildings where timber is exposed or is inadequately protected (e.g. with encapsulating materials). In fires where the timber may contribute, the total fire load will be a combination of the moveable combustible building contents and fittings as well as that contributed by any exposed or inadequately protected timber surfaces in the compartment.

The major consequences of a significant amount of exposed timber on the internal surfaces within a compartment are that in the event of fire, the time to reach flashover will typically be quicker, the fire will be bigger and the duration of burning after flashover will be longer than for an equivalent compartment with no added contribution from the wood surfaces. Furthermore, flames projecting from unprotected openings such as windows and doors may also be larger and persist for a longer period, with a corresponding higher risk of both external vertical fire spread to upper floors and horizontal fire spread to neighbouring buildings.

It is very important for designers to have a good understanding of the fire dynamics in compartments constructed from all types of materials and in the case of timber buildings to be able to address the additional challenges they present. This will depend on the particular characteristics of the building and its occupants, including the compartment size, geometry, height, ventilation, use, the amount and location of exposed wood surfaces as well as the particular performance required or expected and fire safety strategy adopted for its design.

3.2 COMBUSTION OF WOOD PRODUCTS

3.2.1 Effect of temperature and radiant heat

The three main constituents of wood are cellulose, hemicellulose and lignin along with smaller amounts of organic extractives and inorganic species that contribute to ash formation after the fire. The relative proportions of cellulose, hemicellulose and lignin in dry softwoods are typically in the ranges of 40–44%, 20–32% and 25–35%, respectively (Janssens and Douglas, 2004). The thermal decomposition of the different wood constituents typically occurs over different temperature ranges, i.e. hemicellulose 200–260°C, cellulose 240–350°C and lignin 280–500°C. See also Section 5.1.1.

When wood is exposed to external heating, it will decompose to produce a mixture of volatiles and solid carbonaceous residue (char). This means that the material wood (material 1) is transformed into the char layer material (material 2) (see Schmid and Frangi, 2021). In the presence of oxygen, wood can exhibit either flaming combustion or smouldering combustion, depending on the magnitude of the external heat flux. Figure 3.1 illustrates the different degradation zones due to a heat flux applied to the surface. There are four main zones (A–D) that can be demarcated by temperature, and these can be summarised as follows (Browne, 1958):

- Zone A – At temperatures up to 200°C, there is dehydration producing water vapour and small amounts of carbon dioxide, formic and acetic acids and other compounds. These reactions are primarily

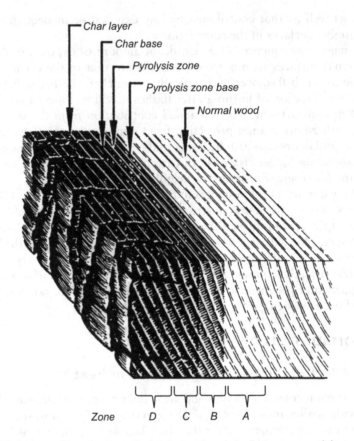

Char layer
Char base
Pyrolysis zone
Pyrolysis zone base
Normal wood

Zone   D   C   B   A

Figure 3.1 Section through charred wood showing degradation zones. Adapted from White (2016) with permission from SFPE.

endothermic and the volatiles produced are non-combustible (Bartlett et al., 2018). There can be a considerable loss of strength (e.g. ~50%) between 100°C and 200°C.

- Zone B – At temperatures between 200°C and 280°C, there is some slow pyrolysis occurring producing water vapour, carbon dioxide and formic and acetic acids as before. Some carbon monoxide may also be produced, along with a slow conversion of the wood to char. A dark brown colour is associated with the onset of pyrolysis.
- Zone C – At temperatures between 280°C and 500°C, the pyrolysis rate increases rapidly producing combustible gases, including carbon monoxide, methane, formaldehyde and formic and acetic acids, along with small amounts of other gases and compounds. Tar droplets are produced as smoke, and the residue is char.
- Zone D – At temperatures above 500°C, the char formed is accompanied by additional reactions involving the gaseous products and

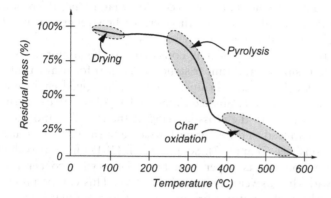

Figure 3.2 Thermal decomposition of timber as a function of temperature in an oxygen-rich environment (illustrative only). Adapted from Law and Hadden (2020) with permission.

tars originating from deeper layers are further pyrolysed to give more highly combustible products, i.e. carbon dioxide and water vapour react with carbon to form carbon monoxide, hydrogen and formaldehyde.

Char development also leads to cracks and fissures being formed that in turn greatly affect the heat and mass transfer between the solid and flame. The combustible volatiles that are released from the heat-exposed surface can mix with the surrounding air/oxygen and burn with a luminous flame. Where flame is not present over the exposed surface, oxygen may diffuse to the surface leading to char oxidation. The exposed surface recedes as combustion progresses due to the char contraction and possible char oxidation (Janssens and Douglas, 2004). Figure 3.2 from Law and Hadden (2020) illustrates where the various thermal decomposition processes occur in terms of the residual mass of the wood as a function of temperature for a piece of wood heated isothermally in an oxygen-rich environment.

During the conversion of structural timber material to the char layer material, a certain amount of the potential chemical energy is released as combustible volatiles. When oxygen is available, these gaseous pyrolysis products are burned and the released heat contributes to the heat release rate (HRR). To describe the combustion behaviour of wood, Schmid and Frangi (2021) consider the energy storage in and the heat release of the char layer.

For engineering design purposes, the temperature within timber elements that demarcates the char from the uncharred wood (i.e. the char depth) is typically assumed to be 300°C, although some pyrolysis is expected at lower temperatures, as previously noted. In addition, for structural calculations, typically a layer to compensate strength and stiffness losses in the virgin wood next to the char layer is considered to reduce the residual cross section, the so-called zero-strength layer, as the strength of wood diminishes quickly at relatively low temperatures.

Typically, char development is considered to be uniform, assuming a homogeneous surface and timber structure typical for heavy timber structures such as columns and beams. However, for glued engineered wood products, such as cross-laminated timber (CLT), observable charring might not be uniform due to the cross-layering of the lamellae and the adhesive used, which can lead to different performance of the CLT structures under fire conditions (see Chapter 7). Schmid et al. (2017) have also shown that char contraction effects depend on both the oxygen concentration and increase with the gas velocity over the surface. This can be important for the decay phase of a natural fire in contrast to a standard fire resistance test (with no decay phase) where oxygen levels are much lower. Variations in thermal exposure within a compartment may also contribute to non-uniform charring behaviour.

3.2.2 Flaming combustion

The heat of combustion of wood is about 15–20 MJ/kg, half to two-thirds of which is released through flaming with cellulose as the main contributor to flaming combustion producing more volatiles than char. Solid timber will not support flaming combustion unless an external heat flux is applied to the surface since the flame heat flux alone is not sufficient to sustain its own burning (Drysdale, 1998). Indeed, the effect of incident heat flux is the most dominant parameter, with an order of magnitude higher influence than the other parameters considered such as material properties, oxygen concentration or surface orientation, over the ranges to be expected in normal design (Bartlett et al., 2018). Flaming combustion will only occur when the rate at which pyrolysis gases are produced is sufficient to sustain a flammable gas–air mixture and below this rate flaming combustion will cease.

The rate of pyrolysis $\dot{m}_p''$ (in kg/m^2·s) as applied to solid wood depends on the heat flux (in kW/m^2) from the flame $\dot{q}_f''$ and from the hot gases and other surfaces in the compartment $\dot{q}_e''$ less the heat losses from the surface $\dot{q}_l''$, which comprises radiative and convective terms as well as the conductive loss into the surface as given by (Bartlett et al., 2018)

$$\dot{m}_p'' = \frac{\dot{q}_f'' + \dot{q}_e'' - \dot{q}_l''}{L_v} \quad \left[\text{kg/m}^2 \cdot \text{s}\right] \tag{3.1}$$

where L_v = heat of gasification. Spearpoint and Quintiere (2000) derived representative values for the heat of gasification of several wood species across the grain in the range 2.5–3.5 kJ/g; also see Table 5.9 for additional data.

The heat flux received by a burning surface also depends on the size and orientation of the surfaces relative to other burning and hot surfaces. For this reason, the burning rate will be higher and sustained for longer where the burning surfaces face each other or are in a wall–corner or wall–ceiling configuration, when compared to a single burning surface in one plane.

3.2.3 Smouldering combustion

Smouldering can be thought of as self-sustained glowing combustion where an external source of heat is not required to sustain the process (although it may be required to start the process). The glowing combustion involves oxidation reactions at the surface of the solid (wood). When glowing combustion occurs, the surface temperature can increase by several hundred degrees over a few seconds (Babrauskas, 2021).

In the context of this guide, with respect to fires in compartments with exposed wood surfaces, the main interest in smouldering is when does it begin (within the decay stage) and when does it end? For the applications of interest here, there are two main possibilities for the conclusion of the smouldering: (1) there is a transition to flaming; or (2) the combustion ceases.

Transition from flaming to smouldering may occur during the decay stage of the fire, when the compartment contents are largely consumed, the rate of burning of exposed wood surfaces is slowing and the compartment temperatures are falling. Generally, for practical purposes, it will be necessary to consider that the possibility of smouldering in wood surfaces in the fire compartment exists after the surface flaming has ceased.

Bartlett et al. (2018) reviewed previous research and the factors affecting the burning behaviour of wood and noted that the critical mass loss rate for the extinction of the flame varied from 2.5 to 5 g/m²·s. Equation 3.1 allows a critical value to be determined where the flaming will not be sustained (Law and Hadden, 2020).

Cessation of flaming combustion is influenced by the oxygen concentration in the environment immediately surrounding the burning surfaces of timber. It is known that a reduction in the oxygen concentration reduces the flame temperature, thereby reducing the heat flux from the flame to the surface. Quintiere and Rangwala (2004) recommended 1,300°C as a critical flame temperature below which flame extinction occurs. Reducing oxygen concentration also reduces the rate of oxidation of the char, increasing the thickness of the char layer but reducing the mass loss rate. This is consistent with Equation 3.1, given a lower heat flux from the flame and reduced conduction into the timber – ultimately resulting in a lower charring rate. Mikkola found that the charring rate in standard fire resistance tests was approximately 20% lower than in oxygen-rich test environments given the same average heat flux over the tests (Mikkola, 1991).

Transition from smouldering to flaming is very complex since heat and mass transfer are not one-dimensional and edge conditions may play a critical role. Flaming is also not likely to erupt within the bulk of a fuel bed unless a cavity is formed but may occur along the boundary of the fuel (Babrauskas, 2021). Airflow or wind may also promote transition to flaming in a smouldering material such that if a smouldering material is disturbed, flaming may occur. Transition to flaming might best be considered a stochastic event.

Cessation of smouldering combustion must eventually occur since at some point all the available fuel will be exhausted. However, the process can stop with unburned fuel left remaining if local circumstances occur that are not favourable for smouldering, such as voids or non-uniformities in the fuel. Airflow could also increase or decrease causing the smouldering to cease.

Crielaard (2015; Crielaard et al., 2019) investigated the self-extinguishment of CLT and concluded that smouldering combustion of CLT ceases when the externally applied heat flux falls below about 5–6 kW/m^2 and the airflow over the surface is below 0.5 m/s. He observed that the cessation of smouldering combustion depends on the airflow across the timber surface. Smouldering is governed by the rate of oxygen diffusion to the reaction zone rather than the amount of oxygen in the surrounding environment. However, to maintain an adequate amount of unburned residual fuel in timber structures, in the context of withstanding a burnout, ultimately it may be necessary to rely on overt extinguishing efforts at the end of the decay stage of the fire to completely halt the smouldering process.

3.3 COMPARTMENT FIRES

3.3.1 Fire development stages

The classic compartment fire has been studied in detail for non-combustible compartments and is described in the literature by authors such as Drysdale (1998) and Torero et al. (2014). There are four different phases of the fire development, as illustrated in Figure 2.3. The fire starts with an incipient period, which occurs prior to flaming, and is followed by ignition and the growth phase, where the type, amount and configuration of fuel determine the burning behaviour. If the fire growth is able to be sustained dependent on the compartment size and ventilation, then flashover may occur followed by a period of relatively steady or fully developed burning. This period is critical for structural design and the gas temperatures are typically high and could reach 1,200°C. Finally, the decay phase is when most of the fuel is consumed and the rate of burning and the gas temperatures inside the compartment decline eventually leading to extinguishment of the fire.

As compartments increase in size or aspect ratio, the assumption that the fire will burn uniformly across the full area of the compartment is less likely. In this case, the seat of the fire (i.e. a localised fire) may be observed to migrate or travel across the floor plate influenced by the location of the fuel and the size and position of the openings. This is called a "travelling fire," discussed later in Section 3.4.4 where alternative methods have been developed to determine the thermal exposure applying to a specific location within the compartment (Rackauskaite et al., 2015; Stern-Gottfried and Rein, 2012).

3.3.2 Fire growth

Early in the fire, reaction to fire characteristics will govern ease of ignition and surface flame spread and the early fire growth behaviour. These topics are covered in Section 5.3, including typical reaction to fire characteristics that can be used to predict time to ignition, rates of surface spread of flame and smoke and toxic combustion products from wood products exposed in a developing fire.

Following ignition, the fire may grow. The growth stage of the fire can range from very fast to very slow, depending on the characteristics of the fuel, the proximity and interactions with the surroundings and the availability of oxygen. The rate of energy release and the rate at which products of combustion are generated are used to describe the fire. Fire growth will be fast when there is flaming combustion of fuels that exhibit rapid surface flame spread, whereas the fire growth will be slow where a lengthy period of smouldering occurs and in some cases the fire may go out with no transition to flaming.

In compartments that are very well-ventilated with large opening areas, or in cases where the surface area of the fuel is small compared to the volume of the compartment, the burning rate may be governed by the surface area of the fuel and this is referred to as fuel-controlled burning.

The burning rate of fuel-controlled fires is dependent upon the nature and surface area of the fuel. In many cases, it is quite difficult to determine the burning rate precisely due to the characteristics and geometry of the fuel packages. For simple, well-defined geometries such as timber cribs, equations exist that allow the fuel pyrolysis rate to be estimated based on the initial fuel mass per unit area and the remaining fuel mass per unit area at a given time (Babrauskas, 2016). Alternatively, "t^2 fires" are commonly used to describe the rate of fire growth as given by Equation 3.2, where α is the fire growth rate coefficient based on the type of fuel load. Typical α values for commonly adopted "t^2 fires" are shown in Table 3.1.

$$\dot{Q}(t) = at^2 \quad [\text{kW}] \tag{3.2}$$

where $\dot{Q}(t)$ = time-dependent heat release rate (kW), and t = time from ignition (s).

The heat release rate for a fuel-controlled fire, $\dot{Q}_f$ (in kW), is generally estimated from either a full-scale test conducted under well-ventilated

Table 3.1 Typical values of α for different
fire growth rates

Growth rate	α (kW/s²)
Ultra-fast	0.19
Fast	0.047
Medium	0.012
Slow	0.003

conditions where the peak heat release rate can be directly measured with an oxygen calorimeter or derived from measurements of the mass loss rate. If the mass loss rate $\dot{m}$ is known, the heat release rate can be calculated as follows:

$$\dot{Q}_f = \dot{m}\Delta H_c \quad [\text{kW}] \tag{3.3}$$

where $\dot{m}$ = mass loss rate of the fuel (kg/s) and ΔH_c = the heat of combustion of the fuel (kJ/kg).

Alternatively, small-scale tests that allow the heat release rate per unit area for the material to be determined allow the maximum heat release rate ($\dot{Q}_{max}$) for a fuel-controlled fire to be determined from

$$\dot{Q}_{max} = \dot{Q}''_{fl} A_f \quad [\text{kW}] \tag{3.4}$$

where $\dot{Q}''_{fl}$ = heat release rate per unit area (kW/m²) and A_f = burning surface area of the fuel (m²). Heat release rate per unit area (or mass loss rate per unit area) is typically measured under well-ventilated free burning conditions where the radiation feedback from the surroundings is negligible. Sometimes, the effect of radiation feedback on the burning rate may need to be considered. Data for the energy release rate per unit floor area $\dot{Q}''_{fl}$ can be found in the literature (e.g. Karlsson and Quintiere, 2000; Babrauskas, 2016).

When multiple fuel packages are present, the respective maximum heat release rates per unit area for all items can be added together assuming all items are burning simultaneously. This provides a conservative estimate of the maximum value for the rate of heat release. Alternatively, fire spread from one item to an adjacent item could also be included if the time for ignition of each item was accounted for keeping in mind the dependency on the spacing and arrangement of the various items within the compartment. This may not be a practical approach for design purposes when the exact spacing and arrangement of fuels is unknown.

During the early fire growth period, the location, size and strength of the fire source in relation to the position of any exposed wood linings is important for estimating the initial fire growth rate where timber surfaces are involved. The scenario where the fire source is positioned in the corner of a room, with flames in close contact with the corner wall surfaces, is commonly recognised as a worst-case scenario for a pre-flashover fire involving surface linings.

When interior finish materials include wood products (or other combustible materials), there is potential for surface flame spread and additional energy release from the finish materials which may increase the fire growth rate compared to the use of non-combustible finishes. This has been demonstrated by Kotsovinos et al. (2022) who conducted an experiment burning wood cribs in a large open-plan compartment with a floor area of 352 m² that included a fully exposed CLT ceiling. They found that the rate of fire spread in this experiment was three to eight times faster than in an equivalent non-combustible compartment due to the presence of timber on the ceiling.

In prescriptive designs, interior finish materials are typically limited in locations within a compartment and in thickness, and their effect on fuel load and fire dynamics is somewhat implicitly considered.

Further information relating to the pyrolysis, burning and heat release rate of wood as well as a description of the applicable regulatory tests can be found in Chapter 5.

3.3.3 Flashover

Flashover is the transition between the fire growth stage and the fully developed stage. It is not a very precise term and commonly criteria such as the gas temperature reaching 500–600°C or an incident heat flux reaching the floor of 15–20 kw/m² are used. This often coincides with flames emerging from the compartment opening as the fire becomes ventilation-controlled.

The energy release rate required to generate a 500°C rise in the gas temperature, i.e. needed to reach flashover, can be estimated by applying the McCaffrey, Quintiere, Harkleroad (MQH) equation assuming a fuel-controlled fire in a conventional room, a simple form of which is given in Equation 3.5 (Karlsson and Quintiere, 2000):

$$\Delta T = 6.85 \left(\frac{\dot{Q}^2}{A_o \sqrt{H_o} h_k A_T} \right)^{\frac{1}{3}} \quad [\text{kW}] \tag{3.5}$$

where $\dot{Q}$ = rate of heat release (kW), h_k = the effective heat conduction coefficient for the solid boundaries (kW/m²·K), A_o = opening area (m²), H_o = opening height (m) and the A_T = boundary surface area (m²) to be used for heat transfer considerations. h_k depends on the duration of the heating and the thermal properties of the compartment. It can be estimated using Equation 3.7 or 3.8 after determining the thermal penetration time t_p, from Equation

3.6, where δ = thickness of the solid (m), k = thermal conductivity (kW/m·K), ρ = density (kg/m³) and c = specific heat (kJ/kg·K):

$$t_p = \frac{\delta^2}{4\left(\dfrac{k}{\rho c}\right)} \ [\text{s}] \tag{3.6}$$

$$\text{For } t < t_p \quad h_k = \sqrt{\frac{k\rho c}{t}} \ \left[\text{kW/m}^2 \cdot \text{K}\right] \tag{3.7}$$

$$\text{For } t \geq t_p \quad h_k = \frac{k}{\delta} \ \left[\text{kW/m}^2 \cdot \text{K}\right] \tag{3.8}$$

Substituting $\Delta T = 500$ K into Equation 3.5 and rearranging allows the heat release rate required to reach flashover $\dot{Q}_{FO}$ (in kW) to be estimated as follows:

$$\dot{Q}_{FO} = 610\left(h_k A_T A_o \sqrt{H_o}\right)^{\frac{1}{2}} \ [\text{kW}] \tag{3.9}$$

Where fires are flush with a wall or in a comer, Mowrer and Williamson (1987) found that the upper-layer temperature could be calculated using Equation 3.5 multiplied by a factor. For fires flush to walls, they recommended a factor of 1.3, and for fires in comers, the equation should be multiplied by 1.7.

Karlsson (1992) found that for the case of combustible lining materials, Equation 3.5 should be multiplied by a factor of 2. In these cases, Equation 3.9 may no longer apply without accounting for the associated change in Equation 3.5.

3.3.4 Fully developed fire

The fully developed stage is characterised by ventilation-controlled burning where the availability of oxygen entering through the openings determines the maximum energy release rate in the compartment. The peak compartment gas temperatures are reached during this stage and typically fall in the range of 700–1,200°C.

The duration of the fully developed phase is strongly influenced by the amount of fuel present. When the compartment surfaces are made from non-combustible materials, i.e. with no combustible finish materials, then it is only the fuel from the compartment contents that need to be considered. However, when the compartment surfaces include combustible materials such as exposed timber walls, floors and ceilings, then it may be necessary to consider the effects of this additional potential fuel on the growth, duration and severity of the compartment fire.

Under ventilation-controlled conditions, the mass flow rate of air through the compartment opening will be approximately proportional to the area of the opening and the square root of the height of the opening. This was confirmed by Kawagoe (1958) in the 1950s based on the burning rate of wood cribs measured inside a small non-combustible compartment considering different sizes of opening, as represented by Equation 3.10:

$$\dot{m}_b = 0.09 A_o \sqrt{h_o} \quad [\text{kg/s}] \tag{3.10}$$

where $\dot{m}_b$ = burning rate (kg/s). This relationship for the burning rate of the wood cribs only applies over a limited range of opening sizes being related to the rate at which air can enter the compartment. For stoichiometric burning of wood cribs in a non-combustible enclosure with all the combustion taking place within the compartment, the empirical mass flow of air $\dot{m}_{in}$ (in kg/s) through a single opening is given by Equation 3.11. This assumes a heat of combustion for wood of 17 MJ/kg; that each kg of oxygen used for combustion produces 13.2 MJ (Huggett, 1980) and comprises 0.23 of the air entering the opening:

$$\dot{m}_{in} \approx 0.09 \times \frac{17}{(13.2)(0.23)} A_o \sqrt{h_o} \approx 0.5 A_o \sqrt{h_o} \quad [\text{kg/s}] \tag{3.11}$$

Equation 3.11 can also be derived from a theoretical analysis of the flow of gas entering and leaving an opening driven by buoyancy forces in a compartment fire where the fire gases are well-stirred. The theoretical analysis assumes that the hot gases leave the compartment above a neutral plane with the cool air from outside entering below the neutral plane with no interaction between the two flow streams. This leads to the simplified expression in Equation 3.12 for the mass flow rate in through the opening (Drysdale, 1998; Karlsson and Quintiere, 2000):

$$\dot{m}_{in} \approx \frac{2}{3} A_o \sqrt{h_o} C_d \rho_o \sqrt{2g} \sqrt{\frac{\frac{\rho_o - \rho_g}{\rho_o}}{\left[1 + \left(\frac{\rho_o}{\rho_g}\right)^{\frac{1}{3}}\right]^3}} \quad [\text{kg/s}] \tag{3.12}$$

where ρ_o = density of ambient gases (1.2 kg/m³), C_d = discharge coefficient for the opening (≈ 0.7) and g = gravitational constant (9.81 m/s²). The square root of the density term on the right of Equation 3.12 is approximately 0.21 where ρ_g = density of the fire gases (kg/m³). Substituting these parameter values into Equation 3.12 gives $\dot{m}_{in} \approx 0.5 A_o \sqrt{h_o}$ reproducing the empirical Equation 3.11 from the Kawagoe experiments.

The ventilation-controlled heat release rate $\dot{Q}_v$ (in kW) for a single opening is then given by Equation 3.13 since the energy released per kilogram of oxygen is 13,100 kJ/kg-O_2 or 3,000 kJ/kg-air for a wide range of fuels (Huggett, 1980). Multiple openings may be considered using the total area and weighted average height of the opening. However, it may underestimate fire severity in compartments with separate ventilation openings at floor and ceiling level and does not apply to fuel-controlled fires.

$$\dot{Q}_v \approx 1,500 A_o \sqrt{h_o} \quad [\text{kW}] \tag{3.13}$$

This simple equation provides an estimate of the maximum possible heat release within the compartment assuming all the oxygen entering through the opening is consumed in the reaction, i.e. the burning process is assumed to be stoichiometric. This means the air is supplied at the exact rate needed to combust the fuel vapours being produced. This is expressed as the stoichiometric air–fuel ratio r (with a value of approximately 5.7 for the combustion of wood). Even if the gases within a compartment form an ideal stoichiometric mixture, some burning gases may still emerge from an opening because the rate of heat release is not instantaneous, and a finite time is needed for the gases to mix and the reaction to be completed. In fully developed fires, the equivalence ratio, i.e. $\dot{m}_{air}/\dot{m}_f$, is typically less than the stoichiometric air–fuel ratio, r. The term Global Equivalence Ratio (GER) is commonly used to describe the degree to which the fuel could burn in a compartment, with a value greater than 1, indicating that unburned fuel must leave the compartment to burn:

$$\text{GER} = r \frac{\dot{m}_f}{\dot{m}_{air}} \quad [-] \tag{3.14}$$

Equation 3.13 can be refined to account for a reduced oxygen utilisation rate or combustion efficiency during the fire. For example, Equation 3.15 is proposed for inclusion in the next version of Eurocode 5 (prEN 1995-1-2, 2021) based on an oxygen utilisation rate of 0.8:

$$\dot{Q}_v \approx 1,260 A_o \sqrt{h_o} \quad [\text{kW}] \tag{3.15}$$

Babrauskas and Williamson (1978, 1979) included a means of allowing for both fuel-controlled and ventilation-controlled burning in a single-zone well-stirred reactor provided the rate of pyrolysis of the fuel is known at all times, so that the rate of heat release of the fire for ventilation-controlled burning can be given by Equation 3.16 (with r = stoichiometric air/fuel ratio) and for fuel-controlled burning as previously given by Equation 3.3.

$$\dot{Q}_v = \dot{m}_{in} \frac{\Delta H_c}{r} \quad [kW] \tag{3.16}$$

For more complicated arrangements (e.g. multiple connected compartments), the ventilation-controlled heat release rate in two-zone models may be estimated from the available oxygen in the gases entrained into the fire plume. The following relationship is commonly adopted in multi-compartment two-zone models with a smoothing function added to account for the lower oxygen limit (Peacock et al., 2017; Wade et al., 2016).

$$\dot{Q}_v \sim \dot{m}_p Y_{O_2} \Delta H_{O_2} \quad [kW] \tag{3.17}$$

where $\dot{m}_p$ = mass flow of gases entrained into the fire plume (kg/s), Y_{O2} = mass fraction of oxygen in the plume flow (–) and ΔH_{O2} = heat of combustion based on oxygen consumption (~13,100 kJ/kg-O_2 for hydrocarbon fuels).

3.3.5 External flame projection

Most fully developed fires also involve some external flaming, i.e. flames projecting from the fire compartment openings. There are various correlations found in the literature such as that from Law (1978) who proposed a simple correlation in Equation 3.18 for estimating the external flame height for a given compartment and opening configuration. This and similar correlations are generally derived from thermal plume data applicable to small non-combustible compartments in the absence of wind.

$$z + H_o = 12.8 \left(\frac{R}{W_o} \right)^{\frac{2}{3}} \quad [kW] \tag{3.18}$$

where z = the flame length from the top of the opening (m), H_o = opening height (m), W_o = opening width (m) and R = mass loss rate of fuel (kg/s).

When a compartment fire is ventilation-controlled and there are more pyrolysis gases generated than able to burn inside the compartment, as discussed previously, some of the excess pyrolysis gases are transported with the outflow of smoke and gases through openings to the outside. Upon mixing with new sources of air/oxygen, these combustible gases can burn generating large flames that project from the openings.

There have been a number of experiments reported in the literature where a large amount of combustible pyrolysis gases have been observed to burn external to the compartment when significant areas of timber on the internal surfaces are exposed to the fire. For example, Hakkarainen reported burning in compartments of protected timber construction where approximately 15% of the burning took place outside the compartment. This compared to another compartment where mass timber construction was fully exposed and where the proportion of external burning was estimated to increase to approximately 50% (Hakkarainen, 2002). More recently, Kanellopoulos et al. (2019) also found that, in contrast to compartments with inert linings, compartments with exposed timber as linings have prolonged external flaming and induce greater heat fluxes above and opposite the opening. Further research is needed to develop calculation methods to quantify the external flaming depending on the GER and the area of the exposed timber in the compartment (e.g. Hopkin and Spearpoint, 2021).

Larger and more intense external flaming has important implications for external fire spread, including the choice of facade-cladding products and their configuration to mitigate vertical fire spread as well as design to prevent fire spread to adjacent buildings. Figure 3.3 illustrates external flaming from a compartment where the CLT ceiling was left exposed.

This behaviour was also confirmed in a recent study involving reduced-scale CLT compartments by Gorska et al. (2021) who found that the burning of timber surfaces resulted in larger flow velocities at the opening being a result of the burning surfaces inducing additional buoyancy and momentum inside the compartment. This also meant that there was less uniform mixing of the pyrolysis gases with oxygen leading to greater external burning. In addition, it was noted that the highest compartment temperatures did not occur directly beneath the ceiling as would be the case in non-combustible compartments, and that the burning rate of the ceiling was lower than for the walls. The findings by Gorska et al. (2021) are illustrated in Figure 3.4.

Depending on both the size of the compartment and the size of the fire, it is possible to have a fire plume that cannot be contained within the compartment, resulting in flame extension out of the openings. Flame extension

Figure 3.3 External flaming from CLT compartment with timber ceiling exposed. Reproduced from Brandon (2021) with permission.

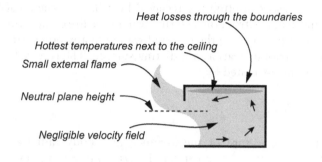

Heat losses through the boundaries

Hottest temperatures next to the ceiling

Small external flame

Neutral plane height

Negligible velocity field

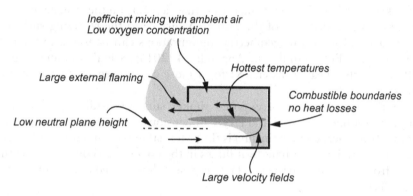

Inefficient mixing with ambient air
Low oxygen concentration

Large external flaming

Hottest temperatures

Combustible boundaries
no heat losses

Low neutral plane height

Large velocity fields

Figure 3.4 Schematic representation of the physical changes that occur when CLT is exposed in a compartment fire. *Top:* Traditional compartment fire with non-combustible linings. *Bottom:* Compartment fire with CLT linings. Adapted from Gorska et al. (2021) with permission from Elsevier.

can occur when the fire plume impinges on the ceiling and the length of the ceiling jet is longer than the distance from the fire plume to opening (Gottuk and Lattimer, 2016). Flame extension is different from the external burning discussed above, but to the observer it may appear to be similar.

Current international practices with respect to mitigating external vertical fire spread rarely consider this difference for compartments of non-combustible versus combustible construction. Nonetheless, many jurisdictions do permit the use of some combustible materials as surface linings or finish materials in buildings where non-combustible construction may be prescribed. These may be limited in thickness, area, location and in the case of timber products may require flame-retardant treatments to meet flame spread indices. Further information is given in Chapter 5.

3.3.6 External fire spread to neighbouring buildings

The permitted separation distance between the external walls of adjacent buildings, or between an external wall and a site boundary is often calculated on the basis of limiting the received heat flux on the adjacent building or at some distance from the external wall of the fire compartment. It is usual to consider openings in the external wall (and any other parts of the wall likely to contribute heat) as a radiating surface. The received heat flux (in kW/m²) can be estimated by

$$\dot{q}'' = \varnothing \varepsilon \sigma T^4 \quad \left[\text{kW/m}^2 \right] \tag{3.19}$$

where T = absolute temperature of the assumed grey-body radiator (K), ε = emissivity of the radiator (–), σ = Stefan–Boltzmann constant (5.67 × 10^{-11} kW/m²·K⁴) and $\varnothing$ = view factor (–) that depends on the size, geometry, orientation and distance of the radiation surface to the receiving surface. View factors for various geometric configurations can be found in the literature, e.g. SFPE Handbook, Appendix 4 (2016). See also Lautenberger (2016) or other heat transfer text for a more comprehensive treatment of the topic.

Another consequence of exposed timber surfaces leading to external flaming of longer duration and larger dimensions (compared to the ejected plumes from inert compartments) is the potential impact on building-to-fire spread, due to larger radiant heat fluxes in the far-field, in comparison with those from inert compartments. Further research is needed to fully quantify this effect.

3.3.7 Species production

The mass of a species product per unit mass of fuel burned is called the yield. For example, the yield of carbon monoxide (CO) is defined as

$$y_{CO} = \frac{m_{CO}}{m_f} \quad [-] \tag{3.20}$$

where m_{CO} = mass of CO (kg) and m_f = mass of fuel burned (kg). The yield is relatively constant for a given fuel for overventilated fires, but increases as the fire becomes underventilated (i.e. GER > 1). Some species yields as a function of ventilation are given in Table 3.2 (Tewarson, 1995; Quintiere, 2017).

Since wood is an oxygenated fuel, it does not require additional oxygen from entrained air to form CO. This enhances the ability of the wood to generate CO in a vitiated atmosphere (Gottuk and Latimer, 2016). CO concentrations greater than 5% have been reported for cellulosic fuels burning in compartments (Tewarson, 1984) and a series of tests were conducted by Lattimer et al. (1998) to evaluate the effect on species production from the addition of wood suspended below the ceiling in the upper layer of a reduced-scale compartment fire. They showed that wood burning in the upper layer resulted in much higher CO concentrations (10.1% vs. 3.2%

Table 3.2 Some species yields for solids

Conditions			Overventilated		Underventilated
Fuel	Y_{CO2} (g/g)	Y_{SOOT} (g/g)	Y_{CO} (g/g)	Y_{CO} (g/g)	
Wood (red oak, pine)	1.27	0.015	0.004	0.138	
Polystyrene (PS)	2.33	0.164	0.060	Not available	
Polyurethane (PU) flexible foam	1.51	0.227	0.031	Not available	
Polyvinyl chloride (PVC)	0.46	0.172	0.063	0.360	

(Tewarson, 2002; Quintiere, 2017)

without wood) with only small increases in the CO_2 concentrations (11.6% vs. 10.4% without wood). A CO yield in underventilated fires of 0.2 g/g is often assumed regardless of the fuel. See Gottuk and Latimer (2016) for further information on this topic and for a description of an engineering methodology for estimating species transported to remote locations based on a compartment equivalence ratio. See also Table 5.10 for additional data for well-ventilated fires.

3.4 COMPARTMENT FIRE TEMPERATURES

This section gives an overview of the different types of fully developed compartment time–temperature curves commonly used. A number of simplified solution techniques for estimating temperatures in fires are summarised by Walton et al. (2016) with more detailed information on the fundamental principles of compartment fire modelling in Quintiere and Wade (2016). Other useful references include the SFPE Standard S.01 for calculating fire exposures to structures (2011), Drysdale (1998), Karlsson and Quintiere (2000) and Wickström (2016).

3.4.1 Energy and mass balance

Compartment fire temperatures can be obtained by solving an energy balance for the compartment. Considering the fully developed stage and treating the compartment as a calorimeter, the energy balance, as illustrated in Figure 3.5, can be described by (Drysdale, 1998)

$$\dot{Q}_F = \dot{Q}_L + \dot{Q}_W + \dot{Q}_R + \dot{Q}_S \quad [kW] \tag{3.21}$$

where $\dot{Q}_F$ = heat release rate due to the combustion (kW); $\dot{Q}_L$ = rate of heat loss due to the convective flows through the opening (kW); $\dot{Q}_W$ = rate of heat loss through the walls, ceiling and floor (kW); $\dot{Q}_R$ = rate of heat loss by radiation through the openings (kW); and $\dot{Q}_S$ = rate of heat storage in the

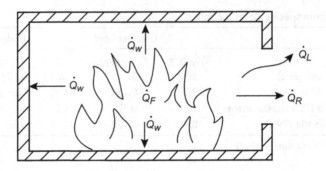

Figure 3.5 Energy balance for a fully developed compartment fire. Copyright © 2000
From Enclosure Fire Dynamics by Karlsson and Quintiere. Reproduced with
permission from Taylor and Francis Group, LLC, a division of Informa plc.

gas volume (kW). The rate of heat storage in the gas volume $\dot{Q}_S$ is small and
often ignored.

The heat release due to the combustion $\dot{Q}_F$ is determined using Equation
3.3 or 3.13 (or 3.15) as applicable. If the fire load L (in kJ) is known and the
ventilation-controlled burning rate is assumed constant, then the duration
of the burning period can be estimated as $L/\dot{Q}_F$ in seconds.

The rate of heat loss due to the convective flows through the opening
$\dot{Q}_L$ can be described by Equation 3.22 (where T_g is the compartment gas
temperature and T_o is the ambient temperature) if we assume that the rate
of air inflow is equal to the outflow and if we ignore the mass contribution
from the fuel.

$$\dot{Q}_L = \dot{m}_{in} c \left(T_g - T_o \right) \quad [\text{kW}] \tag{3.22}$$

The rate of heat loss through the walls, ceiling and floor $\dot{Q}_W$ depends on the
gas temperature inside the compartment and on the surface temperature of
the respective internal surfaces. For the simple case of a semi-infinite solid,
$\dot{Q}_W$ can be written as follows:

$$\dot{Q}_W = \left(A_t - A_o \right) \frac{1}{\sqrt{\pi}} \sqrt{\frac{k \rho c}{t}} \left(T_g - T_o \right) \quad [\text{kW}] \tag{3.23}$$

However, Equation 3.23 assumes the gas temperature is constant and
therefore cannot be used to calculate a changing temperature–time curve.
Consequently, a numerical solution is needed instead. Under transient conditions with constant properties and no internal generation, the appropriate form of the general heat equation should be used. This can be solved
using numerical methods described elsewhere allowing transient heating

conditions and radiative and convective boundary conditions to be considered (Drysdale, 1998; Incropera and DeWitt, 1990; Wickström, 2016).

The rate of heat loss by radiation through the openings $\dot{Q}_R$ (in kW) is calculated as follows:

$$\dot{Q}_R = A_o \varepsilon_f \sigma \left(T_g^4 - T_o^4 \right) \quad [\text{kW}] \tag{3.24}$$

where ε_f = average emissivity of the flames and gases as they radiate out through the opening (–), σ = Stefan–Boltzmann constant (5.67 × 10^{-11} kW/ m$^2 \cdot$K^4), T_g = gas temperature (K), T_o = ambient temperature (K) and A_o = area of the opening (m^2).

Solving the energy balance numerically allows the gas temperature curve to be determined with inputs that include the heat release rate of the fire $\dot{Q}_F$, the thermal properties of the compartment boundary $k\rho c$, the area of the bounding compartment surfaces A_t, the ventilation factor $A_o\sqrt{h_o}$ and the fire load L. Well-known examples of time–temperature curves from an energy balance of this type are those of Magnusson and Thelandersson (1970), with examples shown in Figure 3.6.

Increasing the fuel load in the building will generally increase the duration and overall severity of the fire, as illustrated in Figure 3.6a. In addition to combustible room contents, there may be additional contributions from the building fixtures, fittings and structure which is particularly relevant for timber buildings. During the fire growth period, factors such as the geometric arrangement of the fuel, the exposed surface area (and surface area to mass ratio), thickness and orientation along with the fuel properties (e.g. heat of combustion, heat of gasification) will all affect how quickly the fire grows and the shape of the time–temperature curve. Guidance on selecting fire loads for design is provided in NFPA 557: Standard for Determination of Fire Loads for Use in Structural Fire Protection Design.

Increasing the amount of ventilation (starting with a small opening) in a compartment will lead to a faster and hotter fire, as shown in Figure 3.6b – up to the point where the amount of air and oxygen supplied is enough for complete combustion to take place. However, a continued increase in the ventilation will cool the fire and shorten the duration as more combustion products and heat is lost from the compartment.

3.4.2 Parametric/natural fires

Parametric fires are equation-based expressions of time–temperature curves, with the best-known being those defined in EN 1991-1-2, Appendix A. They are based on Magnusson and Thelandersson's work and later modified and simplified by Wickström (1985).

EN 1991-1-2:2002 Annex A states that the given equations are valid for compartments up to 500 m^2 with a maximum compartment height of 4 m

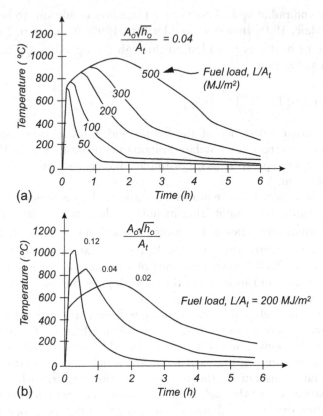

Figure 3.6 (a) Time–temperature curves for varying fuel load and constant ventilation.
(b) Time–temperature curves for varying ventilation and constant fuel load.
Adapted from Buchanan and Abu (2017) with permission from John Wiley &
Sons, Inc.

and assume that the compartment fire load will be completely burned out.
Modifications to these equations have also been proposed by Reitgruber et
al. (2006). Parametric fires, as given in EN 1991-1-2 Annex A, are applicable to timber compartments only if the timber surfaces are encapsulated
such that they do not become exposed during the fire and do not otherwise
contribute fuel to the fire. This requires any glueline failure of CLT and any
failure of the gypsum protection to be avoided (Brandon, 2018a).

As a general note, the maximum temperatures in parametric curves are
often higher than measured in large-scale experiments, which means over-
estimation for those parts of the curves. On the other hand, the decay phase
of the parametric curves is often steeper (and therefore shorter) compared
to experimental results (i.e. an underestimation), suggesting they may not
be appropriate for use with compartments with exposed timber surfaces.
Alternative approaches to overcome these limitations of the current EN

1991-1-2 parametric temperature–time curves are described by Zehfuss and Hosser (2007), and these are currently included in the German National Annex to EN 1991-1-2 (DIN EN 1991-1-2/NA).

See Section 3.8.1 for an engineering approach to account for the contribution of exposed timber surfaces with an example shown in Section 3.9.

3.4.3 Localised fires

Traditional methods for quantifying and modelling compartment fires for structural engineering analysis assume spatially homogeneous temperature conditions. Stern-Gottfried et al. (2010) analysed temperature distributions in a range of post-flashover compartment fires and found that uniform temperature conditions are not present and variation from the compartment average exists. They found peak local temperatures ranged from 23% to 75% higher than the compartment average.

Various localised fire models have been published in the literature. The most widely used are those in EN 1991-1-2, Annex C, giving equations for calculating the heat flux to a specific part of a structure. The heat flux calculations mainly depend on the fire size, the flame length and the relative position of the structural element relative to the fire plume both vertically and horizontally. Different equations are available depending on whether the flames impinge on the ceiling or not.

3.4.4 Travelling fires

In some compartments, a fire has been observed to travel or migrate around the compartment such that the burning is not uniform throughout the compartment. It can also be thought of as a localised fire that moves. This is more commonly associated with large open compartments rather than smaller compartments where traditional post-flashover design fires are often assumed. As a guide, EN 1991-1-2 limits application of uniform parametric fires to floor areas up to 500 m². In travelling fires, fire spread is a function of the fuel load, the size and geometry of the compartment and the location of interest and is less influenced by the thermal properties of the lining materials due to the lesser proximity of the fire to the walls and openings. However, this does not necessarily mean that a travelling fire will be less severe than a uniformly burning fire. Previous studies (Law et al., 2011; Stern-Gottfried & Rein, 2012) have demonstrated that travelling fires can be onerous for the structure as a result of the different thermal and structural responses they produce compared to uniform fires.

As a result, travelling fire methodologies (TFM) for structural fire design purposes have been proposed (Rackauskaite et al., 2015). These methodologies have been developed for non-combustible compartments and are not strictly applicable where timber elements are exposed or partially protected.

TFM considers design fires to be composed of two moving regions: the near-field (flames) and the far-field (smoke). The near-field model represents the flames directly impinging on the ceiling and assumes the peak flame temperatures. The far-field model represents smoke temperatures which decrease with distance away from the fire due to mixing with air. Equations for both localised fires and travelling fires in traditional compartments (without exposed timber on the ceiling) can be found in ISO/TS 16733-2 (2021).

However, it is important to note that the development of the TFM methodology has primarily been limited to non-combustible compartments. There has been very little research conducted in relation to travelling fires in compartments with exposed timber linings, and there is currently no generalised guidance available. In a recent experiment in a large compartment with an exposed CLT ceiling, Kotsovinos et al. (2022) found that the temperature profiles beneath the CLT ceiling were unlike the travelling fire methodologies in the literature. Although there have been few tests of mass timber compartments representative of open-plan layouts, recent experiments suggest that burning on ceiling surfaces will increase the fire spread rate and can result in a fully developed fire rather than a travelling fire (Liu & Fischer, 2022). Therefore, at the current time, existing travelling fire methodologies for non-combustible compartments cannot be used to design large compartments with large areas of exposed mass timber. Given the growing demand for compartments of this type to be constructed, this represents a significant current research need (Rackauskaite et al., 2020).

3.4.5 Standard fire resistance test

The majority of countries require fire resistance of elements and assemblies to be evaluated in accordance with the standard time–temperature curve, as defined in ISO 834 and EN 1363-1 (Equation 3.25). In North America, similar standard test methods are required, such as the ASTM E119 (USA) and CAN/ULC S101 (Canada), as given per Equation 3.26:

$$T_f = 20 + 345\log_{10}\left(8t + 1\right) \quad [°C] \tag{3.25}$$

$$T_f = 20 + 750\left(1 - e^{-0.49\sqrt{t}}\right) + 22\sqrt{t} \quad [°C] \tag{3.26}$$

where T_f = furnace gas temperature (°C) and t = time (min).

Standard fires, in contrast to real or parametric fires, do not incorporate a decay stage, and the standard test is terminated at a specific time during the heating period. By comparing results of a parametric fire curve analysis with the results of room fire experiments, Mikkola et al. (2017) have suggested that when a very large proportion of exposed wood material contributes to the fire, then the hydrocarbon (HC) curve might be used instead of the standard fire resistance curve as a worst-case design fire scenario.

When conducting a standard fire resistance test on a combustible test specimen, the amount of furnace fuel required (typically gas) will be less than for a non-combustible test specimen, to follow the same specified time–temperature curve. However, even if the concept of fire testing in furnaces was developed for non-combustible construction, the actual thermal exposure of combustible assemblies in a standard fire resistance test is similar to that of non-combustible assemblies, because the furnace tests simulate a ventilation-controlled fire development for a predefined duration (Schmid et al., 2019). The lower amount of burned fuel in furnace tests with timber assemblies can be explained by the contribution of combustible gases released from the specimen and the low thermal inertia of the wood. Nevertheless, the fire exposure will conform to the applicable standard fire test requirements.

3.5 FIRE EXPERIMENTS IN CLT COMPARTMENTS

Over recent years, there have been many compartment fire tests incorporating various amounts of exposed and/or protected CLT on the walls and ceiling. These experiments have contributed to the understanding of the fire dynamics in these types of compartments and the influence of various parameters such as adhesive selection, CLT layer configuration and char layer fall-off, amount of exposed CLT, fuel load and the area of openings. The following experiments are a selection of some of the more recent studies.

- Arup conducted an experiment on burning wood cribs in a large open-plan compartment with a floor area of 352 m² that included a fully exposed CLT ceiling (Kotsovinos et al., 2022).
- American Wood Council Project with experiments carried out by the Research Institutes of Sweden (Brandon et al., 2021).
- US Department of Agriculture Forest Products Laboratory experiments on a two-storey mass timber building (Zelinka et al., 2018).
- NFPA Research Foundation Project with experiments on CLT compartments carried out at NIST (Su et al., 2018a).
- National Research Council of Canada, experiments with exposed wood surfaces in encapsulated mass timber construction (Su et al., 2018b).

Brandon and Östman (2016) have published a detailed literature review as part of the NFPA Research Foundation Project that includes a summary of 41 fire experiments in compartments, conducted up until 2016, comprising exposed or protected wood-based construction. Additionally, they give an overview of the relevant test parameters, results and conclusions. Liu and Fischer (2022) also provide a comprehensive review of recent large-scale CLT compartment fire tests, including more recent tests conducted since the Brandon and Östman report.

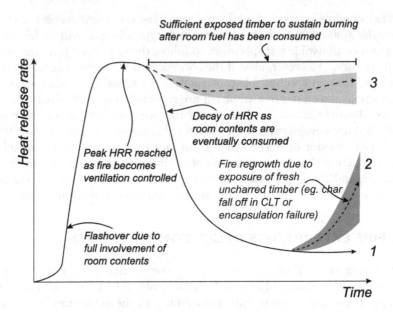

Figure 3.7 Fire development that could result from a compartment fire with exposed mass timber surfaces. Adapted from Barber et al. (2021) with permission.

The major finding from the large number of compartment fire experiments conducted to date is that exposed timber on the compartment internal surfaces impacts the compartment fire dynamics, specifically the fire growth, rate of heat release, fire duration and the fire decay. Figure 3.7 illustrates possible fire development curves for different amounts of exposed wood (Barber et al., 2021). Curve "1" is typical for compartments where the internal surfaces are non-combustible or where the wood is protected. In this case, once the compartment contents (movable fuel load) are consumed, there is a steady and predictable decay. In situations with some exposed wood, there can be fire decay, like curve "1," but then the fire may regrow, due to a possible glueline failure of CLT (char layer fall-off) or the failure of any protective encapsulation (curve "2"). If there are larger amounts of exposed wood or areas of wood exposed to each other promoting re-radiation, then the fire may not decay, even after the contents are consumed, as shown in curve "3."

3.6 OTHER FACTORS FOR TIMBER COMPARTMENTS

3.6.1 Char fall-off

While the char layer in solid wood members may erode, crack and fissure over time due to oxidation of the char including pieces of char falling away, engineered timber in addition may be susceptible to failures associated with the adhesive used to bond the lamellae together. For a given fire scenario, the time to failure will mainly depend on the orientation (vertical or horizontal) and layup (crosswise or in the same direction) of the lamellae. However, if the glueline integrity is not maintained, it will also depend on the thickness of the lamella.

Phenolic-resorcinol-formaldehyde adhesives (RF/PRF) have traditionally performed well in timber structures given their excellent structural performance, long-term durability performance and resistance to fire temperatures. Newer adhesive types such as melamine-(urea)-formaldehyde (MF/MUF), emulsion polymer isocyanate (EPI) and polyurethane (PUR) have become increasingly popular in engineered timber due to their lower cost, handling, shorter hardening times and gap-filling capacity. Some may perform well at fire temperatures, but some may not. In recent times, one-component polyurethane (1C-PUR) adhesives for engineered timber products have also become very popular, but many formulations do not perform well in fire and soften at temperatures before charring of timber occurs. Newer more temperature-resistant formulations are under active development largely driven by imminent code changes for the use of fire rated CLT in North America requiring compliance with ANSI PRG 320-18. In Europe, the next version of Eurocode 5 (prEN 1995-1-2, 2021) will contain a method to quantify if the glueline integrity of a glued engineered wood product is maintained in fire. The mass loss rate and the charring rate of the timber products are key parameters for the evaluation of the performance of engineered glued timber products in fire (see Chapter 7).

In both fire resistance and compartment fire tests of CLT manufactured using adhesives not compliant with ANSI PRG 320-18, fall-off of charred lamellae has been observed (Brandon & Dagenais, 2018), as illustrated in Figure 3.8, followed by an increased charring rate and fire regrowth. Frangi et al. (2009) concluded that CLT with thicker layers showed better fire behaviour compared to that with thinner layers, while Klippel et al. (2014, 2016) reported that ceilings were more prone to fall-off of charring lamellas than exposed walls, likely due to the action of gravity.

A recent series of full-scale compartment experiments were carried out by Su et al. (2018b) with CLT manufactured with a fire-resistant adhesive. All tests in this series showed that the CLT maintaining glueline integrity demonstrated improved resistance to the char layer fall-off. In these experiments, there was no char layer fall-off after the time the char front had passed the adhesive bond and the fire led to self-extinguishment. Other large-scale compartment fires with CLT manufactured with fire-resistant adhesive meeting the North American standard ANSI/APA 320 have been conducted in Sweden for the American Wood Council to further demonstrate the effectiveness of this new generation of CLT panels (Brandon, 2021).

Figure 3.8 Example of char fall-off. (Photo Daniel Brandon with permission.)

There are therefore two main approaches to mitigating the risk of char fall-off in CLT. The first (and more reliable) option is to use an adhesive that is fire-resistant such that the glueline integrity is maintained. The second option is for the thickness of the surface layer of CLT to be greater than the maximum expected depth of charring ensuring that the temperature at the glue line is low enough to avoid a fall-off of charring lamellae for the full duration of the expected fire. However, it is recommended that only adhesives able to resist the expected temperatures reached during a fire are used where a requirement to withstand burnout must be met. Another reason is that currently available verification methods to demonstrate burnout is reached are not able to consider the influence of a char fall-off. This is another area of active research.

3.6.2 Protective coverings

There are generally three methods for protecting engineered timber: (1) use of fire-retardant or intumescent coatings; (2) use of fire-retardant chemicals to pressure-impregnate timber; (3) the use of protective materials, including insulative sheets or other materials applied to the face of the timber. The first two methods are often used to influence the flame spread behaviours (surface flammability) during the early stages of fire development while occupant evacuation is underway. Except for some intumescent coatings, they typically have little effect on fire resistance performance and only improve the reaction to fire class performance. On the other hand, sheet- or board-insulating protective materials can be used to increase the level of fire resistance for timber structures by limiting or delaying the onset of charring.

Sheet or board products can be applied to the surface of mass timber construction, such that some or all of the mass timber is protected either to prevent charring (encapsulation) or to delay charring (partial encapsulation) of the underlying wood, as described in Chapters 2 and 6.

If a strategy of partial encapsulation is used, there is a risk for timber to pyrolyse and contribute to the fire at some stage during the fully developed or decay stages of the fire, and therefore the potential effect on the fire dynamics must be considered. The easiest strategy to fully mitigate the hazards of timber construction is to prevent it from burning or pyrolysing in the first place (i.e. encapsulation). This strategy can be achieved if the surface of the timber is protected so that the maximum temperature of the timber is kept low enough to avoid charring and unacceptable damage prior to burnout. There are standard test methods available that are intended to limit temperatures on structural timber to mitigate against charring or combustion for specified periods of time based on standard fire resistance test exposures. For example, both CAN/ULC S146 and EN 13501-2 (for K-classes with testing to EN 14135) require the average temperature rise at the interface between the timber and encapsulation be no higher than 250°C and also the maximum temperature rise at any single point be no higher than 270°C. Besides the temperature requirement, the onset of local charring or damage on the substrate is considered as additional criterion in Europe. In cases where sufficient protection is to be provided to the underlying structure/substrate to fully mitigate the onset of pyrolysis for the full duration of the compartment fire, the interface temperature between the substrate and lining should remain below 200°C (to avoid the decomposition of hemicellulose and maintain additional strength) as recommended by the Structural Timber Association (2020). Further information on timber protection is found in Chapter 6.

The performance of various encapsulation methods for cross-laminated timber panels, including Type X gypsum board, intumescent coating, rock fibre insulation and spray-applied fire-resistant materials were reported by Hasburgh et al. (2016). Timber strength and stiffness reduction factors due to elevated temperatures are discussed in Chapter 7.

3.6.3 Location of exposed or partially protected timber surfaces

The location and area of exposed timber linings can influence both the fire growth rate during the pre-flashover stage (which may be important for occupant safety during evacuation) and the burning rate in post-flashover fully developed fire, including effects on structural fire behaviour and external flaming. It has been observed that where burning timber surfaces face each other such as opposing surfaces, in corners or across wall/ceiling intersections or are in a corner arrangement, the reradiation between these surfaces may be an important contributor to sustaining higher levels of local heat flux, whereas calculations based on global energy balance for the compartment might not be sufficient to capture these effects. For example, researchers, e.g. Hadden et al. (2017) and Li et al, (2016), have found that radiative exchange between burning wood surfaces was sufficiently high to sustain the flaming and for the burning to continue. This effect can be minimised by including exposed timber on only one wall or ceiling surface, thus avoiding exposed adjacent surfaces in a corner. Hadden et al. (2017) also observed that in compartments with two surfaces of exposed timber burnout could be achieved but was dependent on the char layer remaining attached, i.e. no glueline failure or char fall-off during the burning or the decay period.

In fully developed fires, Gorska et al. (2021) also found that the rate of pyrolysis of an exposed timber ceiling is less than for the walls and therefore exposed timber on the ceiling might be preferred over a wall location. Gorska et al. attributed the lower pyrolysis rate for the ceiling to a less efficient char oxidation process which corresponded to a thicker char layer that diminished the heat flux reaching the pyrolysis front. This reduction in burning rate of the ceiling compared to the other surfaces was approximately 30%. This effect was also seen in a compartment experiment (Test 5) by Su et al. (2018b). However, there are other disadvantages of an exposed timber ceiling such as faster fire spread and higher CO yields, as previously mentioned.

Finally, the top side of a CLT floor panel also requires consideration if it is not encapsulated to prevent onset of charring, such that the contribution of the timber to the fuel load must be considered in any fire severity calculations to verify if burnout is expected to occur.

3.6.4 Wind effects

There is limited information available about the effect of wind and air currents in relation to buildings of all types of structural materials, especially in the post-decay period. With respect to timber buildings, Crielaard found that the airflow did have an influence on the smouldering of CLT, but he only investigated two flow rates (Crielaard et al., 2019).

Wind-driven flows created by openings on both the upwind and downwind sides of a building can be accompanied by higher temperatures and make fire-fighting more difficult, particularly in taller buildings. NIST performed fire experiments in the laboratory and in a seven-storey structure to enable a better understanding of wind-driven fire tactics, including structural ventilation and suppression (Kerber & Madrzykovski, 2009). As part of their laboratory experiments, they found that wind-driven fire behaviour can occur with wind speeds as little as 4.5 m/s. The effect of wind-driven flows on charring rates is difficult to predict and requires further study.

See also Chapter 14 for more information on firefighting in timber buildings.

3.7 DESIGN TO WITHSTAND BURNOUT

3.7.1 Design intent

The fire resistance framework in modern codes and standards based on standardised tests can be traced back to early work on time-equivalence by Ingberg (1928) where the fuel load in compartment fires was related to the fire duration. Codes and standards subsequently formulated fire resistance levels based on characterising the fuel load in a compartment in the expectation that for a given fuel load, the specified fire resistance rating should be sufficient for the building element to withstand burnout without structural failure or fire spread.

Using the current fire resistance framework, it may be possible for designers to obtain the necessary approvals from regulators without explicitly addressing the issue of burnout, i.e. in those buildings not required to be designed for burnout such as some low-rise or mid-rise timber buildings. However, as timber buildings become increasingly taller and more complex, the consequence of failing to design for burnout will also become greater, as does the risk to life, property and the environment (Law & Hadden, 2017). While some regulatory authorities around the world have been reviewing and updating requirements to allow greater use of mass timber structures, there are other countries where the regulatory requirements are much less detailed and where greater reliance is therefore placed on the expertise of the designers to ensure that timber structures achieve the goals and objectives of the local building code.

Since the research being done regarding fire performance of timber structures and the understanding of the fire dynamics in particular is still rapidly evolving, it is essential for structural fire engineers to exercise caution and as much as possible apply the current knowledge regarding design to withstand burnout to those buildings where there is a requirement for the structure to remain stable after the fire. This would likely include almost all tall, and many midrise, buildings with considerable exposed timber surfaces, and likely exclude those considered to be common low-risk. Some countries may have explicit requirements in this regard. For example, in Canada, guidelines for the development of limit states design CSA S408 (2011) stipulate that design for burnout is only required for buildings that are classified as "high buildings." In Canada, this would typically include residential buildings taller than 18 m or office buildings taller than 36 m.

Hybrid buildings using steel frame (or concrete) structures with exposed CLT or other mass timber panels should also include the fuel load from charring of exposed structural timber in any fire severity calculations when verifying the fire performance of the steel frame (or concrete) structure.

3.7.2 Burnout

When designing with the intent of achieving burnout, the response of the timber during the decay stage of the fire may become critically important. This is because temperatures within the timber sections may continue to increase after the peak fire gas temperature in the compartment has been reached and the decay stage commences. This is sometimes referred to as the "thermal wave" and occurs in most structures, including concrete and protected steel. While the higher insulating effects of timber do help to mitigate the impact of this thermal wave, it is still important in timber structures because of the potential reduction of strength and stiffness within the residual cross section that occurs some time after the peak compartment gas temperature has been reached, i.e. during the decay stage of the fire.

When designing for fuel burnout and to ensure structural integrity of timber structures, it is critical to ensure that flaming combustion ceases and smouldering combustion is minimised after the moveable fuel load has been consumed within the compartment. This will typically require a fire watch after the event and facilitating firefighters to manually extinguish any areas of smouldering that may remain. The residual structure can then be assessed for its load-bearing capacity after the fire. This is in contrast to a standard fire resistance test where the test is terminated at the end of the heating period and the specimen is often extinguished with water or left to smoulder and collapse after the test has ended.

The calculation for the rate of pyrolysis (e.g. Equation 3.1) is only applicable for solid timber. Where wood products comprise engineered timber and where layers of timber are bonded together with adhesive, the possibility of char layer fall-off due to glueline integrity failure of the adhesive bond at fire temperatures must be considered. If char fall-off occurs during the fire, fresh timber will be subjected to heating, the pyrolysis rate may increase and the burning may continue.

There have been a number of medium- to large-scale CLT compartment experiments conducted that have shown that burnout can be achieved under certain conditions, including those by Su et al. (2018b) where the CLT was manufactured using a fire-resistant adhesive. Where burnout was achieved, there were typically few exposed CLT faces within the compartment, comparatively low fuel loads and minimal char "fall-off." Conversely, where there have been a larger number of exposed faces and significant char "fall-off," a secondary flashover and regrowth of the fire (or sustained burning) has been observed (Law and Hadden, 2017). A series of five compartment fire experiments (Zelinka et al., 2018) in a two-storey apartment structure conducted by the USDA Forest Products Laboratory at the US Bureau of Alcohol, Tobacco, Firearms and Explosive (ATF) Fire Research Laboratory demonstrated cases where the fire was contained and ceased to burn for a fully protected CLT structure; where only a limited area of ceiling was exposed (20% of the floor area); and also for the case of two perimeter walls with an exposed area equal to 40% of the floor area. This series of compartment experiments has informed proposed changes to the International Building Code (IBC) in relation to mass timber construction. Further compartment fire experiments undertaken by Brandon et al. (2021) using ANSI PRG 320-18 qualifying CLT showed that burnout could be achieved with 100% of the ceiling area exposed with other surfaces protected; or with walls with exposed area 40% of the floor area and at least 4.5 m between any exposed areas on walls that were facing each other. Further proposed changes have been made to the IBC along these lines based on this research (Brandon and Smart, 2021).

If it is possible to calculate the heat losses and the pyrolysis rate with acceptable accuracy, then it may also be possible to show, for a non-fire-resistant adhesive, that burnout could occur such that the depth of char does not reach the first layer of adhesive and that the temperature of the adhesive is sufficiently low that bond failure of the adhesive is unlikely. However, it needs to be recognised that there are considerable uncertainties in calculating the pyrolysis rate, temperature profile and when and how much char fall-off, if any, might occur due to the failure of the adhesive within the timber elements. Therefore, if likely burnout must be verified in projects, a more prudent approach is to specify adhesives that are more resistant to fire temperatures such that the burning behaviour of the engineered timber sections would be expected to be similar to that of solid timber and thus avoiding the prospect of premature failure at the bond line. In Europe, Klippel et al. (2018) proposed a new method using standard fire exposure to assess the adhesive performance in fire with a model-scale fire test where the mass loss rate of a timber panel (such as CLT) is used to draw conclusions on the performance of the timber product and consequently on the bond line performance. The concept compares the mass loss rate of a solid timber panel (serving as a benchmark) with the mass loss rate of the glued engineered wood product. It is recommended in future to use fire tests to determine the charring rate and the mass loss rate of the tested specimen as described in Klippel et al. (2018). This method will likely be adopted in the next version of Eurocode 5 (prEN 1995-1-2, 2021) to define the charring performance of the timber product. It is also important for the construction to be detailed, following specific principles to avoid fire spread via cavities and voids, as discussed in Chapter 9.

The conditions under which burning and smouldering cease in engineered timber is currently an active area of research, and guidance may well be amended and refined as further data becomes available.

3.8 CALCULATION METHODS FOR COMPARTMENTS WITH EXPOSED TIMBER

3.8.1 Methods using parametric fires

Performance-based design with the objective of ensuring a compartment burn-out of contents without collapse or fire spread through compartment boundaries may require an estimate of the ultimate char depth within mass timber construction to be determined. This requires that the contribution from the timber surfaces must be accounted for. Parametric time–temperature relationships developed for natural compartment fires in non-combustible compartments, e.g. EN 1991-1-2 Annex A, are not applicable when the compartment boundaries are combustible and contribute additional fuel to the fire.

However, parametric curves may be modified to account for the contribution from combustible mass timber. Brandon (2018a) proposed an engineering method based on the parametric fire equations in conjunction with an iterative procedure to estimate the char depth by adjusting the fuel density at each iteration. The char rate was based on an empirical model derived from a large number of parametric fire tests. However, Brandon's method is only applicable when glueline integrity of engineered timber lamella is maintained and any protective board encapsulation products used to protect the underlying timber do not fail or fall off. These additional requirements must be separately demonstrated (e.g. using the approaches described in Section 3.6.2).

Barber et al. (2016) proposed a similar two-step engineering methodology for CLT, but included the additional step of checking for smouldering-extinction of CLT. This involved calculating the incident radiant heat flux on the timber surfaces and ensuring it was below a critical value of about 5–6 kW/m^2.

Schmid and Frangi (2021) also proposed a model to estimate the contributions from structural timber to a fire from its fully developed and decay phases until burnout using the energy stored in the char layer as a key characteristic. Their Timber Charring and Heat Storage (TiCHS) model introduces a second material being the "char layer." Schmid and Frangi concluded that the TiCHS model is able to predict burnout and the charring depth. Further, the model allows the determination of the factor α_{st} to describe the combustion behaviour of structural timber. Currently, the model is validated for the gas velocities, which occur in compartments with openings on one side. In the future, it is expected that imposed gas flow by wind can also be considered, a phenomenon that may be important for medium- and high-rise buildings with wind crossflows.

The remainder of this section describes an iterative procedure for determining the amount of fuel contributed by exposed mass timber surfaces and the resulting depth of char for a fire with a parametric temperature–time curve from EN 1991-1-2 Annex A (previously discussed in Section 3.4.2). However, it is important to note that while this type of calculation may be necessary to demonstrate burnout is possible, it is not sufficient on its own and cannot be used to verify the cessation of all smouldering combustion.

Following EN 1991-1-2 Annex A, the gas temperature θ_g (°C) at time t (hours) is given by

$$\theta_g = 20 + 1{,}325\left(1 - 0.324e^{-0.2t\Gamma} - 0.204e^{-1.7t\Gamma} - 0.472e^{-19t\Gamma}\right) \quad [°C] \quad (3.27)$$

where Γ = heating rate factor that depends on the thermal properties of the compartment and the opening factor O, ρ = density (kg/m³), c = specific heat (kJ/kg·K), k = thermal conductivity (kW/m·K) of the compartment's boundaries, A_t = total area of floors, walls and ceiling, including openings (m²), and h_v = weighted average height of the compartment openings (m).

$$\Gamma = \frac{\left(\dfrac{O}{\sqrt{k\rho c}}\right)^2}{\left(\dfrac{0.04}{1160}\right)^2} \quad [-] \tag{3.28}$$

$$O = \frac{A_v}{A_t}\sqrt{h_v} \quad \left[m^{1/2}\right] \tag{3.29}$$

An empirical relationship for the char rate dependent on the heating rate has been proposed by Brandon (2018b). This was based on fire tests in modern furnaces controlled using plate thermometers updating a correlation currently included in Eurocode 5, Appendix A, previously developed by Hadvig (1981). The parametric char rate β_{par} (mm/min) is given by

$$\beta_{par} = \beta\Gamma^{0.25} \quad [mm/min] \tag{3.30}$$

where β is the charring rate corresponding to standard fire resistance tests following ISO 834 and corresponds to either the one-dimensional β_o charring rate for flat surfaces or the notional charring rate β_n for rectangular members, as described in EN 1995-1-2.

Charring is assumed to start reducing at time t_o given by

$$t_o = 0.009\frac{q_{t,d}}{O} \quad [min] \tag{3.31}$$

where $q_{t,d}$ = design fire load per unit area of internal surfaces excluding the openings (MJ/m²).

Charring is assumed to completely stop at time $3t_o$, so the final char depth is given by

$$d_{char} = 2\beta_{par}t_o \quad [mm] \tag{3.32}$$

The gas temperature starts to decline at time t^1_{max} (from EN 1991-1-2 Annex A).

$$t^1_{max} = \max\left[0.0002 q_{t,d} / O; t_{lim}\right] \quad [\text{hour}] \tag{3.33}$$

where $t_{lim} = 0.333$ hour (20 min) assuming a medium fire growth rate (25 min for slow and 15 min for fast fire growth rates).

The contribution of timber is calculated iteratively using the following expression where $q^{i+1}_{t,d}$ is the total fire load at the $(i + 1)$th iteration, including the moveable fire load q_{mfl}, which is the moveable fire load per unit compartment internal surface area, including the openings (MJ/m²). t^1_{max} is constant and does not change for subsequent iterations.

$$q^{i+1}_{t,d} = q_{mfl} + \frac{A_{CLT} \alpha_1 \left(d^i_{char} - 0.7 \beta_{par} t^1_{max}\right)}{A_t} \quad [\text{mm}] \tag{3.34}$$

where A_t = internal compartment surface area, including openings (m²), A_{CLT} = area of exposed timber (m²), and α_1 = ratio between the heat release and char depth and is taken as 5.39 MJ/m² per mm of char depth experimentally determined by Schmid et al. (2016). This was derived from cone calorimeter experiments at an irradiance of 75 kW/m² flux for char depths exceeding 10 mm. The parameter $0.7 \beta_{par} t^1_{max}$ is an estimate of the proportion of the char depth burning outside the compartment during the fully developed stage (of a duration t^1_{max}) in a non-combustible compartment. Equation 3.34 is therefore only valid for compartment fires that reach flashover and become fully developed.

To validate the method based on a selected number of compartment experiments as shown in Table 3.3, Brandon (2018a) produced the comparison of the predicted and experimental char depth, as shown in Figure 3.9. The opening factors applying to the experiments were in the range of 0.03–0.10 m^0.5 and the method might not apply where the opening factor lies outside this range. The method is also applicable only when glueline integrity of engineered timber lamella is maintained and any protective board encapsulation products used to protect the underlying timber do not fail or fall off, and where exposed adjacent wood surfaces do not face each other.

When the char depth converges to a stable value in the calculation, the designer can assume this to be an estimate of the maximum char depth within the exposed wood surfaces within the compartment. However, this excludes any localised effects and hot spots where smouldering combustion may persist requiring additional consideration, as discussed in Section 3.2.3.

Table 3.3 Overview of compartment test properties, results and corresponding predictions

	Compartment test properties					Experimental results				Prediction
Test	Dim (m)	Opening dim (m)	Total unprot surface (m²)	Movable fuel load per floor area (MJ/m²)	Time to flashover (min)	Time of char measured t_{ch} (min)	Char depth at t_{ch} lower limit (mm)	Char depth at t_{ch} upper limit (mm)	Method¹	Calculated charring depth (mm)
I–3	4.6×9.1×2.7	3.6×2	24.6	550	14	69	35	35	A	45
I–4*	4.6×9.1×2.7	1.8×2	41.8	550	11	115	50	50	A	79
A2	9.1×9.1×2.7	7.3×2.4	24.8	550	11	66	23	23	A	49
A3	9.1×9.1×2.7	7.3×2.4	29.9	550	11	39	23	23	A	32
K3	3.5×4.5×2.5	1.1×2	11.3	550	5	68	21	44	B	46
Q1	2.7×2.7×2.8	0.8×1.8	7.4	132	9	51	11	11	N.F.	13
R1*	2.7×5.8×2.4	0.9×1.9	15.9	456	13	179	70	89	B	88
R2	2.7×5.8×2.4	0.9×1.9	15.9	456	13	197	52	70	B	88
R3	2.7×5.8×2.4	0.9×1.9	15.9	456	13	227	49	70	B	88
S1	3.5×4.5×2.5	3×1.5	22.5	600	47	46	35	35	A	39

*Tests involved delamination during the fully developed phase.
¹A = Char depth determined using thermocouple measurements. B = Char depth determined physically after the test.

Source: Reproduced with permission from Fire Protection Research Foundation, Fire Safety Challenges of Tall Wood Buildings – Phase 2: Task 4 – Engineering Methods, Copyright © 2018, Fire Protection Research Foundation, Quincy, MA.

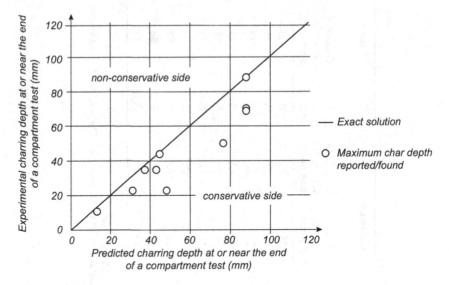

Figure 3.9 Predicted versus experimental char depth at or near the end of the decay phase. Reproduced with permission from Fire Protection Research Foundation, Fire Safety Challenges of Tall Wood Buildings – Phase 2: Task 4 – Engineering Methods, Copyright© 2018, Fire Protection Research Foundation, Quincy, MA. All rights reserved.

3.8.2 Compartment fire models that include wood pyrolysis

Traditional compartment fire models solve the governing equations for mass and energy for discrete control volumes to calculate the fire gas temperatures and heat fluxes within the compartment. Including the pyrolysis of combustible surfaces is a much more complicated calculation and there are relatively few models for this purpose available to the fire engineer.

Fire dynamics simulator (FDS) has been used to determine the response of mass timber structures and the method has been validated against the results from five full-size compartment fire tests with exposed cross-laminated timber (Barber et al., 2018). Inputs for the pyrolysis model include kinetic properties for the timber. In that study, char depths were predicted within 20% based on a fully developed fire. However, CLT char fall-off (due to a failure of the glueline) was not captured, nor any gypsum board fall-off or charring of CLT behind the gypsum board. The computational effort and time required was also very large.

The B-RISK zone model (Wade et al., 2016 and 2018; Wade, 2019) includes optional sub-models for calculating the contribution of exposed mass timber for determining the fully developed fire environment within a compartment where varying amounts of timber are exposed on the walls and ceiling. The wood surfaces are assumed to contribute fuel mass based on the position of the 300°C isotherm within the bounding surfaces. This is a similar approach to that described in Section 3.8.1 where the total fuel available to burn was updated at each time step to account for the additional contribution from the timber surfaces. The model also allows the proportion of burning external to the compartment to be specified. Wade validated the model predictions of gas temperature against 19 full-scale experiment configurations with good estimates of the peak temperature and the duration of burning (Wade, 2019). More recent developments have included a detailed kinetic model for the wood pyrolysis integrated within the zone model framework (Wade, 2019; Wade et al., 2019).

SP-TimFire (Brandon, 2016) is a zone model that calculates the heat release rate of the CLT by assuming a linear relationship with charring depth of 5.39 MJ/m^2 per mm of char depth. The model was used iteratively with heat conduction calculations done using the finite element program SAFIR (Franssen, 2005), and a fall-off of charring lamellae was simulated by removing the exposed lamella from the model when temperatures in the bond line reached a specified temperature.

Schmid and Frangi (2021) presented a simplified engineering model for the consideration of structural timber in compartment fires. Their Timber Charring and Heat Storage model (TiCHS-model) is able to assess the contribution of structural timber to the fire load in the fire compartment. Again, an iterative approach is followed based on the prediction of the compartment environment, i.e. the temperature and the gas characteristics in the compartment. The predictions achieved an overall good agreement unless fall-off of charring layers induce a regrowth of the fire due to the sudden change of the combustion characteristics.

To date, these and similar models and tools are either still under development or have mainly been used within a research environment. While they may also represent a useful advance allowing engineers to quantify some aspects of the fire dynamics in timber buildings, engineers would be well-advised to independently validate the model or tool with data relevant to the application to which it is intended to be used.

There are also other well-validated detailed pyrolysis models decoupled from the fire environment that have been developed in recent times, including by Richter et al. (2020a and 2020b). These are potentially useful to provide greater insight into the charring behaviour of timber structures given the specified boundary conditions in the future, but are not currently in general use today.

Importantly, none of these models can address the cessation of smouldering combustion in localised areas or hot spots, so they are not sufficient on their own. They can only form one part of the overall fire safety strategy for a building.

3.8.3 Time equivalence methods

It was previously seen that compartment time–temperature curves for real fires can be quite different than for the standard fire test. These differences are due to the amount, location and properties of the fuel, the area of openings and the size and properties of the bounding surface materials of the enclosure. While furnace tests could be conducted to more closely follow the expected temperature in a real compartment fire, it is not very practical due to the large number of potential scenarios that could apply, even within a single building. Therefore, methods have been developed to determine the period of time exposed to a standard fire, which would result in the same structural response that would occur when that same structural system is exposed to the real compartment fire. This is referred to as time equivalence. MacIntyre et al. (2021a and 2021b) provide a comprehensive review of the various time equivalence approaches.

In mass timber compartments, the equivalent fire resistance period is the time of exposure to the standard fire at which the char depth equals the maximum char depth reached for the same element exposed to a real fire (Barber et al., 2021). The equivalent standard fire resistance period would be estimated by dividing the maximum char depth in the real fire by the notional char rate for the standard fire (i.e. approximately 0.65 mm/min). While the calculated char depth may be used directly to inform the structural design, in the case of non-structural building elements such as fire doors or fire-stopping systems, estimating the equivalent time in the standard fire test would be more useful.

Where performance-based design of mass timber compartments is being considered, it is important to note that simple time equivalence formulae developed for steel structures such as the formula in EN 1991-1-2 Annex F are not appropriate for compartments with exposed mass timber surfaces, unless the amount and effect of the additional burning timber fuel can be accounted for.

3.8.4 Summary of fire severity models for mass timber buildings with exposed wood

Table 3.4 provides a simple summary of some of the generic methods that can currently be considered to assist designers with the fire design of mass timber elements. All methods will have limits of application and it is necessary for designers to be aware of the capability and validity of any method used for a specific application. In particular, none of the methods can predict

Table 3.4 Summary of fire design methods for mass timber elements

	Model	Burnout*	Notes	Char depth
1	Prescribed fire resistance, using the standard temperature–time curve	No	Typically, 30, 60, 90 or 120 minutes as prescribed by the applicable building code. The decay phase of the fire is not considered, so the structure may continue to degrade after the end of the fire resistance test	Standard fire charring rate
2	Equivalent time of standard fire using the Eurocode formula (EN 1991-1-2 Annex F)	Yes	Should not be used where mass timber surfaces will contribute to the fuel. This formula is based on non-combustible compartments and structural steel assumptions. It is not appropriate for structural timber	Standard fire charring rate
3	Room fire model based on the Eurocode parametric fire (EN 1991-1-2 Annex A), e.g. TiCHS-model, Brandon iterative method.	Yes	Essential to use an iterative method, to account for increasing fuel load from timber surfaces. Concern about the accuracy of the gas temperature decay rate. Need to assume no char fall-off during the fire and uniform heating around the compartment. Floor area limited to max 500 m^2 and 4 m high. Opening factor limits given in EN 1991-1-2 Annex A. The German National Annex to BS EN 1991-1-2 gives a more realistic decay phase for a parametric temperature–time curve and can also be used with the iterative method	Calculated in model
4	Zone model fire incorporating a pyrolysis and boundary heat conduction sub-models, e.g. B-RISK, SP-TimFire	Yes	Users need to demonstrate adequate benchmarking or validation, because of uncertainty about input variables, thermal property assumptions, etc. Need to assume no char fall-off during the fire. Uniform fire throughout	Calculated in model
5	CFD model (field model) incorporating a pyrolysis model, e.g. FDS	Yes	Users need to demonstrate adequate benchmarking or validation, because of uncertainty about input variables, thermal property assumptions etc. Need to assume no char fall-off during the fire	Calculated in model

* None of the methods can predict when local areas of smouldering combustion will cease.

when local areas of smouldering combustion at the end of the decay stage will cease.

3.9 WORKED EXAMPLE

This example is based on the method proposed by Brandon (2018a, 2018b) and estimates the depth of char in an exposed mass timber wall or ceiling resulting from exposure to a fire described using the EN 1991-2 Annex A parametric fire equations where the fire load density also includes the contribution from the exposed mass timber.

3.9.1 Description

The example is for an experiment with gas temperature and char depth data previously recorded being Test 2 in a series of CLT compartment fire experiments conducted by Su et al. (2018b).

Compartment parameters

- Compartment internal dimensions: 2.4 × 4.5 × 2.7 m (width W × length L × height H).
- Ventilation opening: 1 opening with height, $h_v=2.0$ m and area, $A_v=1.52$ m².
- CLT 175-mm thick with five layers and 7% moisture content.
- Exposed CLT surfaces: 33% of walls, 10% of ceiling exposed giving a total of 13.4 m².
- Two layers of 15.9 mm Type X gypsum boards are fixed to the remaining wall and the ceiling. It is assumed that protection boards do not fail and glueline integrity is maintained in exposed CLT. Fire-resistant polyurethane adhesive is used which meets the full-scale fire test requirements by ANSI/APA PRG-320.

Fire parameters:

- As per EN 1991-1-2 Annex A Parametric fire curve.
- Fire growth rate: fast with $t_{lim}=15$ min.
- Fire load density per unit floor area (FLED) excluding CLT is 550 MJ/m². The fuel load comprises wood cribs capable of causing a fully developed fire in a completely non-combustible compartment.
- Thermal parameter, $\sqrt{k\rho c}$ for the compartment boundaries: $\sqrt{k\rho c}=606$ J/s·m·²·K (surface area weighted between plasterboard and timber.

3.9.2 Procedure

First iteration (steps 1–8) for non-combustible or fully protected compartment.

Step	Parameter	Equation and notes	Equation	Value
1	Opening factor	$O = \dfrac{A_v}{A_t}\sqrt{h_v}$ O must be in range of 0.03–0.10 $m^{0.5}$	3.29	0.036 $m^{1/2}$
2	Heating rate factor	$\Gamma = \dfrac{\left(\dfrac{O}{\sqrt{k\rho c}}\right)^2}{\left(\dfrac{0.04}{1,160}\right)^2}$ $\sqrt{k\rho c}$ is assumed to be constant during the fire	3.28	3.05
3	Surface area of compartment boundaries	$A_t = 2\,(L \times W + H(L + W))$		58.9 m^2
4	Movable fire load per surface area of boundaries	$q_{mfl} = FLED \times LW/A_t$ q_{mfl} must be in the range of 50–1,000 MJ/m^2		100.9 MJ/m^2
5	Start time of gas temperature decay (first iteration)	$t'_{max} = \max\left[0.0002 q_{t,d}\,/\,O; t_{lim}\right]$ t'_{max} is a parameter applying to non-combustible compartments and does not change for subsequent iterations	3.33	0.55 hour
6	Initial charring rate	$\beta_{par} = \beta\Gamma^{-0.25}$ assuming $\beta = 0.65$ mm/min	3.30	0.86 mm/min
7	Time at which char rate reduces	$t_o = 0.009\dfrac{q_{t,d}}{O}$ For iteration i, $q_{t,d} = q_{mfl}$	3.31	24.9 min
8	Final char depth (first iteration)	$d_{char} = 2\beta_{par}t_o$	3.32	42.7 mm

For the second (steps 9–12) and subsequent iterations, the following steps are repeated to account for the additional fire load from the charring timber, until the final char depth estimate converges to a value, e.g. when the difference between successive final char depth estimates are below a specified tolerance such at 0.1%.

Step	Parameter	Equation and notes	Equation	Value
9	Total fire load per surface area of boundaries	$q_{td}^{i+1} = q_{mfl}$ $+ \dfrac{A_{CLT}\alpha_i\left(d_{char}^i - 0.7\beta_{par}t_{max}^i\right)}{A_t}$ t_{max}^i is constant between iterations Check $d_{char}^i - 0.7\beta_{par}t_{max}^i > 0$	3.34	128.8 MJ/m² (iteration 2)
10	Initial charring rate	$\beta_{par} = \beta\Gamma^{-0.25}$	3.30	etc.
11	Time at which char rate reduces	$t_o = 0.009\dfrac{q_{t,d}}{O}$	3.31	etc.
12	Final char depth	$d_{char} = 2\beta_{par}t_o$	3.32	etc.
Repeat for as many iterations as required				
	Final char depth (tenth iteration)	$d_{char} = 2\beta_{par}t_o$	3.32	67.2 mm
	Final total fire load per unit surface area of boundaries	$q_{t,d}^{i+1} = q_{mfl}$ $+ \dfrac{A_{CLT}\alpha_i\left(d_{char}^i - 0.7\beta_{par}t_{max}^i\right)}{A_f}$	3.34	158.7 MJ/m²
	Final fire load density per floor area	$FLED = \dfrac{A_t}{A_f}q_{t,d}^{i+1}$		865 MJ/m²

Note in some cases when the exposed surface area of timber is too large, the char depth may not converge, resulting in charring penetrating through the full thickness of mass timber.

Following convergence of the calculation to a stable value (~67 mm), the designer can assume this to be an estimate of the maximum char depth within the exposed wood surfaces within the compartment. However, this excludes any localised effects where smouldering combustion may persist requiring additional consideration (as discussed in Section 3.2.3) such as facilitating fire brigade access and features to ensure manual extinguishment of hot spots following the fire.

For a final char depth of 67.2 mm, an equivalent standard fire resistance duration based upon achieving the same maximum char depth can be determined as 67.2/0.65 = 104 minutes (assuming that the timber member chars with a speed of 0.65 mm/min in the standard fire), which can be used to inform the selection of any fire-rated doors, fire-stopping systems, etc. that may be required (see Section 3.8.4).

3.9.3 Experimental results

Char depths were measured during the experiment using a resistograph device, as shown in Figure 3.10.

Measured and predicted compartment gas temperatures using the iterative method with parametric temperature–time curves are shown in Figure 3.11. Also shown are temperatures predicted using an alternative B-RISK zone model including mass timber pyrolysis, as described by Wade et al. (2020). This latter model includes non-linear thermal properties and

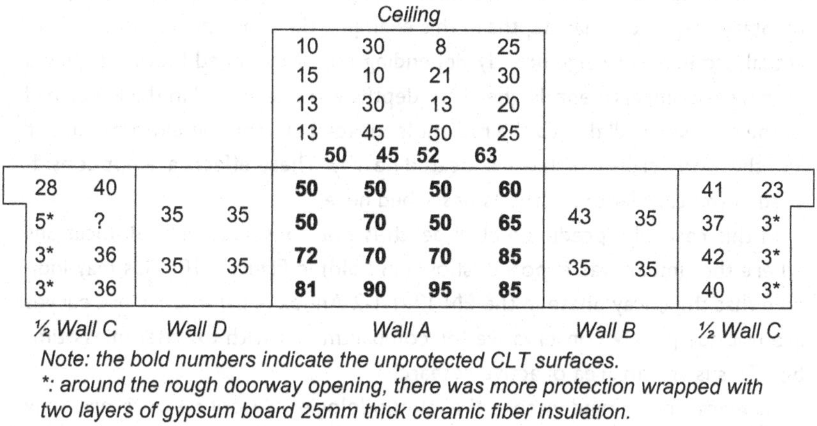

Ceiling
10 30 8 25
15 10 21 30
13 30 13 20
13 45 50 25
50 45 52 63
28 40
5* ?
3* 36
3* 36
35 35
35 35
50 50 50 60
50 70 50 65
72 70 70 85
81 90 95 85
43 35
35 35
41 23
37 3*
42 3*
40 3*
½ Wall C
Wall D
Wall A
Wall B
½ Wall C
Note: the bold numbers indicate the unprotected CLT surfaces.
*: around the rough doorway opening, there was more protection wrapped with two layers of gypsum board 25mm thick ceramic fiber insulation.

Figure 3.10 Measured char depth in experiment with exposed timber measurements in bold. Adapted from Su et al. (2018b) with permission.

Gas temperature (°C)
1400
1200
1000
800
600
400
200
0
0
30
60
90
120
150
180
Time (min)
t^1_{max}
average measured gas temperature
B-RISK zone model with pyrolysis (without calibration) - char depth 41 mm
B-RISK zone model with pyrolysis (calibration 1.6) - char depth 65 mm
B-RISK zone model with pyrolysis (calibration 1.9) - char depth 77 mm
parametric 550 MJ/m² (initial)
parametric 642 MJ/m² (final)

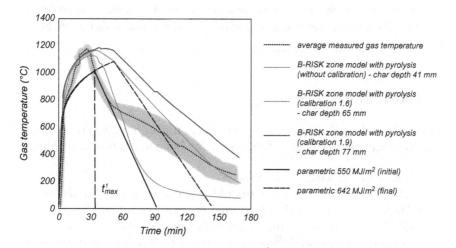

Figure 3.11 Gas temperature measured and predicted using iterative method with parametric temperature–time curves and the B-RISK zone model with pyrolysis (Wade et al., 2020).

a calibration factor applied to the thermal conductivity value. B-RISK with a calibration factor of 1.6 provided a similar final char depth prediction as obtained from the previous calculations.

> ### *NOTE:* LIMITS OF APPLICATION OF CALCULATION METHODS
>
> Additional safety factors may still be required to ensure the calculated char depth is sufficiently conservative for design purposes. As seen in the experimentally measured char depths in this example, the range of char depth in an actual fire may vary significantly depending on location and localised effects within the compartment. Higher char depths were recorded in the lower half of the exposed wall due to the radiant feedback between the exposed timber and the wood cribs which were located nearby. These effects are not considered in the calculation methods described here.
>
> In this case, the predicted char depth is not conservative in all locations where the timber was exposed (shown in bold) in Figure 3.10. This may indicate that the decay phase in the EN 1991-1-2 Annex A parametric fire curves are not adequately conservative for compartments with exposed mass timber. This is still an area of active research.
>
> Readers are reminded that the methodologies described here are only applicable for compartments that reach flashover and fully developed fire. It may not be applicable for large open-plan well-ventilated compartments if travelling fires were to occur. These are areas of active research and highlight the need for caution in the structural fire design of exposed mass timber, especially in higher risk structures such as tall and complex timber buildings.

3.10 RESEARCH NEEDS

Research on the following topics in relation to timber buildings is necessary to encourage improved understanding of fire dynamics in timber buildings:

1. The effect of wind-driven fires.
2. Conditions for cessation of smouldering combustion and burnout.
3. Understanding the risks of external fire exposure of facades.
4. Methods to quantify the external flaming depending on the global equivalence ratio and the area of exposed timber in the compartment.
5. Understanding the contribution of fallen-off charring layers on fire dynamics in compartments.

6. Thermal characteristics of charred timber and its influence on compartments fire dynamics.
7. Fire dynamics in large open-plan compartments.
8. Understanding how travelling fires in timber compartments behave.

REFERENCES

ANSI/APA PRG 320 *Standard for Performance-Rated Cross-Laminated Timber (American National Standard)*. APA – The Engineered Wood Association, Tacoma, WA.

Babrauskas, V. (1979) *COMPF2 – A Program for Calculating Post- Flashover Fire Temperatures*. Technical Note 991. National Bureau of Standards, Gaithersburg, MD.

Babrauskas, V. (2016) Heat release rates, in: Hurley, M.J. (Ed.), *SFPE Handbook of Fire Protection Engineering*. 5th edition, pp. 799–904, Springer, New York.

Babrauskas, V. (2021) *Smoldering Fires*. Fire Science Publishers, New York.

Babrauskas, V. & Williamson, R.B. (1978) Post-flashover compartment fires: Basis of a theoretical model. *Fire and Materials*, 2 (2), 39–53.

Barber, D., Crielaard, R. & Li, X. (2016) Towards fire safety design of exposed timber in tall timber buildings, in: *Proceedings of WCTE 2016 World Conference on Timber Engineering*, Vienna.

Barber, D., Dixon, R., Rackauskaite, E. & Looi, K. (2021) A method for determining time equivalence for compartments with exposed mass timber, using iterative parametric curves, in: *Proceedings of WCTE 2021 World Conference on Timber Engineering*, 24–27 August, Santiago, Chile.

Barber, D., Sieverts, L., Dixon, R. & Alston, J. (2018) A methodology for quantifying fire resistance of exposed structural mass timber elements, in: *Proceedings of the 10th International Conference on Structures in Fire*, Belfast, UK, 217–224.

Bartlett, A., Hadden, R. & Bisby, L. (2018) A review of factors affecting the burning behaviour of wood for application to tall timber construction. *Fire Technology*. https://doi.org/10.1007/s10694-018-0787-y.

Brandon, D. (2016) *Practical Method to Determine the Contribution of Structural Timber to the Rate of Heat Release and Fire Temperature of Post-Flashover Compartment Fires (Report No. 2016:68)*. SP Technical Research Institute of Sweden, Borås.

Brandon, D. (2018a) *Fire Safety Challenges of Tall Wood Buildings – Phase 2: Task 4 Engineering Methods (Report No. FRPF-2018-04)*. Fire Protection Research Foundation, Quincy, MA.

Brandon, D. (2018b) *Engineering Methods for Structural Fire Design of Wood Buildings– Structural Integrity during a Full Natural Fire (RISE Rapport No. 2018:44)*. RISE Research Institutes of Sweden, Stockholm.

Brandon, D. (2021) *Fire Safe Implementation of Visible Mass Timber in Tall Buildings – Compartment Fire Testing (RISE Report No. 2020:94)*. RISE Research Institutes of Sweden, Stockholm.

Brandon, D. & Dagenais, C. (2018) *Fire Safety Challenges of Tall Wood Buildings – Phase 2: Task 5–Experimental Study of Delamination of Cross Laminated*

Timber (CLT) in Fire (Report No. FRPF-2018-05). Fire Protection Research Foundation, Quincy, MA.

Brandon, D. & Östman, B. (2016) *Fire Safety Challenges of Tall Wood Buildings – Phase 2: Task 1 – Literature Review (Report No. FRPF-2016-22).* Fire Protection Research Foundation, Quincy, MA.

Brandon, D., Sjöström, J., Temple, A. Hallberg, E. & Kahl, F. (2021) *Fire Safe Implementation of Visible Mass Timber in Tall Buildings – Compartment Fire Testing (RISE Report No. 2021:40).* RISE Research Institutes of Sweden, Stockholm.

Brandon, D. & Smart, J. (2021) *Expanding mass timber opportunities through testing and code development.* American Wood Council. https://cdn.ymaws.com/members.awc.org/resource/collection/7C07FAE7-6A6B-4CAE-8687-5D905C066CA0/DES612COLOR.pdf.

Browne, F.L. (1958) *Theories of the Combustion of Wood and its Control. A Survey of the Literature. (Report No. 2136).* Forest Products Laboratory, US Department of Agriculture, Madison, WI.

Buchanan, A.H. & Abu A.K. (2017) *Structural Design for Fire Safety.* 2nd Edition. John Wiley & Sons, Ltd, West Sussex, UK.

CAN/ULC-S146 *Standard Methods of Test for the Evaluation of Encapsulation Materials and Assemblies of Materials for the Protection of Structural Timber Elements.* Toronto, ON.

Crielaard, R. (2015) *Self-Extinguishment of Cross-Laminated Timber (Master of Science in Civil Engineering).* Delft University of Technology, Netherlands.

Crielaard, R., van de Kuilen, J.-W., Terwel, K., Ravenshorst, G. & Steenbakkers, P. (2019) Self-extinguishment of cross-laminated timber. *Fire Safety Journal.* https://doi.org/10.1016/j.firesaf.2019.01.008.

CSA S408 (2011) *Guidelines for the Development of Limit States Design Standards.* CSA Group, Toronto, ON.

DIN EN 1991-1-2/NA (2010) *Nationaler Anhang – National festgelegte Parameter – Eurocode 1, Einwirkungen auf Tragwerke 1–2: Allgemeine Einwirkungen Brandeinwirkungen auf Tragwerke.* Beuth Verlag. DIN German Institute for Standardization, Berlin, Germany.

Drysdale, D. (1998) *An Introduction to Fire Dynamics.* 2nd Edition. John Wiley & Sons Ltd., Chichester, Sussex, UK.

EN 1363–1 *Fire Resistance Tests – EN 1363–1.* European Standard. CEN European Committee for Standardization, Brussels.

EN 1991-1-2 *General Actions – Actions on Structures Exposed to Fire.* European Standard. CEN European Committee for Standardization, Brussels.

EN 1995-1-2 (2004) *Eurocode 5 Design of Timber Structures – Part 1–2: General – Structural Fire Design.* European Standard. CEN European Committee for Standardization, Brussels.

EN 13501–2 *Fire Classification of Construction Products and Building Elements – Part 2: Classification Using Data from Fire Resistance Tests, Excluding Ventilation Services.* European Standard. CEN European Committee for Standardization, Brussels.

EN 14135 *Coverings – Determination of Fire Protection Ability.* European Standard. CEN European Committee for Standardization, Brussels.

Frangi, A., Fontana, M., Hugi, E. & Jübstl, R. (2009) Experimental analysis of cross-laminated timber panels in fire. *Fire Safety Journal,* 44, 1078–1087. https://doi.org/10.1016/j.firesaf.2009.07.007.

Franssen, J.-M. (2005) SAFIR: A thermal/structural program for modeling structures under fire. *Engineering Journal*, 42, 143–150.

Gorska, C., Hidalgo, J.P. & Torero, J.L. (2021) Fire dynamics in mass timber compartments. *Fire Safety Journal*, 103098. https://doi.org/10.1016/j.firesaf.2020.103098.

Gottuk, D.T. & Lattimer, B.Y. (2016) Effect of combustion conditions on species production. In *SFPE Handbook of Fire Protection Engineering*, 486–528. Springer, New York.

Hadden, R.M., Bartlett, A.I., Hidalgo, J.P., Santamaria, S., Wiesner, F., Bisby, L.A., Deeny, S. & Lane, B. (2017) Effects of exposed cross laminated timber on compartment fire dynamics. *Fire Safety Journal*, 91, 480–489. https://doi.org/10.1016/j.firesaf.2017.03.074.

Hadvig, S. (1981) *Charring of Wood in Building Fires*. Technical University of Denmark, Kongens Lyngby, Denmark.

Hakkarainen, T. (2002) Post-flashover fires in light and heavy timber construction compartments. *Journal of Fire Science*, 20, 133–175. https://doi.org/10.1177/0734904102020002074.

Hasburgh, L., Bourne, K., Dagenais, C., Ranger (Osborne), L. & Roy-Pourier, A. (2016) Fire performance of mass-timber encapsulation methods and the effect of encapsulation on char rate of cross-laminated timber, in: *Proceedings of the WCTE 2016 World Conference on Timber Engineering*. Vienna University of Technology, Vienna.

Hopkin, D. & Spearpoint, M.J. (2021) A calculation method for flame extension from partially exposed mass timber enclosures. Presented at the *Applications of Structural Fire Engineering*, Ljubljana.

Huggett, C. (1980) Estimation of rate of heat release by means of oxygen consumption measurements. *Fire and Materials*, 4, 61–65. https://doi.org/10.1002/fam.810040202.

Incropera, F.P. & DeWitt, D.P. (1990) *Fundamentals of Heat and Mass Transfer*. John Wiley and Sons, NJ.

Ingberg, S.H. (1928) Tests of the severity of building fires. *National Fire Protection Quarterly*, 22, 43–61.

Janssens, M.L. & Douglas, B. (2004) Wood and wood products, in: Harper, C.A. (Ed.), *Handbook of Building Materials for Fire Protection*. McGraw-Hill, New York, NY.

ISO/TS 16733-2 (2021) *Fire Safety Engineering — Selection of Design Fire Scenarios and Design Fires – Part 2 Design Fires*. International Organization for Standardization, Geneva.

Kanellopoulos, G., Bartlett, A. & Law, A. (2019) Investigation into global equivalence ratio and external plumes from timber lined compartments, in: *Proceedings of the Ninth International Seminar on Fire and Explosion Hazards*, 21–26 April 2019, Saint Petersburg, 468–477. https://doi.org/10.18720/SPBPU/2/k19-68.

Karlsson, B. (1992) *Modeling Fire Growth on Combustible Lining Materials in Enclosures*. Report TVBB-1009. Department of Fire Safety Engineering, Lund University, Sweden.

Karlsson, B. & Quintiere, J.G. (2000) *Enclosure Fire*. CRC Press, Boca Raton, FL.

Kawagoe, K. (1958) *Fire Behaviour in Rooms*. Report No. 27. Building Research Institute, Tokyo.

Kerber, S. & Madrzykovski, D. (2009) *Fire Fighting Tactics under Wind Driven Fire Conditions: 7-Story Building Experiments* (No. TN 1629). National Institute of Standards and Technology, Gaithersburg, MD.

Klippel, M., Leyder, C., Frangi, A. & Fontana, M. (2014) Fire tests on loaded cross-laminated timber wall and floor elements, in: *Proceedings of the Eleventh International Fire Safety Symposium*. International Association for Fire Safety Science, Christchurch, 626–639. https://doi.org/10.3801/IAFSS.FSS.11-626.

Klippel, M., Schmid, J., Fahmi, R. & Frangi, A. (2018) Assessing the adhesive performance in CLT exposed to fire, in: *World Conference on Timber Engineering 20–23 August (WCTE 2018)*. Seoul, Republic of Korea.

Klippel, M., Schmid, J. & Frangi, A. (2016) Fire design of CLT – comparison of design concepts, in: *Proceedings of the Joint Conference of COST Actions FP1402 & FP1404 KTH Building Materials*, 10.3.2016. Presented at the Cross Laminated Timber – A competitive wood product for visionary and fire safe buildings, KTH Royal Institute of Technology, Division of Building Materials, Stockholm, 101–122.

Kotsovinos, P., Rackauskaite, E., Christensen, E., Glew, A., O'Loughlin, E., Mitchell, H., Amin, R., et al. (2022) Fire dynamics inside a large and open-plan compartment with exposed timber ceiling and columns: CodeRed #01. *Fire and Materials*. https://doi.org/10.1002/fam.3049.

Lattimer, B.Y., Vandsburger, U. & Roby, R.J. (1998) Carbon monoxide levels in structure fires: Effects of wood in the upper layer of a post-flashover compartment fire. *Fire Technology*, 34(4), 325–355.

Lautenberger, C. (2016) Radiation heat transfer, in: Hurley, M.J. (Ed.), *SFPE Handbook of Fire Protection Engineering*. 5th edition, Springer, New York, 102–137.

Law, A. & Hadden, R.M. (2017) Burnout means burnout. *SFPE Europe Digital Magazine*, Q1.

Law, A. & Bisby, L. (2020) The rise and rise of fire resistance. *Fire Safety Journal*, 116. https://doi.org/10.1016/j.firesaf.2020.103188.

Law, A. & Hadden, R. (2020) We need to talk about timber: Fire safety design in tall buildings. *Structural Engineers*, 9–15.

Law, A., Stern-Gottfried, J., Gillie, M. & Rein, G. (2011) The influence of travelling fires on a concrete frame. *Engineering Structures*, 33. https://doi.org/10.1016/j.engstruct.2011.01.034.

Law, M. (1978) Fire safety of external building elements – The design approach. *Engineering Journal (American Institute of Steel Construction)*, 2nd Quarter, 59–74.

Li, X., McGregor, C., Medina, A.R., Sun, X., Barber, D. & Hadjisophocleous, G. (2016) Real-scale fire tests on timber constructions, in: *Proceedings of WCTE 2016 World Conference on Timber Engineering*, Vienna University of Technology, Vienna.

Liu, J. & Fischer, E.C. (2022) Review of large-scale CLT compartment fire tests. *Construction and Building Materials*, 318. https://doi.org/10.1016/j.conbuildmat.2021.126099.

MacIntyre, J.D., Abu, A.K., Moss, P.J., Nilsson, D. & Wade, C.A. (2021a) A review of methods for determining structural fire severity—Part I: A historical perspective. *Fire and Materials*. https://doi.org/10.1002/fam.2962.

MacIntyre, J.D., Abu, A.K., Moss, P.J., Nilsson, D. & Wade, C.A. (2021b) A review of methods for determining structural fire severity—Part II: Analysis and review. *Fire and Materials.* https://doi.org/10.1002/fam.2961.

Magnusson, S.E. & Thelandersson, S. (1970) *Temperature-Time Curves for the Complete Process of Fire Development – A Theoretical Study of Wood Fuels in Enclosed Spaces.* Acta Polytechnica Scandinavica, Stockholm.

Mikkola, E. (1991) Charring of wood based materials. *Fire Safety Science,* 3, 547–556. https://doi.org/10.3801/IAFSS.FSS.3-547.

Mikkola, E., Rinne, T. & Granstrom, M. (2017) *Extended Use of Massive Wood Structures without Coverings – Arguments for Revision of Fire Safety Regulations.* KK-Palokonsultti Oy, Espoo.

Mowrer, F.W. & Williamson, R.B. (1987) Estimating room temperatures from fires along walls and in corners. *Fire Technology,* 23(2), 133–145.

NFPA 557 *Standard for Determination of Fire Loads for Use in Structural Fire Protection Design.* National Fire Protection Association, Quincy, MA, USA.

Peacock, R.D., McGrattan, K., Forney, G.P. & Reneke, P.A. (2017) *CFAST – Consolidated Fire and Smoke Transport (Version 7).* Volume 1 Technical Reference Guide (NIST Technical Note No. 1889v1). National Institute of Standards and Technology, Gaithersburg, MD.

prEN 1995-1-2 (2021) *Final Draft Eurocode 5 – Design of Timber Structures, Part 1–2: General – Structural Fire Design.* European Draft Standard. CEN European Committee for Standardization, Brussels.

Quintiere, J.G. (2017) *Principles Of Fire Behavior.* 2nd Edition. CRC Press, Boca Raton, FL.

Quintiere, J.G. & Rangwala, A.S. (2004) A theory for flame extinction based on flame temperature. *Fire and Materials,* 28, 387–402. https://doi.org/10.1002/fam.835.

Quintiere, J.G. & Wade, C.A. (2016) Compartment fire modeling, in: Hurley, M.J. (Ed.), *SFPE Handbook of Fire Protection Engineering.* Springer, New York, 981–995.

Rackauskaite, E., Hamel, C., Law, A. & Rein, G. (2015) Improved formulation of travelling fires and application to concrete and steel structures. *Structures,* 3, 250–260. https://doi.org/10.1016/j.istruc.2015.06.001.

Rackauskaite, E., Kotsovinos, P. & Barber, D. (2020) Letter to the editor: Design fires for open-plan buildings with exposed mass-timber ceiling. *Fire Technology.* https://doi.org/10.1007/s10694-020-01047-0.

Reitgruber, S., Pérez-Jiménez, C., Di Blasi, C. & Franssen, J.-M. (2006) *Some Comments on the Parametric Fire Model of Eurocode 1.* University of Leige, Leige.

Richter, F., Kotsovinos, P., Rackauskaite, E. & Rein, G. (2020a) Thermal response of timber slabs exposed to travelling fires and traditional design fires. *Fire Technology.* https://doi.org/10.1007/s10694-020-01000-1.

Richter, F. & Rein, G. (2020b) A multiscale model of wood pyrolysis in fire to study the roles of chemistry and heat transfer at the mesoscale. *Combustion and Flame,* 216, 316–325. https://doi.org/10.1016/j.combustflame.2020.02.029.

Schmid, J., Brandon, D., Santomaso, A., Wickström, U. & Frangi, A. (2016) Timber under real fire conditions – the influence of oxygen content and gas velocity on the charring behavior, in: *Proceedings of the 9th International Conference on Structures in Fire*. DesTech Publications Inc, Lancaster, Pennsylvania, USA.

Schmid, J. & Frangi, A. (2021) Structural timber in compartment fires – The timber charring and heat storage model. *Open Engineering*, 11, 435–452. https://doi.org/10.1515/eng-2021-0043.

Schmid, J., Santomaso, A., Brandon, D., Wickström, U. & Frangi, A. (2017) Timber under real fire conditions – The influence of oxygen content and gas velocity on the charring behavior. *Journal of Structural Fire Engineering*. https://doi.org/10.1108/JSFE-01-2017-0013.

SFPE (2016) *Handbook of Fire Protection Engineering Appendix 4: Configuration Factors*. 5th Edition. Society of Fire Protection Engineers. Springer, New York.

SFPE Standard S-01 (2011) *Calculating Fire Exposures to Structures*. Society of Fire Protection Engineers, Bethesda, MD.

Spearpoint, M.J. & Quintiere, J.G. (2000). Predicting the burning of wood using an integral model. *Combustion and Flame*, 123 (3), 308–25. https://doi.org/10.1016/S0010-2180(00)00162-0.

Stern-Gottfried, J. & Rein, G. (2012) Travelling fires for structural design-Part II: Design methodology. *Fire Safety Journal*, 54, 96–112. https://doi.org/10.1016/j.firesaf.2012.06.011.

Stern-Gottfried, J., Rein, G., Bisby, L. & Torero, J.L. (2010) Experimental review of the homogeneous temperature assumption in post-flashover compartment fires. *Fire Safety Journal*, 45(4), 249–261. https://doi.org/10.1016/j.firesaf.2010.03.007.

Structural Timber Association (2020) Structural timber buildings fire safety in use guidance. *Volume 6 – Mass Timber Structures; Building Regulation Compliance* B3(1) (No. Version 1.1). Alloa, UK.

Su, J., Lafrance, P., Hoehler, M. & Bundy, M. (2018a) *Fire Safety Challenges of Tall Wood Buildings – Phase 2: Task 2 & 3 – Cross Laminated Timber Compartment Fire Tests (Report No. FPRF-2018-01)*. Fire Protection Research Foundation, Quincy, MA.

Su, J., Leroux, P., Lafrance, P., Berzins, R., Gratton, K., Gibbs, E. & Weinfurter, M. (2018b) *Fire Testing of Rooms with Exposed Wood Surfaces in Encapsulated Mass Timber Construction (Report No. A1- 012710.1)*. National Research Council of Canada, Ottawa, ON.

Tewarson, A. (1984) Fully enveloped enclosure fires of wood cribs, in: *20th Symposium (International) on Combustion*, Combustion Institute, Pittsburgh, PA.

Tewarson, A. (2002) Generation of heat and chemical compounds in fires, Chap. 3-4, in: Dinenno, P.J., Drysdale, D., Beyler. C.L., et al. (eds). *SFPE Handbook of Fire Protection Engineering*. 3rd Edition. National Fire Protection Association, Quincy, MA. 83–161.

Torero, J.L., Majdalani, A.H., Abecassis-Empis, C. & Cowlard, A. (2014) Revisiting the compartment fire, in: *Fire Safety Science – Proceedings of the Eleventh International Symposium*, Christchurch. 28–45.

Wade, C.A. (2019) *A Theoretical Model of Fully Developed Fire in Mass Timber Enclosures* (PhD Thesis). University of Canterbury, Department of Civil and Natural Resources Engineering, Christchurch.

Wade, C.A., Baker, G.B., Frank, K., Harrison, R. & Spearpoint, M.J. (2016) *B-RISK 2016 User Guide and Technical Manual* (Study Report No. SR364). BRANZ, Porirua.

Wade, C.A., Hopkin, D.J., Spearpoint, M.J. & Fleischmann, C.M. (2020) Calibration of a coupled post-flashover fire and pyrolysis model for determining char depth in mass timber enclosures, in: *Presented at the 11th International Conference on Structures in Fire*, Brisbane. 830–841.

Wade, C.A., Hopkin, D., Su, J., Spearpoint, M.J. & Fleischmann, C.M. (2019) Enclosure fire model for mass timber construction – benchmarking with a kinetic pyrolysis submodel, in: *Interflam 2019: 15th International Conference on Fire Science and Engineering. InterScience Communications Limited*, Royal Holloway College, Nr Windsor.

Wade, C.A., Spearpoint, M.J., Fleischmann, C.M., Baker, G.B. & Abu, A.K. (2018) Predicting the fire dynamics of exposed timber surfaces in compartments using a two-zone model. *Fire Technology*, 54, 893–920. https://doi.org/10.1007/s10694-018-0714-2.

Walton, W.D., Thomas, P.H. & Ohmiya, Y. (2016) Estimating temperatures in compartment fires, in: Hurley, M.J. (Ed.), *SFPE Handbook of Fire Protection Engineering*. Springer, New York, 996–1023.

White, R. (2016) Analytical methods for determining fire resistance of timber members, in: Hurley, M.J. (Ed.), *SFPE Handbook of Fire Protection Engineering*. Springer, New York, 1979–2011.

Wickström, U. (1985) Application of the standard fire curve for expressing natural fires for design purposes, in: Harmathy, T.Z. (Ed.), *Fire Safety: Science and Engineering*, ASTM STP 882. American Society of Testing and Materials, Philadelphia, PA, 145–159.

Wickström, U. (2016) *Temperature Calculation in Fire Safety Engineering*. 1st ed. Springer International Publishing, AG Switzerland.

Zehfuss, J. & Hosser, D. (2007) A parametric natural fire model for the structural fire design of multi-storey buildings. *Fire Safety Journal*, 42, 115–126. https://doi.org/10.1016/j.firesaf.2006.08.004.

Zelinka, S., Hasburgh, L., Bourne, K., Tucholski, D. & Ouellette, J. (2018) *Compartment Fire Testing of a Two-Story Cross Laminated Timber (CLT) Building (General Technical Report No. FPL-GTR-247)*. U.S. Department of Agriculture, Forest Service, Forest Products Laboratory, Madison, WI.

Chapter 4

Fire safety requirements in different regions

*Birgit Östman, David Barber, Christian Dagenais,
Andrew Dunn, Koji Kagiya, Eugeniy Kruglov,
Esko Mikkola, Peifang Qiu, Boris Serkov and
Colleen Wade*

CONTENTS

DOI: 10.1201/9781003190318-4

SCOPE OF CHAPTER

This chapter summarises the regulatory control systems for the firesafety design of buildings in different regions around the globe. It is focused on the possibilities of using wood products and timber structures according to prescriptive requirements. The possible use of structural timber elements and visible wood surfaces in interior and exterior applications is reviewed and presented in tables and maps. They apply mainly to residential and office buildings. Performance-based requirements may be used in several countries and can be used to verify further applications of wood (see Chapter 11).

4.1 REGULATORY CONTROL SYSTEMS FOR FIRE SAFETY IN BUILDINGS

The regulatory control systems for fire safety in buildings differ between regions. The main features in Europe, North America, Asia, Australia and New Zealand are summarised below.

4.1.1 Europe

To assure fire safety in buildings, a European system, including performance classes, testing and calculation standards for fire performance, was introduced in 1988 by the Construction Products Directive (CPD). The CPD was replaced by the Construction Products Regulation (CPR) in 2013. The main change is that CPR is mandatory to implement in all European countries. The European standards for fire safety in buildings are concerned mainly with harmonised methods for verification of the fire performance. Products covered by harmonised standards must have a declaration of performance and CE marking.

Six essential requirements were introduced in CPD and remain in CPR, one of which is fire safety.

These essential requirements are implemented and developed by different technical standard committees (TCs) within CEN (European Committee for Standardisation) to European standards (ENs) (see Figure 4.1). Possibilities within EOTA (European Organisation for Technical Assessment) are also included for products without harmonised products standards, mainly new products. For those products, European Technical Assessments (ETA) can be issued based on European Assessment Documents (EAD).

Testing and classification of building products and structures are specified in two European standards (EN 13501-1, EN 13501-2).

The five parts of the essential requirements for fire safety are that structures must be designed and built such that in the case of fire:

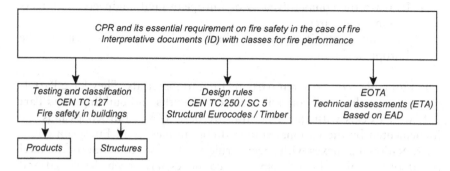

Figure 4.1 Systems for developing European fire standards (ENs) and Technical Assessments (ETAs) for building products.

1. Load-bearing capacity can be assumed to be maintained for a specific period of time
2. The generation and spread of fire and smoke is limited
3. The spread of fire to neighbouring structures is limited
4. Occupants can leave the building or be rescued by other means
5. The safety of rescue teams is taken into consideration

For building products, a system with reaction to fire classes A to F are defined for products except floorings, where classes A1 and A2 are for non-combustible products, which cannot be reached by wood products. For floorings, a similar system is defined with classes A_{fl} to F_{fl}, where "fl" means floorings.

For structures, the classification is based on the parameters for stability R, integrity E and insulation I, without including requirements on non-combustibility, which earlier were used in many countries and formed an obstacle to building higher buildings with a wooden structure.

Harmonised product standards specify the requirements for different building products and form the basis for using the CE-mark to declare conformity with the European legislation for specified products. Fire properties are mandatory to declare for all building products. There are about ten harmonised product standards for wood and timber building products.

For structural engineering, a set of European design standards for structures have been published, called Eurocodes, to standardise design rules within Europe. The Eurocodes aim to:

• Provide common design criteria and calculation methods to merge necessary requirements
• Establish a common understanding of the design of structures
• Enable the exchange of construction services within Europe
• Provide a common basis for research and development in the construction industry

- Increase the competitiveness of European civil engineers, architects and manufacturers
- Contribute significantly to single-market activities within the European Union

The Eurocodes comprise ten parts relating to materials. They have Part 1, which covers the design of civil engineering works and buildings, and Part 1-2, which deals with the structural fire design. The Eurocodes must be implemented by the national standard committees in all European countries. National annexes with specific rules and values to maintain the level of safety prevailing in the respective countries have been developed and form essential documents to enable Eurocodes to be used. The following appropriate information must be included in the annexes:

- Values or classes where alternatives are given in the Eurocode
- Values to be quantified where only a symbol is given in the Eurocode
- Specific data, e.g., for material properties, wind or snow load
- The procedure to be used when alternative procedures are given in the Eurocode
- Decision on the application of informative annexes

Eurocodes allow the calculation and verification of load-bearing capacity of components and structures for different materials, based on a semi-probabilistic design concept with partial safety coefficients. It is also possible to design structures or components for desired behaviour in the case of fire, based on tabular values and simplified or general calculation methods, and to optimise the design of fire protection. Application of the Eurocodes fire parts permits the integration of parametric temperature-time curves and natural fire curves to represent real-fire scenarios as an alternative to the standard time/temperature curve in evaluating the fire resistance of components, which can be useful, especially in the evaluation of existing structures. However, the use of extended methods requires an increased level of expertise from the user.

The present Eurocode 5 for timber 5, EN 1995, was published in 2004. An extensive revision is ongoing, and a new version is planned to be published in 2025. The new design models that have been developed since 2004 are included in this global guideline (Chapters 6, 7 and 8).

Fire test and classification methods are harmonised across Europe, but regulatory requirements applicable to building types and end users remain on national bases. The European standards exist on the *technical level*, but fire safety is governed by national legislation and is thus on the *political level*. National fire regulations therefore remain, but the new European harmonisation of standards is intended to provide means of achieving common national regulations (Dimova et al., 2019).

4.1.2 Canada

The Canadian requirements are given in the model National Building Code of Canada (NBCC, 2015), which is then adopted by the Canadian provinces with or without modifications. Since its 2005 edition, the NBCC is an objective-based code where compliance can be achieved by using prescriptive solutions (called "acceptable solutions") or by using alternative solutions that will achieve at least the minimum level of performance required by the prescriptive solutions in the areas defined by the objectives and functional statements attributed to the applicable acceptable solutions. Depending on the scale and scope of an alternative solution, its approval is typically provided by the provincial or municipal authorities.

Prescriptive (acceptable) solutions can be found in Division B of the NBCC, with Part 3 providing the requirements for fire protection, occupant safety and accessibility. In the NBCC, building construction systems are classified into two categories: 1) combustible construction and 2) non-combustible construction. This division is based on the non-combustibility characteristic of materials, when tested in the standard test method (CAN/ULC S114). The required type of construction depends on the building's major occupancy, height, area, and whether it is equipped with an automatic sprinkler system.

In the NBCC, a non-combustible construction is a type of construction in which a degree of fire safety is attained using non-combustible materials for structural elements and other building assemblies. The objective of requiring non-combustible materials is to limit the probability that materials will contribute to the growth and spread of fire. However, other aspects of material behaviour when exposed to fire conditions such as structural performance, thermal expansion, spalling, etc., are not intended to be addressed through the requirement for non-combustible materials (Ni and Popovski, 2015).

Combustible construction relates to a type of construction not meeting the requirements for non-combustible construction and as such implies a risk to fire growth and spread. The types of timber structures, as presented in Chapter 1, are classified as combustible, according to the NBCC.

The 2021 edition of the NBCC incorporates a new type of construction called "Encapsulated Mass Timber Construction" (EMTC), defined as a type of construction in which a degree of fire safety is attained using encapsulated mass timber elements with an encapsulation rating and minimum dimensions for structural members and other building assemblies. Prescriptive provisions for using EMTC up to 12 storeys for residential and office occupancies are provided. This new type of construction acknowledges the enhanced fire performance of mass timber construction.

4.1.3 USA

Each state within the US adopts one or more model building codes. All 50 states adopt the International Code Council's (ICC) International Building

Code (IBC) (ICC, 2018), with some states also adopting NFPA 101 "Life Safety Code" (NFPA). Each state adapts and amends the model codes to provide the basis for construction compliance. The adoption process may take several years. Some cities have their own building code, such as New York City and Chicago. Also required to be met are other relevant codes and standards that will impact aspects of construction, fire protection system design, maintenance and firefighting operations, including the International Fire Code (IFC, 2018) and numerous referenced standards. The IBC has fire protection requirements that provide for occupant life safety in fire, access and equipment for firefighters and to prevent fire spread to neighbouring buildings. Protection of the building structure from fire varies with height and area. The IBC requires buildings with an occupied floor above 22.9 m (75 feet), defined as high-rise, to have an increased level of fire protection and structural performance.

Timber construction is referred to as combustible construction in the IBC. Concrete and steel construction is referred to as non-combustible construction. Within the IBC, timber construction can be utilised within Types III, IV and V construction. Types III, IV and V have been limited to low- and medium-rise buildings up to 25.9 m (85 feet), with limited building area. From 2021 the IBC (ICC, 2021) will change significantly and will allow for high-rise mass timber construction.

ICC 2021 introduces three new construction types, with Type IV-A and IV-B allowing mass timber buildings to be built up to 12 storeys with limited areas of exposed mass timber or up to 18 storeys with all the mass timber protected (encapsulated) (Breneman et al., 2018). High-rise mass timber will require fire resistance ratings of 120 minutes for the structure for buildings up to 12 storeys and 180 minutes for buildings up to 18 storeys. All mass timber buildings are to be fully sprinkler protected. The code changes include additional fire safety measures for improved firefighting and sprinkler water supplies, protection for concealed spaces, specifications for non-combustible gypsum board protection to mass timber and measures for fire safety during construction.

The IBC changes were based on research carried out by the Fire Protection Research Foundation and USDA FPL (Su et al., 2018; Zelinka et al., 2018), in which full-scale fire tests were carried out on exposed CLT panels to determine how large areas of exposed CLT will perform in real building fires. The lessons from those full-scale fire tests, and tests undertaken in Canada (McGregor, 2013; Medina, 2014; Taber et al., 2014) have influenced how CLT is manufactured in North America for high-rise buildings. This change in building code requirements has resulted in a significant boost to mass timber construction and for the use of CLT.

4.1.4 China

To satisfy the requirements of market economy and the TBT (Technical Barriers to Trade) Agreement of WTO (World Trade Organisation), China has been revising its regulatory system since 2001. In 2015, the

Standardization Administration of China released "The Plan for Furthering the Standardization Reforms" to standardise the national governance system and economic and social development. The plan reformed the regulatory system and standardisation management system. The new regulatory system includes government-leading and market-leading standards.

Government-leading standards include mandatory national standards, recommended national standards, recommended industry standards, and recommended local standards. Market-leading standards include group standards and enterprise standards. The government-leading standards focus on ensuring safety, health, environmental protection, etc. The market-leading standards focus on improving market competitiveness.

For fire safety in buildings, the regulatory system has evolved from completely prescriptive to being more and more performance-based. Figure 4.2 sets out the regulatory system for fire protection of buildings.

Basic standards refer to the terminology, symbols, measurement units, graphics, modulus, basic classification, basic principles etc., which are the basis for other standards within a certain professional scope and are commonly used, for example, "Fire protection vocabulary – Part 1 general terms" (GB/T 5907.1-2014).

General standards have greater coverage for a certain type of standardised object. Such standards can be used as the basis for formulating special standards, such as general requirements for safety, health and environmental protection, general quality requirements, general design, construction requirements and test methods etc. "Code for fire protection design of buildings" (GB 50016) belongs to this category.

Dedicated standards refer to special standards formulated for a specific standardisation object or as a supplement or an extension to a general standard. Its coverage is generally not large. For example, the requirements and methods for the survey, planning, design, construction, installation and quality acceptance of a certain project, the safety, health, and environmental protection requirements of a certain range, a certain test method, the application and management technology of a certain type of product, etc. For example, "Technical standard for multi-story and high-rise timber buildings" (GB/T 51226-2017) and "Code for fire protection design of civil airport terminal" (GB 51236-2017).

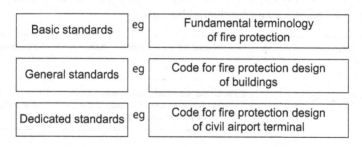

Figure 4.2 Regulatory system for fire protection of buildings in China.

There is another classification in terms of the hierarchy of the components of the regulatory system. Figure 4.3 sets out the hierarchy of the regulatory system for fire protection of buildings.

"General code for fire protection of buildings and constructions" is the primary legislation governing the fire safety design of buildings and constructions in China. It sets the minimum requirements that buildings must meet. Three objectives for the fire safety of buildings are specified in the code:

1. Health, life and property
2. Continuity of business and operation of important facilities
3. Environmental protection, energy saving and public interests

To achieve these objectives, buildings must meet certain functional requirements; for example, the performance of the load-bearing elements shall be able to withstand the fire within a certain period of time, evacuation design shall ensure that people who use the building can escape from the building in case of fire, the fire separation elements shall be able to prevent fire spread, and building materials and decorations shall not contribute to the fire severity or fire spread.

There are two ways to demonstrate compliance with the requirements of "General code for fire protection of buildings and constructions". One way is to design a new building by following the requirements in the "Code for fire protection design of buildings" (GB 50016), which provides solutions for achieving those objectives and functional requirements. The other way is by performance-based design.

"Code for fire protection design of buildings" (GB 50016) is the mother code for fire safety of buildings in China and is applicable to most buildings and constructions. It not only specifies requirements for the fire safety design of civil buildings but also for factory and storage buildings. It gives detailed requirements for building classifications, building height, number of storeys, fire compartmentation, fire separation distance, evacuation and fire extinguishing facilities, etc. In terms of timber buildings, it is expected (under approval 2022) that the number of storeys can be eight and the maximum building height 32 m in "Code for fire protection design of buildings" (GB 50016). Timber buildings with the number of storeys not less than four shall be fully sprinklered. Timber buildings with five or more storeys can only be used as office and residential buildings.

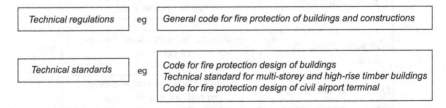

Figure 4.3 Hierarchy of the regulatory system for fire protection of buildings in China.

4.1.5 Japan

The Japanese regulatory system for building fire safety is based on two laws: Building Standard Law and Fire Service Law:

- Building Standards Law (BSL) stipulates
  - Fire resistance
  - Fire protection equipment such as fire door
  - Smoke control system, evacuation facilities such as staircase
  - Materials etc.
- Fire Service Law (FSL) stipulates
  - Suppression system such as sprinkler
  - Alarm system
  - Emergency equipment such as ladder, guidance system
  - Water source
  - Facilities for fire service such as heat and smoke exhaust system, standpipe etc.

The overall system of BSL is illustrated in Figure 4.4. In 2000, performance-based code was introduced into the fire regulatory system in Japan.

The Japanese building regulation classifies the fire safety performance of buildings into "Fireproof Buildings", "Quasi-fireproof Buildings", "Building with Fire-rated Envelope" and others. The fireproof buildings are defined as those that can stand even after a fire, while quasi-fireproof buildings have to stand for only the required period during fire.

The conventional fire regulations of fire-resistive buildings may be either fireproof (FP) buildings or quasi-fireproof (QFP) buildings.

Principal structural parts (columns, beams, floors etc.) in FP buildings should continuously support themselves and not collapse even after a normal fire ends. Requirement for the FP buildings is harder than the "fire-resistant buildings" in most other countries due to the need of the consideration of earthquake fire in Japan, where fire service and automatic sprinklers may not work. For fire resistance rating tests, the structural members will be left loaded for three times the prescribed fire resistance time in a furnace, i.e., 3–9 hours for the cooling phase after a 1–3 hours of fire resistance test. Combustible structural members such as timber are left loaded for 24 hours after testing in the furnace (see Figure 4.5).

Many of the FP structural members are covered with non-combustible material (fire-resistive insulation). However, an exposed wood surface of an FP structural member is used as a sacrificed layer, and its underlying layer is designed to stop charring. Wood-based fireproof construction has been developed and put to practical use since the introduction of performance-based regulation in 2000. Examples of wooden fireproof members are given in Figure 4.6.

Regardless of materials used for structural parts, the number of storeys of FP building that can be built depends on its fire-resistive performance.

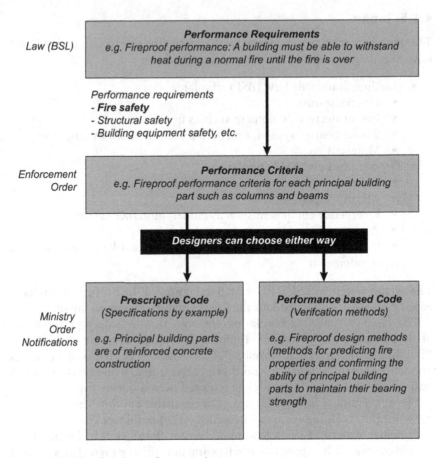

Figure 4.4 Building Regulatory System in Japan.

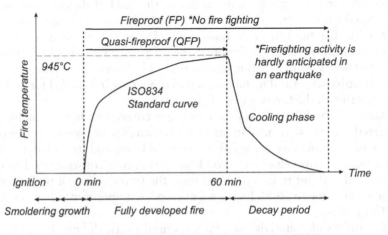

Figure 4.5 Fireproof and Quasi-fireproof testing in Japan (Kagiya, 2021).

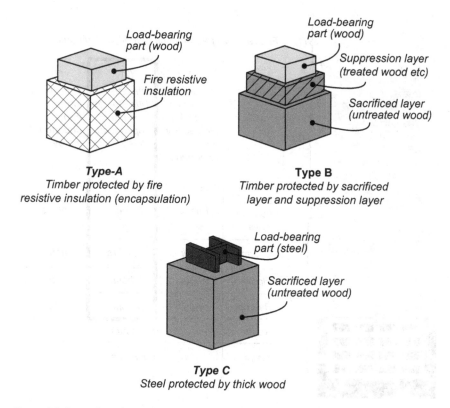

Figure 4.6 Examples of wooden fireproof structural members (Hasemi et al., 2016).

Figure 4.7 illustrates the fire-resistive performance requirement for FP buildings depending on the number of storeys.

However, mid-rise FP building is costly to be built with timber compared with other buildings, especially satisfying the demand for "visible" wood construction in the Japanese market.

In recent years, alternative solutions of FP buildings have been developed, called "advanced" QFP buildings, considering the time for firefighting and evacuation of all occupants, including rescue service. Legally, "advanced QFP" buildings are equivalent to FP buildings, with no limit of building height if fire resistance rating is satisfied. A verification method will be notified soon.

Performance-based requirements for alternative solutions for FP buildings include:

- Not to collapse causing damage to the surrounding
  - Keep time for firefighting activity by compartmentation, etc.
  - Stipulated by BSL article 21

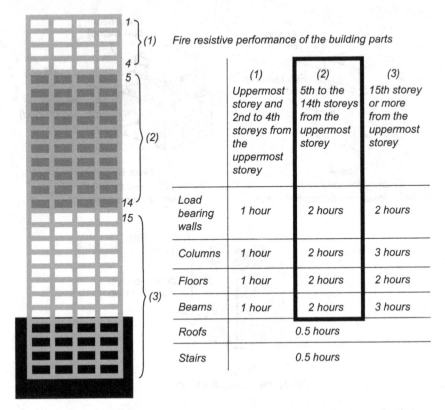

	(1) Uppermost storey and 2nd to 4th storeys from the uppermost storey	(2) 5th to the 14th storeys from the uppermost storey	(3) 15th storey or more from the uppermost storey
Load bearing walls	1 hour	2 hours	2 hours
Columns	1 hour	2 hours	3 hours
Floors	1 hour	2 hours	2 hours
Beams	1 hour	2 hours	3 hours
Roofs	0.5 hours		
Stairs	0.5 hours		

Fire resistive performance of the building parts

Figure 4.7 Fire-resistive performance requirement for structural parts of FP building (Kagiya, 2021).

- Not to collapse until all occupants escape from the building, including search and rescue service
  - Stipulated by BSL article 27
- Not to cause rapid fire spread that leads to city fire
  - Stipulated by BSL article 61.

4.1.6 Russian Federation

An overview of fire behaviour and fire protection in timber buildings from a Russian perspective is available (Aseeva et al., 2014). The present legal requirements are reviewed below.

The main requirements for fire resistance and fire hazard of buildings, structures and fire compartments, as well as requirements for building structures, are contained in Federal Law No. 123 (2018). The law regulates fire safety and establishes general fire safety requirements for protected objects (products), including buildings and structures, production facilities, fire-technical products and general-purpose products.

Each object of protection must have a fire safety system, which is aimed at preventing fire, ensuring the safety of people and protecting property in case of fire. The system for ensuring the fire safety of the facility must exclude the possibility of exceeding the values of the permissible fire risk established by the Federal Law and aimed at preventing the danger of harm to third parties by the fire safety system, which can be a fire prevention system, a fire protection system, or a set of organisational and technical measures to ensure fire safety.

Fire safety of the protected object is considered to be ensured if one of the following conditions for the compliance with fire protection objectives is met with fire safety requirements in accordance with the Federal Law on Technical Regulation:

1. The fire risk does not exceed the permissible values *or*
2. The fire safety regulations are fulfilled

Fire safety requirements for objects of protection are presented based on identification, which is established according to the features shown in Figure 4.8.

The classification of buildings, structures and fire compartments by functional fire hazard depends on their purpose, age, physical condition and the number of people in the building, and whether they are awake or asleep. The functional purpose of buildings is divided into:

F1 – buildings intended for permanent and temporary residence of people
F2 – buildings of entertainment, cultural and educational institutions
F3 – buildings of public service organisations
F4 – buildings of educational organisations, scientific and design organisations, management bodies of institutions
F5 – buildings for industrial or warehouse purposes

Requirements are imposed on building structures in two main parameters: fire resistance and fire hazard. When analysing the fire safety of buildings, structures and fire compartments, the concepts of "required" and "actual" fire resistance and fire hazard are used. The "required" fire resistance and fire hazard of structures depend on the required degree of fire resistance (I–V) and the building's structural fire hazard class (C0–C3). The "required" fire resistance of a building structure defines the requirements for standard limit states, depending on the functional purpose of the building structure (REI), and is expressed by its fire resistance limit in minutes, as shown in Table 4.1.

The required fire hazard class determines the degree of participation of the building structure in the development of a fire and its ability to form hazardous fire factors (K0, K1, K2, K3) as shown in Table 4.2.

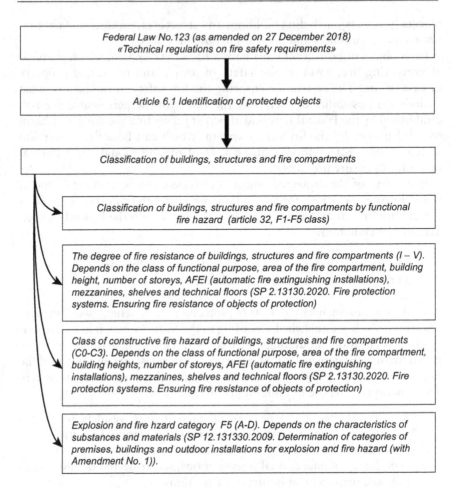

Figure 4.8 Fire-technical classification of buildings, structures and fire compartments in the Russian Federation (Federal Law 123, 2018).

The "actual" values of the parameters of the fire resistance limits and the classes of the fire hazard of building structures must be determined under standard test conditions according to the methods established by the fire safety regulations. The actual fire resistance limits and fire hazard classes of building structures similar in shape, materials, and design to building structures that have passed fire tests can be determined by calculation and analytical methods established by fire safety documents, as shown in Figure 4.9.

Table 4.1 The limit of fire resistance of building structures and building components in Russia (SNiP-II-2-80)

Fire resistance degree of buildings and structures	Fire resistance limit of building structures						
	Load-bearing walls, columns and other load-bearing elements	Exterior curtain walls	Inter-floor ceilings (including the attic and above basements)	Building structures of attic roofs		Staircase building structures	
				Decks (including with insulation)	Trusses, beams, purlins	Internal walls	Marches and landings of stairs
I	R 120	E 30	REI 60	RE 30	R 30	REI 120	R 60
II	R 90	E 15	REI 45	RE 15	R 15	REI 90	R 60
III	R 45	E 15	REI 45	RE 15	R 15	REI 60	R 45
IV	R 15	E 15	REI 15	RE 15	R 15	REI 45	R 15
V	not standardised						

Table 4.2 The required constructive fire hazard class for buildings and structures in
Russia

	Fire hazard class of building structures				
Building constructive fire hazard class	Bearing bar elements (columns, beams, trusses)	External walls from the outside	Walls, partitions, ceilings and roofs	Staircase walls and fire barriers	Marches and landings of stairs in stairwells
C0	K0	K0	K0	K0	K0
C1	K1	K2	K1	K0	K0
C2	K3	K3	K2	K1	K1
C3	not standardised			K1	K3

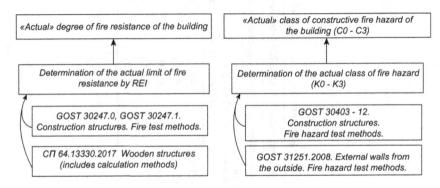

Figure 4.9 Scheme for determining the "actual" indicators of fire resistance and fire
hazard of building structures in Russia.

4.1.7 Australia

The Australian requirements are given in the National Construction Code
– Building Code of Australia (NCC) and its latest version 2019, which
was amended in 2020. The NCC is a performance-based code containing all Performance Requirements for the construction of buildings. It is
built around a hierarchy of guidance and code-compliance levels, with the
Performance Requirements being the minimum level that buildings and
building elements must meet. A building solution will comply with the
NCC if it satisfies the Performance Requirements, which are the mandatory
requirements of the NCC.

The key to the performance-based NCC is that there is no obligation to
adopt any particular material, component, design factor or construction
method. This provides for a choice of compliance pathways. The Performance
Requirements can be met using either a Performance Solution (Alternative
Solution) or using a Deemed-to-Satisfy (DTS) Solution (see Figure 4.10).

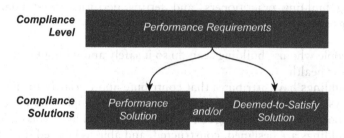

Figure 4.10 Performance Solution (Alternative Solution) and Deemed-to-Satisfy (DTS) Solution to meet the Performance Requirements in Australia (NCC, 2020).

Performance Solution

A Performance Solution is unique for each individual situation. These solutions are often flexible in achieving the outcomes and encouraging innovative design and technology use. A Performance Solution directly addresses the Performance Requirements by using one or more of the Assessment Methods available in the NCC.

Deemed-to-Satisfy Solution (DTS)

A DTS Solution follows a set recipe of what, when and how to do something. It uses the DTS Solutions from the NCC, which include materials, components, design factors, and construction methods that, if used, are deemed to meet the Performance Requirements. The form of the Australian DTS Solution is similar to the prescriptive solution in many other countries.

Often, building solutions are not just a Performance Solution or deemed-to-satisfy, but a combination of both. Performance solutions may only be used for solutions that can't meet the DTS.

Regarding fire safety, the Performance Requirements relate to:

- Structural stability during a fire
- Spread of fire within the building or to another building
- Spread of fire and smoke in health and residential care buildings
- Safe conditions for evacuation
- Fire protection of service, emergency equipment and openings and penetrations
- Fire brigade access

4.1.8 New Zealand

New Zealand has a performance-based building regulatory system. The Building Act 2004 specifies four purposes that regulation of building work, licensing building practitioners, and setting performance standards for buildings should achieve, so that:

- People who use buildings can do so it safely and without endangering their health
- Buildings have attributes that contribute appropriately to the health, physical independence, and well-being of the people who use them
- People who use a building can escape from the building if it is on fire, and
- Buildings are designed, constructed, and able to be used in ways that promote sustainable development.

To achieve these objectives, building work must meet certain requirements set out in legislation and regulations. The Ministry of Business Innovation and Employment (MBIE) is the government agency responsible for administering the New Zealand Building Code (NZBC). Figure 4.11 sets out the key components of the building regulatory system in New Zealand and methods for demonstrating compliance. This includes:

- Building Act 2004 – the primary legislation governing the building and construction industry
- Building Regulations – detail for particular building controls
- Building Code – contained in Schedule 1 of the Building Regulations 1992, setting the minimum performance standards that buildings must meet

There are different ways to demonstrate compliance with the NZBC, and these are summarised in Figure 4.12. For the design of a new building, the usual methods would be to follow an Acceptable Solution or Verification Method or to develop an Alternative Solution to demonstrate that the Code objectives, functional requirements and performance criteria are achieved.

Where an Acceptable Solution is used for establishing compliance with the Protection from Fire clauses of the NZBC, C/AS2 (MBIE, 2020b) is

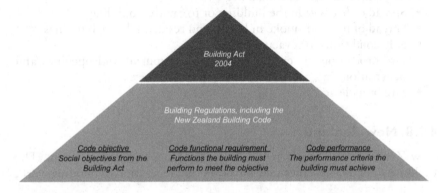

Figure 4.11 Components of the New Zealand Regulatory System (MBIE, 2017).

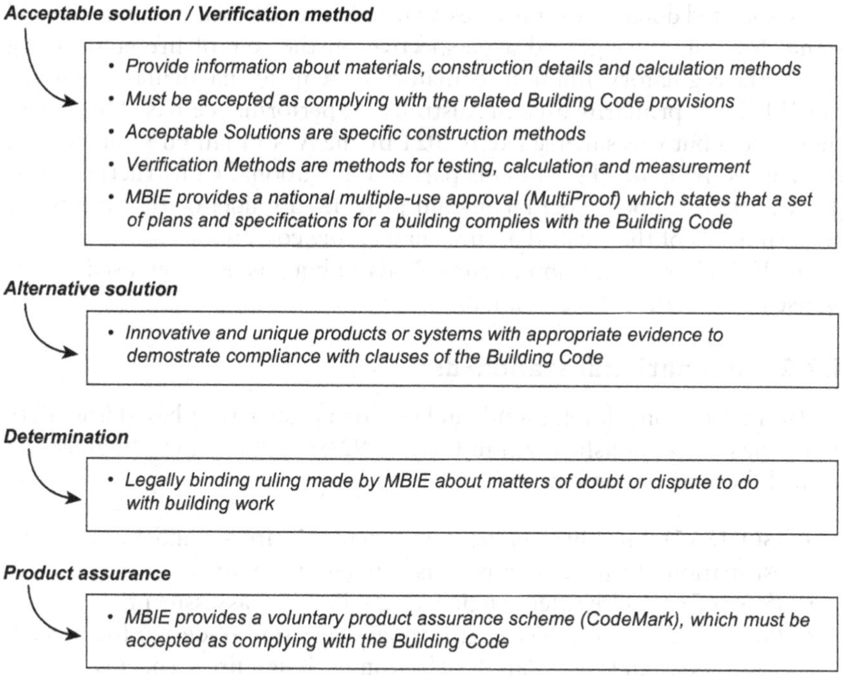

Acceptable solution / Verification method

- Provide information about materials, construction details and calculation methods
- Must be accepted as complying with the related Building Code provisions
- Acceptable Solutions are specific construction methods
- Verification Methods are methods for testing, calculation and measurement
- MBIE provides a national multiple-use approval (MultiProof) which states that a set of plans and specifications for a building complies with the Building Code

Alternative solution

- Innovative and unique products or systems with appropriate evidence to demostrate compliance with clauses of the Building Code

Determination

- Legally binding ruling made by MBIE about matters of doubt or dispute to do with building work

Product assurance

- MBIE provides a voluntary product assurance scheme (CodeMark), which must be accepted as complying with the Building Code

Figure 4.12 Methods of Demonstrating Compliance with the New Zealand Building Code (MBIE, 2017).

used, and this is applicable to most buildings, except low-rise simple residential buildings where C/AS1 (MBIE, 2020a) applies instead. However, C/AS2 is not applicable to buildings with complex features (e.g. atria, prisons, hospitals or some spaces with large numbers of people). In those cases, the use of a Verification Method or Alternative Solution is necessary. Any building of more than 20 storeys need an Alternative Solution.

4.1.9 Other regions

The regulatory systems in other regions of the world are usually different, and no overviews have been found. Timber structures and wood products are usually not widely used in larger or higher buildings in these regions.

4.2 INTERNATIONAL GUIDES AND STANDARDS

4.2.1 International Fire Engineering Guide (IFEG)

The International Fire Engineering Guidelines (IFEG) were developed 2005 and made available for use in Australia, Canada, USA and New Zealand. This IFEG guide references national and international standards, guides and associated documents, provides an insight into the issues that go beyond actual fire engineering, and a perspective on the role of fire engineering within the regulatory and non-regulatory systems for particular countries. The IFEG was primarily used in Australia for performance-based fire safety engineering but was superseded in 2021 by the Australian Fire Engineering Guidelines (AFEG). The AFEG is part of the National Construction Code (NCC, 2020) support documents and provides a guideline that meets the modern needs of the Australian fire engineering community.

The IFEG has recognition in New Zealand but has not been used or recognised within the USA or Canada.

4.2.2 International standards

On the international level, useful guidance on performance-based fire safety design has been published within ISO TC 92/SC 4 Fire Safety Engineering in the following documents:

- ISO 13571 Life-threatening components of fire – Guidelines for the estimation of time to compromised tenability in fires
- ISO 16732-1 Fire safety engineering – Fire risk assessment
- ISO 16733-1 Fire safety engineering – Selection of design fire scenarios and design fires – Part 1: Selection of design fire scenarios
- ISO 16733-2 Fire safety engineering – Selection of design fire scenarios and design fires – Part 2: Design fires
- ISO 19706 Guidelines for assessing the fire threat to people
- ISO 23932-1 Fire safety engineering – General principles
- ISO 24679-1 Fire safety engineering – Performance of structures in fire
- ISO/TR 20413 Fire safety engineering – Survey of performance-based fire safety design practices in different countries

The International Fire Safety Standards Coalition (IFSS) has recently published the 1st edition of International Fire Safety Standards – Common Principles (IFSS).

On the European level, a review of national requirements and applications has recently been published (CEN/TR 17524). Nordic INSTA standards (InterNordicSTAndards) have been developed to support the transition to more performance-based fire safety design (INSTA 950, INSTA 951 and INSTA 952). They are now being considered for inclusion within the European technical committee CEN TC 127 Fire safety in buildings and internationally within ISO TC 92/SC4 fire safety engineering.

4.2.3 European guideline

A European guideline on fire safety in timber buildings was published in 2010 (Östman et al.). It was the very first Europe-wide technical guideline on the fire safe use of wood products and timber structures in buildings.

The aim of this global guideline is to update and extend the European guideline as far as possible.

4.3 NATIONAL AND REGIONAL DIFFERENCES FOR THE USE OF WOOD

National and regional differences between countries have been reviewed, both in terms of the number of storeys permitted in timber structures and amounts of visible wood surfaces in interior and exterior applications, as an update to an earlier review (Östman and Rydholm, 2002). A later review has also been presented (Mikkola and Pilar, 2015). The results of the recent review 2020 are presented below in tables and maps (Östman, 2022).

This type of information must be treated with great caution, as there are often very detailed requirements and conditions that are difficult to simplify as a fair comparison between countries. There are also lesser requirements in certain countries that might be interpreted to indicate that the requirements in other countries are too strict, where in fact the opposite may be true; i.e. the requirements in some countries may be inadequate (or silent in respect of tall timber buildings).

The information below is therefore just an indication of current regulatory differences in prescriptive requirements. For real building projects, the full regulations must be consulted, and performance-based alternatives may be available, especially for larger and more complex buildings.

4.3.1 Residential buildings

4.3.1.1 Load-bearing timber elements

The maximum number of storeys allowed with load-bearing timber elements in multi-unit residential buildings is summarised in Table 4.3 and illustrated in maps in Figure 4.13. Data for both unsprinklered and sprinklered buildings are included.

4.3.1.2 Visible wood surfaces

The maximum number of storeys allowed with visible wood surfaces, both as interior linings and as exterior façade claddings, in residential buildings are summarised in Table 4.4 and illustrated in maps in Figures 4.14 and 4.15. Data for both unsprinklered and sprinkled buildings are included.

4.3.2 Office buildings

Data for office buildings are available for several countries are similar to residential buildings in most cases (Östman, 2022).

Table 4.3 Maximum number of storeys/maximum height and fire resistance requirements on *load-bearing elements* in *residential timber buildings* – prescriptive/pre-accepted requirements

Country	Max number of storeys *		Max height (m)		Same for all materials	Additional req. for wood	PFB design allowed	Valid since	Fire resistance requirements (minutes) Residential buildings Max number of storeys			
	Unspr.	Spr.	Unspr.	Spr.					1–2	3–4	5–8	> 8
Australia	2–3	(8)	–	25	Yes	Yes	Yes	2019	30–60	60–90	90	–
Austria	(7)[2]	(7)	22	–	Yes	Yes	Yes	2019	30–60	30–90	60–90	–
Belarus	2	2	–	–	No	No	No	2018	–	–	–	–
Belgium	NL[2]	NL	NL	NL	Yes	No	Yes	2020	30	30–60	60–120	120
Bulgaria	1–2	(4)	–	12	–	No	No	2010	–	30	60	120
Canada	3	12	–	42	No	Yes	Yes	2020	45	45–60	60[1]/120	120
China	3	5	10	–	Yes/No	Yes	Yes	2017	30	60	120	–
Croatia	(7)	(7)	22	22	Yes	Yes	Yes	2015	30	60	90	–
Czech Rep.	(3–4)[2]	(3–4)	9–12	9–12	No	Yes	Yes	1980+	15[2]/30	30[2]/60	45[2]/60	–
Denmark	(3–4)[2]	(3–4)	12[2]	12	No	Yes	Yes	2020	60	60	–	–
Estonia	4	8	–	–	No	Yes	Yes	2017	30	60–180	60[2]/120	–
Finland	2	8	9	28	No	Yes	Yes	2011	30	60[2]	60[2]	–
France	(16)	(16)	50	50	No	No	Yes	1986	15–30	30–60	60	90–120
Germany	(7–8)	(7–8)	22	22	Yes	Yes	Yes	2021	30	60	90	–
Greece	NL	NL	NL	NL	Yes	No	No	2018	30	60	60–90	90–120
Hungary	3	3	14	14	Yes	Yes	Yes	2020	15	30	60	90–120

(Continued)

Table 4.3 (Continued) Maximum number of storeys/maximum height and fire resistance requirements on load-bearing elements in residential timber buildings – prescriptive/pre-accepted requirements

Country	Max number of storeys *		Max height (m)		Same for all materials	Additional req. for wood	PFB design allowed	Valid since	Fire resistance requirements (minutes) Residential buildings Max number of storeys			
	Unspr.	Spr.	Unspr.	Spr.					1–2	3–4	5–8	>8
Iceland	8	NL	23	NL	Yes	No	Yes	2012	30/90[3]	60[4]/90	60[4]/90	90[4]/120
Ireland	3	4	10	10	No	Yes	Yes/No	2006	30	30–60	–	–
Italy	NL	NL	NL	NL	Yes	No	Yes	2006	30–60	–	60	90–120
Japan	4	4	16	16	Yes	Yes	Yes	2019	30	60/75	–	–
Latvia	(7)[2]	(7)[2]	21[2]	21[2]	Yes	Yes	(Yes)	2018	30	30[2]–60	60[2]	60[2]–180
Lithuania	(3)	(3)	10	10	Yes	No	No	2010	NL	45	60–120	60–120
Netherlands	NL	NL	NL	NL	Yes/No	No	Yes/no	2012		60	90	120
New Zealand	20	20	25	–	Yes	No	Yes	2020	60/30[4]	60/30[4]	60/30[4]	60/30[4]
Norway	4	4	–	–	Yes	No	Yes	2007	30	60	–	–
N Macedonia	1–2	1–2	6–9	6–9	Yes	No	No	1984	120	120	120	120
Poland	8	>8	25	>25	Yes	No	No	2017	30	30	30	120
Portugal	NL	NL	NL	NL	Yes/No	No	No	2009	30	30	60	90
Romania	3	4	–	–	No	Yes	Yes	1999	–	–	–	–
Russia	NL	NL	75	75	Yes	Yes	Yes	2012	0–30[2]	45	45	90[4]–120
Serbia	1–2	1–2	6–9	6–9	Yes/No	Yes	Yes	2019	15–30	30–60	–	–

(Continued)

Table 4.3 (Continued) Maximum number of storeys/maximum height and fire resistance requirements on load-bearing elements in residential timber buildings – prescriptive/pre-accepted requirements

Country	Max number of storeys*		Max height (m)		Same for all materials	Additional req. for wood	PFB design allowed	Valid since	Fire resistance requirements (minutes) Residential buildings Max number of storeys			
	Unspr.	Spr.	Unspr.	Spr.					1–2	3–4	5–8	> 8
Slovakia	3	3	–	–	No	Yes	No	2019	15–30	30–60	–	–
Slovenia	6	(7)	–	22	No	Yes	Yes	2019	–	60	–	–
Spain	NL	NL	NL	–	Yes	No	Yes	2019	30	60	90	120
Sweden	NL	NL	NL	NL	Yes	No	Yes	2012	60	60	60–90[2]	90
Switzerland	(33)	(33)	100	100	Yes	No	Yes	2015	0[1]/30	0[1]/30	30[1]/60	60[1]/90
Turkey	10	NL	30.5	NL	No	No	No	2007	30[2]/60	60[2]	60[2]	90/120[2]
Ukraine	NL	NL	NL	NL	Yes	Yes	Yes	2016	30	30	60	120–180
UK	3–4	NL	11	NL	Yes	No	Yes	2020	30	60	60[4]/90	90[4]/120
US	0	(18)	0	83	No	Yes	Yes	2021	0	0[1]/30	60[1]/120	120[4]/180

*Storey height estimated to 3 m, if only building height specified in national answers (estimimated number of storeys in brackets).
NL = No Limit for wood
PBD = Performance-Based Design
[1] For five to six storeys.
[2] Additional details apply.
[3] If different storeys and, in some countries, different fire compartments.
[4] With sprinklers.

Maximum number of storeys with *load-bearing timber structure* in residential buildings

Without sprinklers

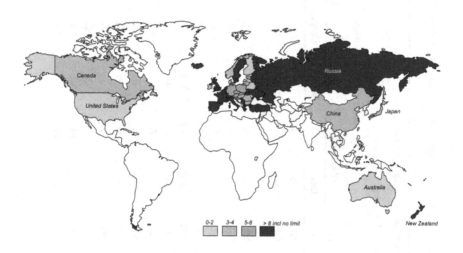

With sprinklers

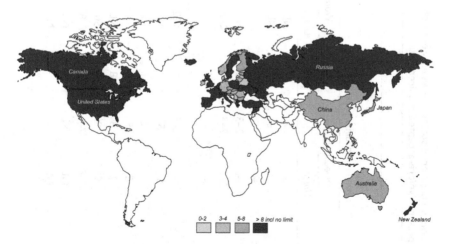

Figure 4.13 Maximum number of storeys allowed with load-bearing timber elements in *residential* buildings (prescriptive requirements).

Table 4.4 Maximum number of storeys for visible wood surfaces, exterior and interior – reaction to fire requirements in residential buildings – prescriptive/pre-accepted requirements

Country	Façade claddings – exterior			Maximum number of storeys generally						Floorings	
				Wall and ceiling linings – interior							
				Flats			Escape routes			Flats	Escape routes
	Wood, untr.[1]		FRT wood[1]	Wood, untr.		FRT wood[1]	Wood, untr.		FRT wood[1]	Wood, untr.	Wood, untr.
	Unspr.	Spr.		Unspr.	Spr.		Unspr.	Spr.		Unspr.	Unspr.
Australia	2	2	2	NL	NL	NL	0	0	0	NL[2]	NL[2]
Austria	6[3]	6	6	NL	NL	NL	3-4	3-4	4	NL	3-4[1]
Belarus	–	–	–	–	–	–	–	–	–	–	–
Belgium	2-3	2-3	8	NL	NL	NL	0	0	–	NL	0
Bulgaria	2-3	2-3	7-8	2-3	–	2-3	0	0	0	NL	–
Canada	3	6	6	3	6	NL	0	0	NL	3	3
China	–	–	–	0	0	–	–	–	–	0	3
Croatia	2-3	2-3	≤7	≤22	NL	≤22	2-3	2-3	2-3	–	2-3
Czech Rep.	5[3]	5[3]	5[3]	NL	NL	NL	4[3]	4[3]	4[3]	NL	4[3]
Denmark	1	2	–	1-7[3]	1-7[3]	–	0	0	–	NL	NL[1]
Estonia	8	8	NL	NL	NL	NL	0	0	8	NL	NL
Finland	2	8	8	NL	NL	NL	0	0	–	NL	NL
France	9	9	9	50	50	50	0	0	3	NL	0
Germany	7-8	7-8	7-8	7-8[3]	7-8[3]	7-8[3]	0	0	0	NL	7-8
Greece	NL	NL	NL	NL	NL	NL	NL	NL	NL	NL	NL
Hungary	NL	NL	–	NL	NL	–	NL	NL	–	NL	NL

(Continued)

Table 4.4 (Continued) Maximum number of storeys for visible wood surfaces, exterior and interior – reaction to fire requirements in residential buildings – prescriptive/pre-accepted requirements

	Façade claddings – exterior			Maximum number of storeys generally						Floorings	
				Wall and ceiling linings – interior							
				Flats			Escape routes			Flats	Escape routes
	Wood, untr.[1]		FRT wood[1]	Wood, untr.		FRT wood[1]	Wood, untr.		FRT wood[1]		Wood, untr.
Country	Unspr.	Spr.		Unspr.	Spr.		Unspr.	Spr.		Unspr.	Unspr.
Iceland	–	NL	NL	1–2	NL	NL	0	0	NL	NL	≤23 m
Ireland	4	–	–	0	0	–	0	0	–	NL	NL
Italy	5–8[1]	5–8[1]	5–8[1]	NL	NL	NL	0	0	NL	NL	0
Japan	NL	NL	NL	NL[3]	NL[3]	NL[3]	0	0	0	NL	NL
Latvia	3	3	9	NL	NL	NL	3	3	–	NL	NL[1]
Lithuania[3]	2	2	4	4	4	4	2	2	4	4	2
Netherlands	4	4	NL	NL	NL	NL	0	0	–	NL	0
New Zealand	3–4	3–4	3–4[6]	20	20	20	0	0	20	8	8
Norway	1–2	1–2	–	3–4	3–4	–	0	0	NL	NL	NL
N Macedonia	–	–	–	–	–	–	–	–	–	–	–
Poland	8	8	8	NL	NL	NL	0	0	NL	NL	0
Portugal	10	10	–	NL	NL	NL	0	0	0	NL	0
Romania	3	4	–	3	4	–	3	4	–	3	4
Russia	2	2	–	NL	NL	NL	2	2	2	NL	2
Serbia	–	–	–	–	–	–	–	–	–	–	–

(Continued)

Table 4.4 (Continued) Maximum number of storeys for visible wood surfaces, exterior and interior – reaction to fire requirements in residential buildings – prescriptive/pre-accepted requirements

Country	Façade claddings – exterior			Maximum number of storeys generally							Floorings	
	Wood, untr.[1]		FRT wood[1]	Wall and ceiling linings – interior							Flats	Escape routes
				Flats			Escape routes				Wood, untr.	Wood, untr.
				Wood, untr.		FRT wood[1]	Wood, untr.		FRT wood[1]			
	Unspr.	Spr.		Unspr.	Spr.		Unspr.	Spr.			Unspr.	Unspr.
Slovakia	5	5	5	NL	NL	NL	3	3	3		NL	NL
Slovenia	3–4	3–4	–	NL	NL	–	0	0	–		NL	0
Spain	3	3	–	NL	NL	NL	0	–	NL		NL	0
Sweden	2	2	8	2	8	NL	0	0	NL		NL	NL[1]
Switzerland	10	10	10	10	30	30[3]	10[3]	10[3]	30[3]		30	30[3]
Turkey	–	–	–	–	–	–	–	–	–		–	–
Ukraine	2	2	2	NL	NL	NL	0	0	0		NL	0
UK	3–4	6	6	0	0	NL	0	0	NL		–	–
US	3	3	6[1]	NL[1,3]	NL[1,3]	NL	0	NL[1]	NL		NL	NL

NL = No Limit for wood.
[1] Only if meeting required class.
[2] Minimum Critical Radiant Flux applies; not all timber species can comply.
[3] Additional details apply.
[4] If > 10 m to other buildings.
[5] With sprinklers.
[6] Up to 20 if passes full-scale façade test.

Maximum number of storeys with *wooden façade claddings* in residential buildings

Without sprinklers

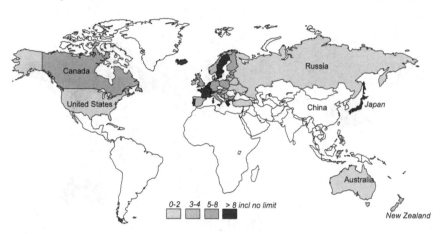

With sprinklers

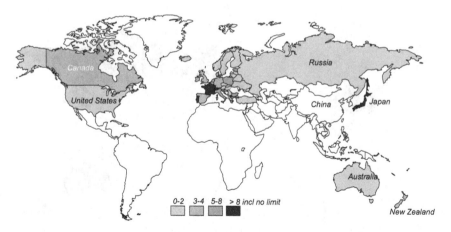

Figure 4.14 Maximum number of storeys allowed with wooden façade claddings in *residential* buildings (prescriptive requirements).

Maximum number of storeys with *visible interior wood* surfaces in residential buildings

Without sprinklers

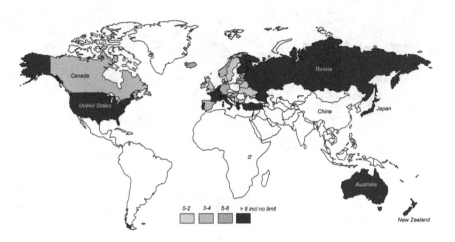

With sprinklers

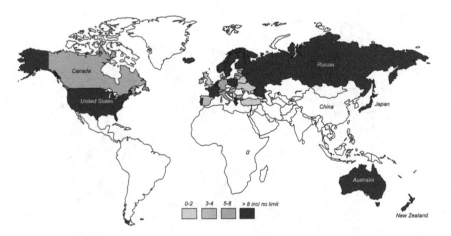

Figure 4.15 Maximum number of storeys allowed with visible wood in *interior* applications in *residential* buildings (prescriptive requirements).

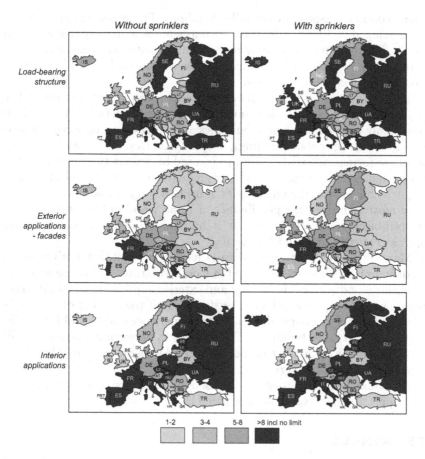

Figure 4.16 Possibilities to use wood in *residential buildings* in different applications in Europe.

4.3.3 Differences between European countries

The differences between European countries are further detailed in Figure 4.16 and in Östman, 2022.

4.4 CONCLUSIONS

The regulatory control systems for fire safety design of buildings differ between regions around the globe, but it is based on the same principles of saving life and property and specifying requirements for structural and non-structural applications.

The possibilities for building in wood have gradually increased in recent decades in many countries, mainly due to the environmental benefits of using wood. But there are still restrictions in terms of fire regulations in many countries, especially for taller buildings. The situation has therefore been mapped in 40 countries on four continents as an update to a survey in 2002. The main issues are how high buildings with load-bearing wooden frames may be built and how much visible wood may be used both inside and as façade claddings (Östman, 2022).

The requirements shown in this chapter apply primarily to prescriptive fire design according to so-called simplified design with detailed rules, which are mainly used for residential buildings and offices. For more complicated construction e.g. public buildings, shopping centres, arenas and assembly halls, performance-based design can be used by fire safety engineers using, for example, engineering methods for predicting evacuation and smoke filling, which increases the possibilities of using wood in buildings.

In most countries, the possibilities of using wood in buildings increase if sprinklers are installed, which is highlighted. More information on sprinklers is presented in Chapter 10.

Major differences between countries have been identified, both in terms of the number of storeys permitted in wood structures and of the amounts of visible wood surfaces in interior and exterior applications. Several countries have no specific regulations or do not limit the number of storeys in wooden buildings, mainly due to limited experience and lack of interest in using wood in taller buildings. The differences between countries are still large, and many countries have not yet started to use larger wood buildings despite supplies of forest resources.

Performance-based design may be used in several countries to verify further applications of wood (see Chapter 11).

REFERENCES

General

IFEG (2005) *International Fire Engineering Guide. Australian Building Codes Board.* ISBN 1741 614 562. IFEG.

IFSS (2020) *International Fire Safety Standards: Common Principles. International Fire Safety Standards Coalition,* 1st edition. IFSS.

ISO/TR 20413 (2020). *Fire Safety Engineering: Survey of Performance-based Fire Safety Design Practices in Different Countries.* International Standards Organization.

Mikkola, E., Pilar, M. G. (2015) Bio-based products and national fire safety requirements. In *COST Action FP1404.* Workshop Berlin.

Östman, B. (2022) *National Fire Regulations for the Use of Wood in Buildings: Worldwide Review 2020.* Report Linnaeus University.

Östman, B., Rydholm, D. (2002) *National Fire Regulations in Relation to the Use of Wood in European and Some Other Countries 2002.* Trätek Publication, 0212044.

Östman, B. et al. (2010). *Fire Safety in Timber Buildings: Technical Guideline for Europe.* SP Report 2010:19. Stockholm, Sweden.

Australia

NCC (2020) *National Construction Code: Building Code of Australia, Amendment 1.* NCC, Australia.

Canada

CAN/ULC-S114 (n.d.) *Standard Method of Test for Determination of Non-Combustibility in Building Materials.* Underwriters Laboratories of Canada, Toronto, ON.

NBCC (2015) *National Building Code of Canada.* National Research Council, Canada.

Ni, C. and Popovski, M. (2015) *Mid-Rise Wood-Frame Construction Handbook (SP-57E).* FPInnovations, Pointe-Claire, QC.

China

GB 50016 (2018) *General Code for Fire Protection of Buildings and Constructions.* China Planning Press, Beijing.

GB/T 51226 (n.d.) *Technical Standard for Multi-story and High Rise Timber Buildings.* China Architecture Publishing & Media Co., Ltd, Beijing.

GB 51236 (n.d.) *Code for Fire Protection Design of Civil Airport Terminal.* China Planning Press, Beijing.

GB/T 5907.1 (n.d.) *Fire Protection Vocabulary: Part 1 General Terms.* China Quality and Standards Publishing & Media Co., Ltd, Beijing.

Europe

CEN/TR 17524 (2020). *Fire Safety Engineering in Europe: Review of National Requirements and Application.* Technical Report. CEN European Committee for Standardization, Brussels.

CPR Construction Products Regulation (2011). *Official Journal Council Directive.* 89/106/EEC OJ L 88 of 4. European Commission, Brussels.

Dimova, S., Pinto, A., Athanasopoulou, A., Sousa, L. (2019) Introduction of fire engineering approach in building regulations. *European Commission, DG JRC.* SFPE Europe Conference Fire Safety Engineering.

EN 13501-1 (n.d.) Fire classification of construction products and building elements – Part 1: Classification using test data from reaction-to-fire tests. *European Standard.* European Committee for Standardization, Brussels.

EN 13501-2 (n.d.) Fire classification of construction products and building elements - Part 2: classification using data from fire resistance tests, excluding ventilation services. *European Standard.* European Committee for Standardization, Brussels.

INSTA TS 950 (n.d.) *Fire Safety Engineering: Comparative Method to Verify Fire Safety Design in Buildings.* Nordic INSTA (InterNordicSTAndards) Technical Specification. Standards Norway, Oslo.

INSTA 951 (n.d.) *Fire Safety Engineering: Guide to Probabilistic Analysis for Verifying Fire Safety Design in Buildings*. Nordic INSTA (InterNordicSTAndards) Standard. Standards Norway, Oslo.

INSTA 952 (n.d.) *Fire Safety Engineering: Review and Control in the Building Process*. Nordic INSTA (InterNordicSTAndards) Standard. Standards Norway, Oslo.

Japan

BSL (2019) *Building Standards Law*, Article 21, Article 27, Article 61

Hasemi Y. et al. (2016) *Development of Wood-based "Fireproof" Buildings in Japan, World Conference on Timber Engineering (WCTE)*.

Kagiya, K. (2021) *Fire Safety Engineering for Timber Buildings: From a Perspective of Japan*. Timber Workshop, IAFSS International Symposium on Fire Safety Science.

New Zealand

MBIE (2017) *Ministry of Business Innovation and Employment. Regulatory Charter – Building Regulatory System*. Wellington, New Zealand. https://www.mbie.govt.nz/dmsdocument/3083-regulatory-charter-building-pdf.

MBIE (2020a) *C/AS1 Acceptable Solution for Buildings with Sleeping (Residential) (Risk Group SH). Including Amendment 5*. Wellington, New Zealand: The Ministry of Business, Innovation and Employment.

MBIE (2020b) *C/AS2 Acceptable Solution for Buildings other than Risk Group SH (including Amendment 2)*. Wellington, New Zealand: The Ministry of Business, Innovation and Employment.

Russian Federation

Aseeva, R., Serkov, B., Sivenkov, A. (2014) *Fire Behaviour and Fire Protection in Timber Buildings*. Springer Series in Wood Science. Dortrecht, Heidelberg, New York, London. DOI 10.1007/978-94-007-7460-5.

С П 64.13330 (n.d.) *Wooden Structures (Includes Calculation Methods)*. Russian standard, GOST, Moscow.

Federal Law No 123 (2018) *Technical Regulations on Fire Safety Requirements*. Russian Federation, Moscow.

GOST 30247.0 (n.d.) *Elements of Building Constructions. Fire-resistance Test Methods. General Requirements*. Russian standard, GOST, Moscow.

GOST 30247.1 (n.d.) *Elements of Building Constructions. Fire-resistance Test Methods. Load Bearing and Separating Constructions*. Russian standard, GOST, Moscow.

GOST 30403 (n.d.) *Building Structures. Fire Hazard Test Method*. Russian standard, GOST, Moscow.

GOST 31251 (n.d.) *Facades of Buildings. Fire Hazard Test Method*. Russian standard, GOST, Moscow.

SNiP II-2 (n.d.) *A Guide to Determining the Fire Resistance Limits of Structures, the Fire Propagation Limits for Structures and the Combustibility Groups of Materials.* Russian norms and rules, Moscow.

USA

Breneman, S., Timmers, M., Richardson, D. (2018) *Tall Wood Buildings in the 2021 IBC Up to 18 Stories of Mass Timber, Woodworks.* Washington DC, USA.

ICC (2018) *International Building Code.* International Code Council, Washington DC, USA.

ICC (2021) *International Building Code.* International Code Council, Washington DC, USA.

IFC (2018) *International Fire Code.* International Code Council, Washington DC, USA.

McGregor, C. (2013) *Contribution of Cross Laminated Timber Panels to Room Fires.* Thesis, Carleton University, Ottawa, Canada.

Medina, A. (2014) *Fire Resistance of Partially Protected CLT Rooms.* Thesis, Carleton University, Ottawa, Canada.

NFPA (2018) *National Fire Protection Association "Life Safety Code" NFPA 101,* Quincy, Massachusetts, USA.

Su, J., LaFrance, P., Hoehler, M., Bundy, M. (2018) *Fire Safety Challenges of Tall Wood Buildings – Phase 2: Task 2 & 3 Cross Laminated Timber Compartment Fire Tests. Fire Protection Research Foundation.* Report no. FPRF-2018-01.

Taber, B. C., Lougheed, G., Su J. Z., Bénichou, N. (2014) *Solutions for Mid-rise Wood Construction: Apartment Fire Test with Encapsulated Cross Laminated Timber Construction: Test APT-CLT Report to Research Consortium for Wood and Wood-Hybrid Mid-Rise Buildings, National Research Council of Canada.* Technical Report no. A1-100035-01.10.

Zelinka, S., Hasburgh, L., Bourne, K., Tucholski., D, Ouellette, J. (2018) *Compartment Fire Testing of a Two-Story Mass Timber Building.* General Technical Report FPL–GTR–247.

Chapter 5

Reaction to fire performance

Marc Janssens and Birgit Östman

CONTENTS

DOI: 10.1201/9781003190318-5

SCOPE OF CHAPTER

This chapter presents the reaction to fire performance of wood products used in buildings as internal surface finishes, exterior wall claddings and roof coverings. It describes the systems used for compliance with prescriptive regulations in different regions, and it also covers the characteristics of wood products for performance-based design and methods for improving the reaction to fire performance of wood products.

5.1 WOOD PRODUCTS USED AS INTERIOR FINISH, EXTERIOR CLADDING OR ROOF COVERING

This section briefly describes the different types of wood products that are used as interior finish, exterior cladding, and roof covering in buildings. It supplements the information in Chapter 1 of this guide.

5.1.1 Sawn timber

Logs are converted to rectangular-shaped sawn timber in sawmills. Sawn timber is used primarily for structural applications, and a wide range of shapes are readily available. Cladding and decking timber are the more important applications in terms of reaction to fire. Sawn timber is mostly produced from softwood trees, but hardwoods can be used as well, e.g., for exterior claddings.

It is important to note that the reaction to fire characteristics of a wood product is affected by the composition of the wood because its three main components (cellulose, hemi-celluloses and lignin) have quite different thermal degradation characteristics. This is evident from the results of thermogravimetric analyses (TGA), which show that the constituents decompose to release flammable volatiles over different temperature ranges (Roberts, 1970). Typical decomposition ranges are 240–350°C for cellulose, 200–260°C for hemi-celluloses, and 280–500°C for lignin. Consequently, the thermal degradation characteristics of wood shift towards higher temperatures with increasing lignin content. This explains why the surface temperature at ignition is significantly higher for softwoods than for hardwoods (see below). Moreover, only about 50% by mass of the lignin (which typically accounts for 18–35% by mass of the wood) decomposes to volatiles and a higher lignin content therefore results in an increased char yield.

In addition to the three principal components, wood also contains removable extraneous organic compounds, referred to as "extractives" (typically between 4% and 10% of the wood) and small amounts of inorganic minerals (less than 1%). Extractives, which are a collection of various organic compounds, adversely affect the flammability of the wood. Petterson (1984) compiled detailed chemical composition data of wood species found in the U.S. and other parts of the world.

Wood products can be treated in a variety of ways to increase durability, and previous toxic treatment methods are being replaced by more environment-friendly methods. This includes different types of wood modifications, for example, acetylation (treatment with acetic acid), furfurylation (treatment with furfuryl alcohol) and thermal treatment (Gérardin, 2016). These treatments change the chemistry of the wood and reduce the amount of water that the cell walls can absorb. This suggests that the fire performance of the material can be altered, which has been confirmed by a few studies (Morozovs & Bukšāns, 2009; Dong et al., 2015).

5.1.2 Panel products

Wood is the principal component in the production of a variety of engineered panel products for use in structural or decorative applications. The most common of these products are briefly described below. Additional discussion of panel products can be found in Chapter 1 of this guide and Chapter 11 of the Wood Handbook (Wood Handbook, 2010):

- Hardboard is manufactured primarily from inter-felted lignocellulosic fibres (usually wood), mixed with a synthetic resin and additives to improve specific properties, and consolidated under heat and pressure in a hot press. Density range ≥ 900 kg/m^3.
- Medium-density fibreboard (MDF) is manufactured from lignocellulosic fibres in a similar fashion to hardboard but has a lower density than hardboard. Density range: 400–900 kg/m^3.

- Oriented strandboard (OSB) is an engineered structural-use panel manufactured from thin wood strands (in North America, primarily aspen, which is a hardwood species) bonded together with water-resistant resin. The resin may have adverse effects on the reaction to fire performance. Density range: 500–800 kg/m³.
- Particleboard is produced by mechanically reducing the wood raw material into small particles, applying an adhesive to the particles, and consolidating a loose mat of the particles with heat and pressure into a panel product. Density range: 600–800 kg/m³.
- Plywood is a glued wood panel made up of an odd number (usually) of relatively thin layers of veneer (also referred to as plies) with the grain of adjacent layers at right angles. In North America, structural plywood generally is made from softwood veneers, while hardwood plywood is used primarily for decorative purposes. Density range: 350–800 kg/m³.
- Waferboard is a particle panel product made of wafer-type flakes. It is usually manufactured to possess equal mechanical properties in all directions parallel to the plane of the panel. Density range: 470–640 kg/m³.

A variety of adhesives are available for the manufacture of engineered panel products, and it has been demonstrated that the type of adhesive does not influence the reaction to the fire performance of wood-based panels (Östman and Mikkola, 2010), contrary to the structural fire performance of some engineered wood products containing adhesives (see Chapter 7).

5.1.3 Engineered structural wood products

Engineered structural wood products such as cross-laminated timber (CLT), glued laminated timber (glulam), laminated veneer lumber (LVL), parallel strand lumber (PSL) etc., may be used with visual wood surfaces and need evidence of their declared reaction to fire performance based on testing. These products are described in Chapter 1 of this guide.

5.2 ASSESSING REACTION TO FIRE PERFORMANCE OF WOOD PRODUCTS FOR COMPLIANCE WITH PRESCRIPTIVE REGULATIONS

This section provides an overview of the reaction to fire tests and classification systems that are used in prescriptive building regulations in different parts of the world, summarises how wood products perform and explains how this affects their use in different geographical regions. The systems are different for wall and ceiling linings, floor coverings, roof coverings and façade claddings, as described below. The prescriptive requirements are usually based on fire testing, and the main test methods used are therefore described.

Further information on fire safety requirements in different regions is presented in Chapter 4.

5.2.1 Wall and ceiling linings

Most of the available information exists for wall and ceiling linings.

International methods

Common internationally standardised methods to assess the reaction to fire performance of wall linings and ceiling materials are the Room/Corner test standard ISO 9705-1 and the Cone Calorimeter ISO 5660-1.

The ISO 9705 Room/Corner test consists of a room measuring 3.6 m deep by 2.4 m wide by 2.4 m high, with a single ventilation opening (doorway) approximately 0.8 m wide by 2 m high in the front wall. In the standard configuration, the interior surfaces of all walls (except the front wall) and the ceiling are covered with the test product. The product is exposed to a propane-burner ignition source located on the floor in one of the rear corners of the room opposite the doorway.

At the start of a test, the propane gas burner is ignited and the material system is exposed to a 100 kW flame. After 10 minutes of exposure to 100 kW, the gas flow to the burner is increased to 300 kW and maintained at that level for an additional 10 minutes. The products of combustion emerging through the doorway are collected in a hood and extracted through an exhaust duct by a fan. A gas sample is drawn from the exhaust duct to measure the concentrations of oxygen, carbon dioxide and carbon monoxide in the fire effluents. The gas temperature and differential pressure across a bi-directional probe are measured to determine the mass flow rate of the exhaust gases. The gas concentrations and duct flow rate measurements are used to calculate the heat release rate based on the oxygen consumption technique (Janssens, 1991a). The smoke production rate is determined based on the measured light opacity in the duct using a white-light extinction photometer located close to the gas sampling point. The primary measurements are the heat release rate, smoke production rate and heat flux to the floor in the room. The test is generally terminated when flashover occurs during the 20-minute test period. Flashover is assumed to have occurred when the total heat release rate reaches 1000 kW. The Room/Corner test apparatus is shown schematically in Figure 5.1.

The Room/Corner test has been used for classification in Australia, see below, and as a reference scenario for the European reaction to fire classification system (Sundström, 2007).

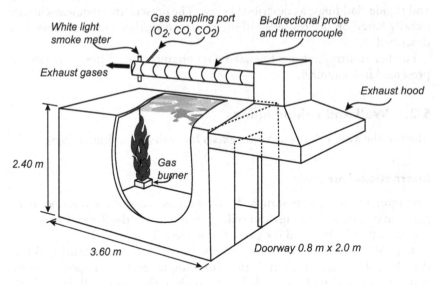

Figure 5.1 Room/Corner test ISO 9705-1. Specimen size 23 m² (walls only) or 31.6 m² (incl. ceiling).

The Cone Calorimeter described in ISO 5660-1 is the most commonly used test method to assess the reaction to the fire performance of building products. It is a sophisticated small-scale test apparatus, which is capable of measuring the heat release rate of materials and products under a wide range of thermal exposure conditions based on the oxygen consumption technique. Other useful information obtained from Cone Calorimeter tests includes time to ignition, mass loss rate, smoke production rate and effective heat of combustion. At the start of a test, a square specimen of 100×100 mm is placed on a load cell and exposed to a pre-set radiant heat flux from an electric heater. The heater is in the shape of a truncated cone and can provide heat fluxes to the specimen in the range 0–100 kW/m². An electric spark ignition source is used for piloted ignition of the pyrolysis gases produced by the heated specimen. The products of combustion and entrained air are collected in a hood and extracted through a duct by a blower. A gas sample is drawn from the exhaust duct and analysed for oxygen (and often for carbon dioxide and carbon monoxide as well). Smoke production is determined based on the measured light obscuration in the duct using a laser photometer located close to the gas sampling point. Gas temperature at and differential pressure across an orifice plate are used for calculating the mass flow rate of the exhaust gases. The oxygen concentration and mass flow rate measurements are used to calculate the heat release rate based on oxygen consumption calorimetry (Janssens, 1991a). A schematic sketch of the Cone Calorimeter is shown in Figure 5.2.

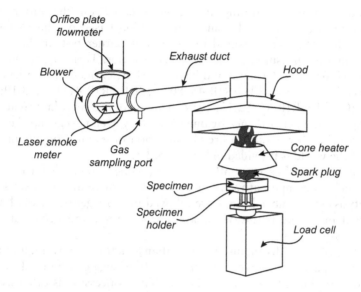

Figure 5.2 Cone Calorimeter ISO 5660-1. Specimen size 0.01 m².

Table 5.1 Australian building code BCA (NCC, 2019) classification based on ISO 9705

Classification	Time to flashover, $t_{flashover}$ (s)
Group 1	No flashover
Group 2	601–1200
Group 3	121–600
Group 4	0–120

Australia and New Zealand

In Australia, the prescriptive (or "deemed-to-satisfy") reaction to fire requirements for building materials and components are covered under the "Fire Hazard Properties" provisions in the Building Code of Australia or BCA (NCC, 2019). The Fire Hazard Properties specification has two parts: one part that specifies the reaction to fire requirements and a separate part that prescribes the minimum fire performance requirements for any material used in the construction of buildings, except single-family homes.

As far as reaction to fire is concerned, the BCA classifies wall and ceiling linings into four groups based on the time to flashover measured in the ISO 9705 Room/Corner test (see Table 5.1). An alternative approach to determine the group classification of a lining involves the use of a calculation method developed by Kokkala et al. (1993) to predict ISO 9705 room/corner test performance, based on the Cone Calorimeter data.

In addition to controlling fire growth based on the group number, the BCA also limits the smoke production based on the SMOGRA (SMOkeGrowthRate) measured in the Room/Corner test or the specific extinction area measured in the Cone Calorimeter. Typically, wood products are classified in Group 3 and have a specific extinction area less than (and well below) 750 m^2/kg (RIR 45980.10, 2018; RIR 41117.9, 2019; RIR 45981.10, 2019; RIR 45982.13, 2019). The use of fire-retardant treatments can improve the performance by one or possibly even two group classes. A comparison with reaction to fire classification of wood products in other parts of the world is provided in Section 5.2.5.

Group 4 materials are not allowed at all, while Group 1 is for restricted areas, such as escape pathways or areas where there are occupants with mobility issues. Timber linings are allowed in most general areas but are restricted in escape pathways or areas where there are occupants with mobility issues.

One aspect that distinguishes Australian reaction to fire requirements from other parts of the world is that timber linings can be used in most parts of a building to any storey height. This concession is often used to explain why the Australian fire-resistance rating requirements are generally higher than in other developed countries.

In addition to reaction to fire, all materials used in a building, except for housing, have to meet minimum fire-performance requirements. These requirements are based on performance in the test method described in AS/NZS 1530.3. The test results are used to calculate Ignitability, Flame Propagation, Heat Evolved and Smoke Developed Indices. The Ignitability Index is an integer between 0 and 20 and the other three indices vary between 0 and 10, where 0 is best and 20 or 10 is worst. AS/NZS 1530.3 performance data for a range of wood products are published (WoodSolutions, 2021). In practice, the requirements are based on the Spread-of-Flame and Smoke-Developed Index only. For solid sawn wood, the Spread-of-Flame Index for wood species varies between 0 (for merbau) and 10 (for Western red cedar), but for most species it ranges from 7 to 10. The Smoke-Developed Index for solid sawn wood ranges from 2 (for jarrah) to 5 (for merbau).

Finally, the reaction to fire-performance requirements for building materials and components in Australia are specified in Volumes 1 and 2 of the National Construction Code (NCC, 2019). These two volumes of the NCC constitute the Building Code of Australia (BCA). The NCC is a performance-based code. The performance requirements can be met by adopting one of the deemed-to-satisfy solutions, which are provided in the BCA in the form of Acceptable Solutions and Verification Methods. Acceptable Solutions are deterministic in nature, while Verification Methods prescribe another way to comply with the BCA performance requirements based on tests and/or calculations. The performance requirements also can be met by developing an alternative solution, which typically involves testing and/

or engineering analyses to demonstrate that the material or system meets or exceeds the pertinent objective(s) and level of safety implicit in the code.

In New Zealand, the New Zealand Building Code or NZBC (NZ legislation) requires wall linings and ceiling materials to be classified in one of four groups based on the measured or calculated flashover time in the ISO 9705-1 room/corner test, similar to Australia, as shown in Table 5.1. As in Australia, an alternative approach to determine the group classification of a product using the calculation method developed by Kokkala et al. (1993) based on Cone Calorimeter data can also be used in some instances. Untreated wood products are usually classified in Group 3 and generally can be used as wall linings in all occupied spaces, except those in specific locations and certain types of buildings (e.g., exit ways, sleeping areas in buildings where care or detention is provided, buildings that must be operational following an earthquake, etc.). Additional restrictions apply to the use of untreated wood products used as ceiling materials in crowd/assembly spaces.

Smoke production limits for wall linings and ceiling materials only apply in unsprinklered buildings where a group number of 1 or 2 is required. The average smoke production rate over the period 0 to 10 minutes in the ISO 9705-1 test must not be greater than 5.0 m²/s, or the average specific extinction area must not be greater than 250 m²/kg when ISO 5660-1 is used.

The NZBC (NZ legislation) Acceptable Solution C/AS2 allows solid wood or wood products at least 9 mm thick and a density of at least 400 kg/m³ (or 600 kg/m³ for particleboard) to be assigned group number 3 without further testing. This also applies where waterborne or solvent-borne paint coating, varnish or stain is applied to the surface, provided it is not more than 0.4 mm thick and not more than 100 g/m².

Products with European Classifications using EN 13501-1 of Class B (or better), C and D are also treated as equivalent to group number 1, 2 and 3, respectively. See also Table 5.2.

Europe

In Europe, common test methods to evaluate the reaction to fire of construction products have been agreed. For construction products other than floor coverings, the main method is the Single Burning Item (SBI) test EN 13823. Two specimens of the material to be tested are positioned in a specimen holder frame at a 90° angle to form an open corner section. Both specimens are 1.5 m high. One specimen is 1 m wide and is referred to as the long wing. The other specimen is 0.5 m wide and is referred to as the short wing. During a test, the specimens are exposed for 20 minutes to the flame of a triangular diffusion propane gas burner operating at 30 kW. The products of combustion are collected in a hood and are extracted through an exhaust duct. Instrumentation is provided in the duct

Table 5.2 European reaction to fire classification system for building products, except floorings

Euroclass	Smoke class*	Burning droplets class§	Requirements according to				Example products
			Non comb.	SBI	Small flame	FIGRA (W/s)	
A1	–	–	✓	–	–	–	Stone wool with limited binder content
A2	s1, s2 or s3	d0, d1 or d2	✓	✓	–	≤ 120	Gypsum boards (thin paper), mineral wool
B	s1, s2 or s3	d0, d1 or d2	–	✓	✓	≤ 120	Gypsum boards (thick paper), FRT wood
C	s1, s2 or s3	d0, d1 or d2	–	✓	✓	≤ 250	Wall coverings on gypsum board, FRT wood
D	s1, s2 or s3	d0, d1 or d2	–	✓	✓	≤ 750	Wood, wood-based panels
E	–	– or d2	–	–	✓	–	Some synthetic polymers
F	–	–	–	–	–	–	Do not fulfil class E

* s1: $SMOGRA \leq 30$ m²/s² and $TSP_{600s} \leq 50$ m²; s2: $SMOGRA \leq 180$ m²/s² and $TSP_{600s} \leq 200$ m²; and s3: not s1 or s2

§ d0: No flaming droplets/particles in EN 13823 within 600 s; d1: no flaming droplets/particles persisting longer than 10 s in EN 13823 within 600 s; and d2 = not d0 or d1

to measure temperature, velocity, gas composition (oxygen, carbon dioxide and carbon monoxide concentrations) and light opacity. The velocity and gas composition data are used to determine the heat release rate on the basis of the oxygen consumption technique (Janssens, 1991a). Smoke production rate is determined based on the measured flow rate and light opacity in the exhaust duct. During the test, observations are made of lateral flame spread (LFS) over the specimen surface and the presence of flaming droplets or particles. Classification is based primarily on a fire-growth rating (FIGRA, FIreGrowthRAte), total heat released over the first 10 minutes of the test (THR$_{600s}$), and lateral flame spread (LFS) across the long-wing specimen. A smoke development index (SMOGRA), as well as visual observations of flaming droplets and/or particles are used for additional classification. The FIGRA index is equal to the maximum value of (heat release rate)/(elapsed time). To reduce the noise, the FIGRA is calculated based on the 30-second running average heat release rate. In addition, only heat release rates that exceed a class-dependent minimum value are considered in the calculations. The SMOGRA is equal to the maximum value of (smoke production rate) / (elapsed time). The smoke production rate is based on a 60-second running average.

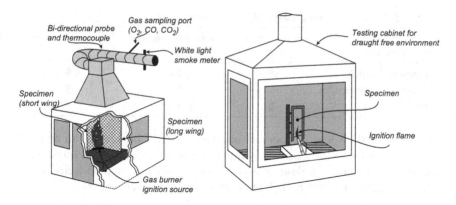

Figure 5.3 SBI Single Burning Item test EN 13823, specimen size 0.0225 m² (left) and small flame test EN ISO 11925-2 (right), specimen size 0.00225 m².

In addition to the SBI test, the small flame test EN ISO 11925-2 has to be used. Both test methods are illustrated schematically in Figure 5.3.

Table 5.2 provides the European reaction to fire classification system for construction products, except flooring materials (EN 13501-1). It is important to note that the official Euroclasses B, C and D of a combustible product are based on performance in EN 13823, as well as in EN ISO 11925-2. Most wood products with density > 300 kg/m³ and thickness > 9–12 mm, depending on mounting, fulfil class D (Östman et al., 2010). Fire-retardant treated (FRT) wood products may fulfil Class B (see Section 5.4.1). Classes A1 and A2 are for non-combustible products and classes E and F, which require testing only according to the small flame test EN ISO 11925-2, are seldom used in buildings. Some results for wood products and comparisons with classification in other countries are presented in Section 5.2.1.6.

Japan

The building regulations in Japan were updated in 1998 to facilitate the adoption of internationally accepted fire-test methods and implementation of performance-based requirements for compliance. This resulted in the development of a new reaction to fire classification system that came into effect in June of 2000 (Hakkarainen and Hayashi, 2001). The system recognises three classes of interior finish materials based on heat release rate measurements in the Cone Calorimeter, ISO 5660-1 (described earlier in Section 5.2.1). The criteria are identical for the three classes, but the test duration is different, as shown in Table 5.3. The peak heat release rate can exceed the limit for a maximum period of 10 seconds.

To obtain a classification, building products and materials also need to pass a small-scale smoke toxicity test. The model box test described in ISO/TS 17431 is similar to a reduced-scale version (~ 40% in the linear dimension)

Table 5.3 Classification based on ISO 5660-1 used in Japan

Classification	Test duration (min)	Peak heat release rate (kW/m²)	Total heat released (MJ/m²)
Non-combustible	20	≤ 200	≤ 8
Quasi non-combustible	10	≤ 200	≤ 8
Fire retardant	5	≤ 200	≤ 8

of the ISO 9705 Room/Corner test but uses a 40 kW burner. It can be used as an alternative to the Cone Calorimeter to qualify materials as quasi non-combustible or fire-retardant. Wood products can be treated with fire retardants to achieve the fire-retardant classification (see Section 5.4).

North America

In the **United States**, the Steiner tunnel test is the most common material-flammability test method prescribed by building codes to limit flame spread over wall and ceiling finishes. The apparatus and test procedure are described in ASTM E84. The test specimen is 7.6 m long and is mounted in the ceiling position of a long tunnel-like enclosure. It is exposed at one end to a 79 kW gas burner. There is a forced draft through the tunnel from the burner end. The measurements consist of flame spread over the surface and light obscuration by the smoke in the exhaust duct of the tunnel. Test duration is 10 min. A flame spread index (FSI) is calculated based on the area under the curve of flame tip location versus time. The FSI is 0 for a cement board and is normalised to approximately 100 for red oak. The smoke-developed index (SDI) is equal to 100 times the ratio of the area under the curve of light absorption versus time to the area under the curve for a heptane pan fire.

The test standard ASTM E2768 is an extended duration version of ASTM E84 used to qualify FRT wood for use in buildings of non-combustible construction. To pass this test, FRT wood needs to achieve an FSI of 25 or less during the first 10 minutes and the flame shall not progress beyond 3.2 m from the centreline of the burners during the entire 30-minute test.

The classification of linings in the model building codes is based on the FSI. There are three classes: Class A for products with FSI ≤ 25; Class B for products with 25 < FSI ≤ 75; and Class C for products with 75 < FSI ≤ 200. In addition, all three classes require that the product have an SDI of 450 or less. Class A products are generally permitted in enclosed vertical exits. Class B products can be used in exit access corridors, and Class C products are allowed in other rooms and areas.

The American Wood Council (AWC) has published a list of FSI and SDI values for a large number of solid sawn wood species and panel products (AWC, 2019). Most solid sawn wood specimens achieve Class B while some are Class C. The reverse is the case for wood panel products, i.e., most are Class C while some are Class B.

The Room/Corner test described in NFPA 286 can be used as an alternative to the Steiner tunnel test to qualify wall and ceiling linings for use in areas where Class A materials are required. The NFPA 286 test apparatus is similar to ISO 9705 but has some differences. It consists of a room measuring $3.66 \times 2.44 \times 2.44$ m high, with a single ventilation opening (doorway) measuring approximately 0.76 m $\times$ 2.03 m high in the front wall. Typically, only the interior surfaces of the sidewalls and the back wall are covered with the test material. The test material is exposed to the flame of a 0.3×0.3 m propane diffusion "sand box" burner, located with the top surface 0.3 m above the floor in one of the rear corners of the room opposite the doorway. Propane is supplied at a specified rate so that a net heat release rate of 40 kW is achieved for the first 5 minutes of the test, followed by 160 kW for the remaining 10 min. The products of combustion generated in the fire are collected in a hood and extracted through an exhaust duct, which is instrumented to measure the heat release rate based on oxygen consumption calorimetry, and smoke production rate using a white light or laser photometer. The primary pass/fail criteria are the occurrence of flashover at any time during the test, and the total amount of smoke produced exceeding 1000 m^2 at the end of the 15-minute test.

Wood products cannot pass NFPA 286 unless treated with fire retardants. For most untreated wood panel products, flashover occurs between 5 and 7 minutes, i.e., within two minutes from the increase of the burner output from 40 to 160 kW, but flashover occurred prior to the burner increases for OSB treated with a water repellent (Tran and Janssens, 1991).

In Canada, the Acceptable Solution for reaction to fire performance of wall lining and ceiling materials is based on the surface burning characteristics determined according to CAN/ULC-S102. The apparatus and test procedure are nearly identical to those in ASTM E84, except that the windows are installed flush mounted to the outside face of the tunnel furnace, creating cavities which provide turbulence and mixing of the air and combustion gases. As a consequence, the ULC versions of the tunnel test do not require turbulence bricks. However, the test results are evaluated differently.

The net result of the differences is that the highest allowable flame spread rating for interior wall and ceiling finishes in the National Building Code of Canada (NBCC) is 150 versus 200 in the U.S. CAN/ULC-S102 Flame Spread Ratings and Smoke-Developed Classifications for various wood products are published by the Canadian Wood Council (CWC, 2020).

The NBCC is objective-based and provides deemed-to-satisfy solutions, which can, but do not have to, be used to meet the code objectives.

Comparison of reaction to fire classification of surface linings in different countries

Table 5.4 compares the reaction to fire classification for six wood products in four countries (Janssens et al., 2006). Small open-flame testing was

Table 5.4 Comparison of reaction to fire classification in different countries

Product	United States	Australia/NZ	Europe	Japan
Douglas fir plywood	C	Group3	D	Unclassified
FRT Douglas fir plywood	A	Group 1 or 2	C	Unclassified
Oriented strandboard 1	C	Group 3	D	Unclassified
Oriented strandboard 2	C	Group 3	D	Unclassified
White pine planks	C	Group 3	D	Unclassified
White oak planks	C	Group 3	D	Unclassified

Table 5.5 Flashover times for different wood products in Room/Corner tests

		Flashover time (s)*	
Wood product	Thickness (mm)	NFPA 286	ISO 9705
FRT plywood N	13	NFO[+]	640
FRT plywood F	13	NFO[+]	633
FRT plywood R	13	NFO[+]	NFO[+]
Spruce plywood	13	372	186
Oak veneered plywood	13	330	78
Particleboard N1	13	306	156
Particleboard N2	13	335	140
Hardboard with stucco coating	10	324	174

* Flashover is defined based on the time when flames emerge through the door or heat flux to the floor reaches 20 kW/m², whichever occurs first.
[+] Flashover did not occur prior to the end of the test (15 minutes for NFPA 265 and 20 minutes for ISO 9705).

not conducted for these materials, so the reported Euroclass in Table 5.4 is based solely on EN 13823 test results. However, it is well known that wood products with a density over about 300 kg/m³ will pass the EN ISO 11925-2 test (Östman et al., 2010).

White et al. (1999) published a comparison of the flashover times for a range of wood products evaluated according to different protocols in the standard Room/Corner test apparatus. Table 5.5 provides a subset of the data for wood products that were tested according to NFPA 286 with material on walls only and ISO 9705 with material on walls and ceiling. Flashover was not observed for the three FRT plywoods in the NFPA 286 test. Depending on the treatment, flashover in the ISO 9705 Room/Corner test is either delayed until after the change of the burner output from 100 to 300 kW (products N and F) or does not occur within the 20-minute test duration. For untreated wood products, flashover occurs in the NFPA 286 test within 1 to 2 minutes after the burner increases from 40 to 160 kW. The flashover time in the ISO 9705 Room/Corner test exceeds 2 minutes for all untreated wood products, except oak veneer plywood, which quickly delaminated during the test.

5.2.2 Floor coverings

International method

The main test method for flooring coverings is the Radiant Flooring Panel test, EN ISO 9239-1. The test apparatus consists of a premixed gas-fired radiant panel inclined at 30° to and directed at a horizontally mounted floor covering specimen. The radiant panel generates a heat flux distribution along the length of the test specimen from a nominal maximum of 10 kW/m² to a minimum of 1 kW/m². The test is initiated by open-flame ignition from a pilot burner. The heat flux at the location of maximum flame propagation is reported as the critical heat flux, CHF. A smoke photometer is used to measure light attenuation as a function of time in the exhaust stack. The area under the light attenuation curve expressed in %·min is referred to as the SDR (Smoke Development Rate).

A schematic of the Radiant Flooring Panel test apparatus is shown in Figure 5.4.

Australia and New Zealand

In Australia, the Radiant Flooring Panel Test described in AS ISO 9239.1 is used, which is functionally identical to the international test standard EN ISO 9239-1. The deemed-to-satisfy solution for flooring materials in the Australian code BCA (NCC, 2019) is based on acceptance criteria for

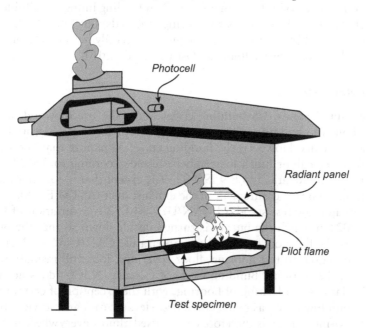

Figure 5.4 Radiant flooring panel test EN ISO 9239-1. Specimen size 0.24 m².

the CHF and the SDR. In general, solid wood flooring 12 mm or thicker achieves a CHF of 2.2 kW/m² or higher and an SDR below 750%·min (RIR 21419-05, 2018; RIR 41117.9, 2019). As a result, solid wood flooring can be used in nearly all locations and buildings. Examples of exceptions where a CHF of 4.5 kW/m² or greater is required are fire-isolated exits and fire control rooms in most types of buildings and patient care areas in unsprinklered health care facilities. Some high-density Australian and Asian hardwoods exceed the CHF requirements for these restricted areas.

In New Zealand, similar to Australia, the NZBC (NZ legislation) requires that timber or other combustible flooring materials, when tested to AS ISO 9239.1, exceed specified minimum CHF values depending on the occupancy, use of sprinklers and location in the building. Generally, a minimum CHF of either 1.2 or 2.2 kW/m² applies and these are typically achieved by solid wood flooring. New Zealand does not regulate the SDR of flooring materials.

For the purposes of compliance with Clause C3.4(b) of the NZBC, wood products, plywood or solid timber, if not less than 12 mm thick and not less than 400 kg/m³ in density, can be assigned a CHF of 2.2 kW/m² without further evidence of testing to AS ISO 9239.1.

Europe

In Europe, the Radiant Flooring Panel test EN ISO 9239-1 is used. The classification system is similar to that for wall and ceiling linings (see Table 5.2), with classes B_{fl} to E_{fl} (fl refers to flooring materials) but EN ISO 9239-1 is used instead of EN 13823. Wood floorings are mainly in Class D_{fl}, but some may reach C_{fl}, e.g. spruce flooring (Östman et al., 2010).

North America

In the United States, US building codes allow the use of wood flooring throughout the building without testing, except for some essential areas such as fire exits. However, the building codes do require that "non-traditional" interior floor finish materials be tested according to ASTM E648, which is nearly identical to EN ISO 9239-1, except that it does not include light transmission measurement in the exhaust duct (ASTM E648).

In Canada, the test standard CAN/ULC S102.2 is a variation of CAN/ULC S102, in which the burner is turned upside down so that the burner flame hits the floor of the tunnel test apparatus (CAN/ULC S102.2). This variation is used to test flooring, floor coverings, loose fill insulation, etc. However, the National Building Code of Canada (NBCC) does not regulate the flame spread rating of flooring, with the exception of certain areas in high buildings such as exits, corridors, elevator cars and service spaces. Wood flooring materials therefore can be used almost everywhere in buildings of any type of construction.

5.2.3 Roof coverings

International method

There is an international standard ISO 12468-1 External exposure of roofs to fire – Part 1: Test method, but it is unclear if it is used in any country.

New Zealand

In New Zealand, the reaction to fire properties of roof coverings is not regulated.

Europe

In Europe, there are four alternatives for testing and verifying the external fire performance of roofs in the European system according to CEN/TS 1187:

1. Method with burning brands
2. Method with burning brands and wind
3. Method with burning brands, wind and supplementary radiant heat
4. Two-stage method incorporating brands, wind and supplementary radiant heat

The four test methods originate from different European countries, and no harmonisation has been possible so far. One or both of the following hazard conditions are considered:

- Fire spread over the surface and/or immediately below the roof covering
- Penetration of fire through the roof

The classification of external fire performance of roofs is specified in EN 13501-5. When using wood products in roofs, the main concern is the possibility of using wood as the substrate for the roof covering, because the use of wood products as the exterior roof covering is not very common.

North America

In the United States, the test standard ASTM E108 covers the measurement of the relative fire characteristics of roof coverings under simulated fire originating outside the building. The following fire test methods are included:

- Spread of flame test
- Intermittent flame exposure test
- Burning brand test
- Flying brand test

When a roof covering is not restricted for use on non-combustible decks, the spread of flame, intermittent flame and burning brand tests are required. The burning brand test is required for roof coverings that have the potential of generating embers that continue to burn or smoulder after reaching the floor of the test facility. Three classes of fire test exposure are described:

- Class A tests are applicable to roof coverings that are effective against severe test exposure
- Class B tests are applicable to roof coverings that are effective against moderate test exposure
- Class C tests are applicable to roof coverings that are effective against light test exposure

The procedures measure the surface spread of flame and the ability of the roof-covering material or system to resist fire penetration from the exterior to the underside of a roof deck under the conditions of exposure. The tests are conducted with a gas burner flame or with cribs of Douglas fir. Class A tests use larger cribs than Class B or C tests, in this order.

Each specimen consists of the roof covering mounted on a 1.0×1.3 m deck. FRT *wood shingles* and shakes first need to be subjected to rain and weathering tests described in ASTM D2898. Roof assemblies that have wood shingles and shakes can be rated Class A if the shingles or shakes are treated with fire retardants and a specific water-tight underlayment is installed. Without the underlayment, the maximum rating is Class B.

In **Canada,** the test standard CAN/ULC-S107 is used to evaluate the fire performance of roof coverings. The Canadian method is conceptually similar to that described in ASTM E108. In addition, in certain unsprinklered one-story buildings, the NBCC permits the use of a roof deck construction system using FRT wood that meets the flame-spread performance standard CAN/ULC-S126, originally developed for non-combustible roof assemblies (CWC, 2015).

5.2.4 Façade claddings

There are several ways of assessing the fire performance of facades and exterior wall systems. This section is focussed on the systems for the exterior façade claddings and large-scale methods. Some countries use also small or medium scale reaction to fire test methods for wall and ceiling linings (see Section 5.2.1).

International methods

International fire test methods for façade claddings are summarised in Table 5.6 (White and Delichatsios, 2014).

Table 5.6 International full-scale façade fire tests

Test standard	Country used	Test scenario	Geometry of test rig	Fire source
ISO 13785 Part 1:2002	International	Flames emerging from a flashover compartment fire via a window	One fire source opening. Two walls in a corner "L" arrangement	Series of large perforated pipe propane burners. Total peak output 120 g/s (5.5 MW) within standard fire enclosure
NFPA 285	USA	Flames emerging from a flashover compartment fire via a window	One fire source opening One wall	Rectangular pipe gas burner in fire compartment (room burner). 1.52 m long pipe gas burner near opening soffit (window burner). Room burner increases from 690 kW to 900 kW over 30 min test period. Window burner ignited 5 min after room burner and increases from 160 kW to 400 kW over remaining 25 min test period
CAN/ULC S134	Canada	External (or internal) pellet fire located directly against the base of a re-entrant wall corner	One fire compartment opening One wall	Four 3.8 m long linear propane burners. Total output 120 g/s propane (5.5 MW)
FM 25 ft high corner test	US/ International	External (or internal) pellet fire located directly against the base of a re-entrant wall corner	Two walls in a corner "L" arrangement. Ceiling over top of walls	340 ± 4.5 kg crib constructed of 1.065 by 1.065 m oak pallets, max height 1.5 m. Located in corner 305 mm from each wall. Ignited using 0.24 L gasoline at crib base
FM 50 ft high corner test	US/ International	Flames at base of small section of façade	Same as FM 25 ft test	Same as FM 25 ft test

White and Delichatsios, 2014

Australia and New Zealand

In Australia, the Australian code BCA (NCC, 2019) recognises three types of construction: A, B and C, of which type A is the most fire resistant. According to the deemed-to-satisfy solution in the BCA, exterior wood claddings can be used in type A apartments, large-scale boarding houses, guest houses and hostels up to three or four stories depending on whether specific concessions and additional requirements are met, and up to one or two stories in other types of buildings. Moreover, exterior wood cladding systems are not permitted in buildings 25 m or higher because, even when treated with fire retardants, they cannot meet the acceptance criteria of any of the four full-scale fire tests that can be used to demonstrate compliance with the reaction to fire performance requirements in the code.

In bushfire-prone areas, the BCA has specific requirements for the fire performance of external construction elements when exposed to radiant heat, burning embers and debris. The pertinent standard AS 3959 provides a method to calculate the "Bushfire Attack Level" (BAL) for a building in a bushfire-prone area. The BAL class of a construction product used in an exterior building component is determined through testing according to AS 1530.8.1 for BAL 12.5 to BAL 40 (where the number refers to the incident radiant heat flux in the test). The BAL classification for different wood species can be found on the WoodSolutions (2021) website. For example, the wood species blackbutt, merbau and spotted gum achieved BAL 29 for all applications. Jarrah and radiate pine achieved BAL 19 for all applications, although the latter achieved BAL 29 for decking used in conjunction with non-combustible wall cladding (RIR 30930800). Western ash and white cypress achieved BAL 19 for door and window joinery.

In New Zealand, exterior cladding materials are tested in the Cone Calorimeter according to AS/NZS 3837 or ISO 5660-1 (see Section 5.2.1 for a description of the test method). Wood specimens treated with fire retardants need to be subjected to accelerated weathering, according to ASTM D2898, prior to testing in the Cone Calorimeter. When using the Cone Calorimeter, the fire performance of exterior cladding material is based on the peak heat release rate and the total heat released in a 15-minute test at a heat flux level of 50 kW/m^2. Type A cladding materials have a peak heat release rate of 100 kW/m^2 or less and a total heat released of 25 MJ/m^2 or less. The corresponding limits for Type B materials are 150 kW/m^2 and 50 MJ/m^2, respectively. Wood products have to be treated with fire retardants to have a chance of obtaining a Type A or Type B classification.

For buildings less than 10 m high, the Acceptable Solution in the NZBC (NZ legislation) requires the use of Type A or non-combustible claddings when the exterior wall is within 1 m from the boundary. If the exterior wall is at a greater distance from the boundary, the Acceptable Solution allows Type B claddings in buildings where care or detention is provided and the occupants need help from others but has no requirements for other types of buildings and uses. Type A claddings are allowed in buildings between 10 and 25 m. For buildings 25 m or higher, the exterior wall system has to pass a full-scale façade test. Exterior wall systems with untreated wood claddings cannot meet the acceptance criteria for the full-scale façade test. BS 8414 is now most commonly used in New Zealand, where a full-scale façade test is required, but this is not yet formally included in the Acceptable Solution C/AS2, which currently refers to NFPA 285. The topic is under current research and review and some interim guidance is available from the regulator.

Europe

Some European countries use the reaction to fire classes also for facades, while other countries have different requirements (Östman et al., 2010), but there is at present no European harmonised solution to assessing and quantifying their fire performance. Development work is ongoing (Anderson et al., 2021). The goal is a European approach to assess the fire performance of façades, but this might take 5–10 years to achieve. A review of the present situation in Europe is available (Östman and Mikkola, 2018) and methods (full or medium scale) used in Europe are summarized in Table 5.7 (Boström et al., 2018)

Structural fire protection and fire stops are fundamental requirements when façades are used as the outer surface of external walls of multi-storey buildings (regardless of the material used). The goal is to prevent uncontrolled fire spread on the surface and in ventilation cavities (if present) of the external wall for a required time period (see Chapter 9).

Japan

Japan has a very long tradition of using exterior wood siding on homes. Since the eighteenth century, the "shou sugi ban" or "charred cedar board" technique has provided a cost-effective way to make exterior wood siding resistant to the weather and to fire. However, until recently, the Building Standards Law (BSL) in Japan only regulated the fire resistance of exterior walls in multi-story buildings but did not have any requirements to limit fire propagation over the surface of a façade and fire spread between adjacent buildings in the densely populated residential areas in Japan. In 2015, the Japanese Standards Association published JIS A 1310. Exterior wall systems with wood claddings can pass this full-scale fire test that is used to evaluate fire propagation over building façades, provided the wood is treated with fire retardants. However, prior to conducting the fire test, such claddings have to be subjected to accelerated weathering according to JIS A 1326 (see also Section 5.4.2).

Table 5.7 Fire test methods for facades used in Europe

Country	Test method	Scope	Application	Scale	Configuration
Austria, Switzerland	ÖNorm B 3800-5	Simulates a fire from a window burnout of an apartment and the flame height in the second floor above the fire floor	Ventilated and non-ventilated facades, ETICS etc.	Full scale	Vertical wall and a right angle wing
Belgium	Lepir 2, BS 8414-1, BS 8414-2 or DIN 4102-20	See other countries	All façade systems for high-rise and mid-rise buildings	Full or Medium scale	See other countries
Czech Republic	ISO 13785-1	Reaction to fire test		Medium scale	Right angle, return wall
France	Lepir 2	Fire behaviour of building facades with windows. Includes classification criteria	All façade systems incl. windows	Full scale	Single vertical wall
Germany, Switzerland	DIN 4102-20	Complementary test of cladding systems (each part has to be of low flammability)	For classification as a low flammability system	Medium scale	Two wings (corner)
Germany	Technical regulation A 2.2.1.5	Test for ETICS and EPS insulation	Fire performance for fire outside the building	Full scale	Two wings (corner)
Hungary	MSZ 14800-6:2009	Fire propagation test for building facades	No provisions for extending the test results	Full scale	Single wall with two openings
Poland	I. PN-B-02867:2013	Determination of fire behaviour of facades without windows	All façade systems	Medium scale	Single vertical wall without openings
Slovakia	ISO 13785-2	Reaction to fire test	All external thermal insulation systems	Full scale	Right angle, return wall
Sweden, Norway, Denmark	SP Fire 105	Reaction to fire of external wall assemblies	External wall assemblies and façade claddings added to an existing wall	Full scale	Single vertical wall
Switzerland	Prüfbestimmung für Aussenwand-bekleidungssysteme	Fire behaviour of external wall covering systems when exposed to fire from a simulated apartment fire with flames out of a window	Linings and surface coating used on exterior walls, incl. elements with a limited application area	Full scale	Single vertical wall, no wing
UK, Republic of Ireland	BS 8414-1 and BS 8414-2	Fire performance of external cladding systems	The system tested	Full scale	Right angle, return wall

Andersson et al., 2017; Boström et al., 2018

North America

In the United States, the standard NFPA 285 describes a test method intended to evaluate the capability of exterior wall assemblies constructed using combustible materials to resist vertical and, to some extent, lateral flame propagation over the exterior and interior faces and within the core of the assembly. The apparatus consists of a two-story structure with an open window in the lower compartment. Two gas burners are used to create the exposing fire. The main burner is located inside the first-floor burn room and is used to develop a temperature-time curve that is comparable to that prescribed in the fire resistance test standard ASTM E119. A second burner is located inside the window opening so that flames hit the window head, which is the most vulnerable part of the exterior wall assembly for flame penetration into the core of the wall. In 2019 the scope of the standard was expanded from non-combustible buildings (steel and concrete) to all construction types, including mass timber buildings. However, the building codes do not allow the use of combustible (wood) claddings and components in mass timber buildings over four storeys.

The building codes specify a minimum separation distance between adjacent buildings based on the assumption that exterior walls are covered with wood siding and that the minimum heat flux for piloted ignition of the siding is 12.5 kW/m². The test standard NFPA 268 is used to determine the piloted ignition threshold of alternative siding materials and verify that it does not exceed 12.5 kW/m². Wood siding is assumed to meet these requirements and does not need to be tested.

In Canada, the standard CAN/ULC-S134 is the test method used to evaluate the fire performance of exterior wall systems. The Canadian method is conceptually similar to that described in NFPA 285. However, the NFPA 285 test wall is shorter 4×5.33 m versus 5×7 m, and the heat flux to the exterior wall from the flame and plume above the window is significantly higher but of shorter duration in the Canadian test (average 45 kW/m² over 15 minutes at 0.5 m above the window versus average 25 kW/m² over 30 minutes at 0.6 m). Exterior wall designs with FRT wood claddings can pass the test, but the wood first needs to be exposed to accelerated weathering according to the methods described in ASTM D2898. For example, the deemed-to-satisfy exterior wall assembly EXTW-1 in section D-6 of the NBCC has 12.7 mm FRT plywood cladding.

5.3 REACTION TO FIRE CHARACTERISTICS OF WOOD PRODUCTS FOR PERFORMANCE-BASED DESIGN

This section provides typical reaction to fire characteristics that can be used to predict time to ignition, rate of the surface spread of flame, heat release and mass loss rate, and generation rate of smoke and toxic combustion products from wood products exposed in a developing fire. In North America, reaction to fire is also referred to as material flammability. Additional information on fire dynamics is provided in Chapter 3.

5.3.1 Ignitability

Ignition is defined as "the initiation of combustion". For a solid material such as wood, a distinction is made between combustion that takes place in the gas phase versus that which occurs at the surface of the solid. Initiation of the former is referred to as "flaming ignition" because combustion is visually manifested by the formation of a luminous flame. Initiation of the latter is called "glowing ignition" because combustion progresses at a much slower rate and is evident from glowing at the surface.

Flaming ignition

When a combustible material is exposed to the heat flux from an external heat source (radiative, convective or a combination), its temperature will rise. If the net heat flux into the material is sufficiently high, the surface temperature will eventually reach a level at which the material starts to pyrolyse. The fuel vapours generated emerge from the exposed surface and mix with air in the gas phase. This mixture may ignite when the fuel vapour concentration exceeds the lower flammability limit. Sustained flaming initiated by a local heat source in the gas phase, such as a small flame or a hot spark, is referred to as piloted ignition. Auto-ignition occurs if there is no pilot present, and flaming is initiated at the hot surface of the heated solid.

Piloted ignition of wood

Piloted ignition of wood has been studied extensively since the 1950s. These studies usually involved laboratory-scale experiments to measure the time to ignition at different levels of incident heat flux from a radiant heater. Janssens (1991b) showed that the following expression is suitable for correlating piloted ignition data of "thermally thick" wood products and other solid materials:

$$\dot{q}_e'' = \dot{q}_{cr}'' \left[1 + 0.717 \left(\frac{k\rho c}{h_{ig}^2 t_{ig}} \right)^{0.55} \right] \tag{5.1}$$

where

$\dot{q}_e''$ = Incident radiant heat flux (kW/m^2)
$\dot{q}_{cr}''$ = Critical heat flux for ignition (kW/m^2)
$k\rho c$ = Apparent thermal inertia (kJ/m^4·K^2·s)

h_{ig} = Total heat transfer coefficient from the surface at ignition (kW/m²·K)

t_{ig} = Time to ignition at heat flux $\dot{q}''_e$ (s)

A specimen is considered thermally thick under specified thermal exposure conditions, if ignition occurs before the substrate starts to have an effect on the ignition time. In other words, the specimen behaves as if it were a semi-infinite solid. A specimen may behave as a thermally thick solid at high heat fluxes (short ignition times) and as thermally thin at low heat fluxes (long ignition times).

The critical heat flux for ignition, $\dot{q}''_{cr}$, in Equation 5.1 is just sufficient to heat the material surface to the ignition temperature, T_{ig}, for very long exposure times (theoretically ∞). The relationship between $\dot{q}''_{cr}$ and T_{ig} therefore follows from a steady-state heat balance at the specimen surface:

$$\dot{q}''_{cr} = h_c\left(T_{ig} - T_\infty\right) + \sigma\left(T_{ig}^4 - T_\infty^4\right) \equiv h_{ig}\left(T_{ig} - T_\infty\right) \tag{5.2}$$

where

ε = Surface emissivity/absorptivity (~0.88 for wood (Janssens, 1991b))

h_c = Convection coefficient (kW/m²·K)

T_{ig} = Surface temperature at ignition (K)

T_∞ = Ambient and initial temperature (K)

σ = Boltzmann constant (5.67·10⁻¹¹ kW/K⁴·m²)

The convection coefficient for the Cone Calorimeter in the horizontal orientation and the Lateral Ignition and Flame Spread Test apparatus is estimated at 12 and 15 W/m²·K, respectively (Janssens, 2013; ASTM E1321). The practical significance of Equation 5.1 is that for a thick material, ignition data points plotted as $(1/t_{ig})^{0.55}$ versus $\dot{q}''_e$ should fall on a straight line. The intercept with the abscissa of a linear fit through the data is $\dot{q}''_{cr}$. Once $\dot{q}''_{cr}$ is found, T_{ig} and h_{ig} can be obtained from Equation 5.2. Finally, $k\rho c$, which is the product of the thermal conductivity (k in kW/m·K), density (ρ in kg/m³) and specific heat capacity (c in kJ/kg·K) of the solid, can then be calculated from the slope of the straight-line fit. Note that $k\rho c$ estimated from an analysis of piloted ignition data is an apparent value over the temperature range between T_∞ and T_{ig}.

Janssens (1991b) used this method to estimate T_{ig} and $k\rho c$ for a range of wood products. The results of this work can be summarised as follows:

- T_{ig} for oven dry tested softwoods varied between 350 and 365°C
- T_{ig} for oven dry tested hardwoods varied between 300 and 310°C
- T_{ig} increased by approximately 2°C per percent increase in moisture content

- Apparent $k\rho c$ values ranged approximately from 0.09 to 0.4 kJ/m^4·K^2·s for dry wood with densities between 330 and 810 kg/m^3 and increased by 30–40% for specimens conditioned to equilibrium at 23°C and 50% relative humidity prior to testing
- The apparent $k\rho c$ obtained from analysis of piloted ignition data of wood was approximately equal to the product of ρ and literature values of k and c at a temperature halfway between ambient and T_{ig}

The ignition properties (T_{ig} and $k\rho c$) can then be used to predict the time to ignition of a material under time-varying heat flux conditions in an actual fire. The net heat flux over a specified area of an exposed material is equal to the incident heat flux received over the area minus the convective and radiative heat losses from the surface:

$$\dot{q}''_{net}(t) = \left[\dot{q}''_e(t) - \sigma T_s(t)^4\right] + h_c\left[T_f(t) - T_s(t)\right] \tag{5.3}$$

where

$\dot{q}''_{net}$ = Net heat flux into the solid at time t (kW/m^2);
t = Time (s); and
T_f = Temperature of the fluid in contact with the surface of the solid (K).

Because the heat losses are a function of the surface temperature, the net heat flux is not only a function of time but varies with surface temperature as well. For a thermally thick solid, the surface temperature can then be calculated by applying Duhamel's superposition theorem, which leads to the following integral equation:

$$T_s(t) = T_\infty + \sqrt{\frac{1}{\pi k\rho c}} \int_0^t \frac{\dot{q}''_{net}(\tau)}{\sqrt{t-\tau}} d\tau \tag{5.4}$$

If the $k\rho c$ of the material is known and the net heat flux is specified as a function of time and surface temperature, Equation 5.4 can be solved to calculate T_s as a function of time. The material will ignite when $T_s = T_{ig}$. However, because $\dot{q}''_{net}$ is usually a function of t and T_s (which in turn is a non-linear function of t), it is necessary to solve the equation numerically.

A limitation of Janssens' approach is that it assumes that the material behaves as a thermally thick solid. Dietenberger (2004) used a weighted average of Equation 5.1 with a similar equation for thermally thin materials backed by insulation, i.e., as tested in the Cone Calorimeter ISO 5660-1 and LIFT apparatus ASTM E1321, to correlate piloted ignition data that cover the thermally thick (high heat fluxes leading to short ignition times) and thermally thin (long ignition times at low heat fluxes) regimes. Table 5.8 provides a comparison between ignition properties for various wood products estimated from Janssens's and Dietenberger's methods. The

Table 5.8 Ignition properties for various wood products obtained according to two methods

Material	δ (mm)	Janssens method			Dietenberger method		
		ρ (kg/m³)	$k\rho c$ (kJ²/m⁴·K²·s)	T_{ig} (°C)	$k\rho c$ (kJ²/m⁴·K²·s)	$k/\rho c$ (10⁻⁷m²/s)	T_{ig} (°C)
FRT Douglas fir plywood	11.8	563	0.515	350	0.261	1.37	374
Oak veneer plywood	13.0	479	1.103	304	0.413	1.77	290
FRT plywood	11.5	599	0.412	376	0.346	1.31	377
Douglas fir plywood	11.5	537	0.282	335	0.221	1.37	332
FRT Southern pine plywood	11.0	606	1.209	342	0.547	2.26	399
Douglas fir plywood	12.0	549	0.231	361	0.233	1.38	346
Southern pine plywood	11.0	605	0.256	367	0.290	1.38	347
Particleboard	13.0	794	1.195	251	0.763	2.72	290
Oriented strandboard	11.0	643	0.244	348	0.342	1.54	326
Hardboard	6.0	1,026	0.400	351	0.504	0.90	320
Redwood lumber	19.0	421	0.165	380	0.173	1.67	365
White spruce lumber	17.0	479	0.286	342	0.201	1.67	348
Southern pine boards	18.0	537	0.328	354	0.260	1.63	371
Waferboard	13.0	631	0.793	236	0.442	2.69	290

table indicates that T_{ig} is generally lower for wood panel products compared to solid sawn wood.

The advantage of Dietenberger's method is that, in addition to T_{ig} and apparent kρc, it also provides an estimate of an apparent value of the thermal diffusivity k/ρc. Because density ρ is known, apparent values of k and c can then be calculated from the thermal inertia and diffusivity estimates for use in heat conduction calculations for wood surfaces and ignition targets in compartment fire models such as B-RISK, CFAST and Fire Dynamics Simulator (FDS) (Wade et al., 2016; Peacock et al., 2021; McGrattan et al., 2021).

Auto-ignition of wood

Auto-ignition of wood exposed to radiant heat is similar to piloted ignition, except that the hot surface triggers ignition of the flammable mixture of volatiles and air in the boundary layer. Consequently, T_{ig} for auto-ignition is much higher than for piloted ignition. Abu-Zaid (1988) measured 510°C and 550°C for Douglas fir with 0% and 17% moisture content respectively, i.e., about 150–200°C higher than for piloted ignition. Abu-Zaid's data indicate that the corresponding critical heat flux is between 30 and 40 kW/m², which is consistent with the value of 33 kW/m² based on an earlier work by Simms (1960).

Glowing ignition

The following conditions are necessary and favourable for the initiation of glowing combustion of wood:

- The incident heat flux is too low to generate combustible vapours at a sufficient rate to create a flammable mixture in the gas phase
- The incident heat flux is high enough and is applied for a sufficient duration to promote self-accelerating exothermic reactions at the surface with oxygen in the surrounding air
- The reactions can be sustained because the surface temperature rises to about 600°C following ignition and conduction heat losses into the solid are low due to the porous nature of wood

Babrauskas (2002) reported that smouldering ignition occurred when wood exposed to a minimum heat flux of 4.3 kW/m² reached a temperature of 250°C. After some time, glowing combustion may, but not always does, transition to flaming combustion, depending on the specific conditions.

5.3.2 Surface spread of flame

Flames can propagate over a solid surface in two modes. The first mode is referred to as the wind-aided flame spread. In this mode, flames spread in the same direction as the surrounding airflow or are driven by buoyancy. The second mode is referred to as opposed-flow flame spread, which occurs when flames spread in the opposite direction of the surrounding airflow. Upward and downward flame propagation over the vertical surface of a wall are examples of wind-aided and opposed-flow flame spread, respectively.

The rate of wind-aided flame spread is a function of the external heat flux distribution over the surface and can be calculated based on the ignition source characteristics and ignition, heat release rate properties of the solid, e.g. Kokkala et al. (1997). Quintiere (1981) developed the following equation to predict the opposed-flow spread rate of a turbulent flame over thick fuel sheets based on earlier work by deRis (1969) for laminar flames, after whom the equation is named:

$$V_p = \frac{\phi}{k\rho c \left(T_{ig} - T_s\right)^2} \tag{5.5}$$

where

ϕ = Flame heating parameter (kW^2/m^3)
T_{ig} = Surface temperature at ignition (K)
T_s = Surface temperature of the time of flame front arrival (K)

The ignition properties $k\rho c$ and T_{ig} can be estimated according to the procedures described in Section 5.3.1.1. The flame heating parameter ϕ can be estimated from flame spread data obtained in the LIFT apparatus according to a procedure described in ASTM E1321. The standard also includes a procedure to estimate $T_{s,min}$, i.e., the minimum surface temperature at which the opposed-flow flame spread front ceases to advance. Janssens (1991b) performed tests on specimens of six solid sawn wood products and five wood panel products in the LIFT apparatus, and reported values of ϕ and $T_{s,min}$ are in the range of 1.7–8.8 kW^2/m^3 and 73–183°C, respectively.

5.3.3 Burning rate

Heat release rate

A typical heat release rate curve for wood measured in the Cone Calorimeter (or similar device) is bimodal. Shortly after ignition, the heat release rate rises rapidly to the first peak. A protective char layer builds up as the pyrolysis front moves inward. The char layer forms an increasing thermal resistance between the exposed surface and the pyrolysis front, resulting in a decrease of the heat release rate after the first peak. At some point, the surface recedes at approximately the same rate as the pyrolysis front and the heat release rate becomes relatively steady. Standard Cone Calorimeter specimens are backed by high-temperature ceramic fibre insulation. This causes the heat release rate to start rising again when the pyrolysis front approaches the back surface. After the second peak, the heat release drops again until flaming ceases and the char residue continues to smoulder.

Janssens (1991c) and Tran (1992) published extensive surveys of the heat release rate data of wood. Janssens, for example, reported a first peak heat release rate for a number of conditioned wood specimens in the range of 180 to 230 kW/m². For wood treated with fire retardants to obtain a Class A rating in the Steiner tunnel test (see Section 5.2.4), this value can be well below 100 kW/m². The heat release of a wood product measured in the Cone Calorimeter or a similar device depends on many factors (species, density, moisture content, heat flux, specimen orientation, oxygen concentration, etc.). The results of thousands of tests have been published, but it is not practical to include them in this chapter. Instead, the reader is referred to the aforementioned surveys by Janssens (1991c) and Tran (1992) and the public domain repository of Cone Calorimeter data for wood products tested at the Forest Products Laboratory in Madison, Wisconsin, USA (USDA).

Pyrolysis models

Models to predict the pyrolysis rate of wood range from simple approximate analytical equations to detailed numerical solutions of the conservation equations of mass, momentum and energy supplemented with algebraic equations to predict thermal properties of wood and char as a function of temperature, evaporation rate of free and bound water, thermal decomposition rate of the active components of wood, etc. A detailed review of wood pyrolysis models was made by Moghtaderi (2006). A simple approach to estimating the pyrolysis rate of wood relies on the concept of heat of gasification:

$$\dot{m}'' = \frac{\dot{q}''_{net}}{\Delta h_g} \quad \text{and} \quad \dot{Q}'' = \Delta h_{c,eff} \dot{m}'' = \Delta h_{c,eff} \frac{\dot{q}''_{net}}{\Delta h_g} \tag{5.6}$$

where
$\dot{m}''$ = Pyrolysis rate per unit area (g/m²·s)
$\dot{q}''_{net}$ = Net heat flux into the solid at the exposed surface (kW/m²)
Δh_g = Heat of gasification (kJ/g)
$\dot{Q}''$ = Heat release rate per unit area (kW)
$\Delta h_{c,eff}$ = Effective heat of combustion (kJ/g)

Equation 5.6 can be used to calculate the burning rate of liquid and thermoplastic pool fires, which, after an initial transient following ignition, reach and remain at a steady state until burnout. This is because, in this case, the surface temperature and heat of gasification are relatively constant. However, for char-forming materials, such as wood exposed to a constant heat flux in the Cone Calorimeter or a similar device, neither the burning rate nor the surface temperature is constant. Consequently, Δh_g

of wood varies over time. Janssens (1993) developed a method to calculate Δh_g of wood in the Cone Calorimeter as a function of char depth. Table 5.9 gives average values for $\Delta h_{c,eff}$ (which, unlike Δh_g, is relatively constant and can be measured directly in the Cone Calorimeter), Δh_g and $\Delta h_g / \Delta h_{c,eff}$ for various wood products. The $\Delta h_{c,eff}$ and Δh_g values in Table 5.9 are based on the data reported by Janssens (1993) and can be used in conjunction with Equation 5.6 to obtain an estimate of the pyrolysis rate or heat release rate of wood under quasi-steady thermal exposure conditions. $\Delta h_g / \Delta h_{c,eff}$ is a measure of the flammability of the material, where a lower ratio corresponds to increased flammability. The ratio varies between 0.22 and 0.27 for solid sawn wood and is somewhat lower (~0.18) for three of the four panel products. Table 5.9 also includes $\Delta h_{c,eff}$ and Δh_g for some common plastics (Lyon, 2004) for comparison purposes. Δh_g for untreated plastics is comparable to but generally somewhat lower than that of wood products.

Over the past decade, the pyrolysis model in Fire Dynamics Simulator (McGrattan et al., 2021) has become one of the most commonly used methods to simulate the thermal degradation of solid materials in general and wood in particular. Often, the modelling approach involves testing of specimens of the material in the Cone Calorimeter (or similar device), thermogravimetric analysis (TGA) apparatus and sometimes other small-scale test apparatuses. The test data are used in conjunction with the model to estimate the kinetic parameters for the thermal degradation reactions, apparent thermal properties of the material and its char and other model parameters. In addition, part of the Cone Calorimeter data is used for model validation. A detailed example of this approach to model pyrolysis of wood was

Table 5.9 Effective heat of combustion and heat of gasification values for various wood products and common plastics

Material	$\Delta h_{c,eff}$ (MJ/kg)	Δh_g (MJ/kg)	$\Delta h_g / \Delta h_{c,eff}$
Western red cedar	13.1	3.27	0.25
Redwood	12.6	3.14	0.25
Radiata pine	11.9	3.22	0.27
Douglas fir	12.0	2.64	0.22
Victorian ash	11.7	2.57	0.22
Blackbutt	10.6	2.54	0.24
Douglas fir plywood	12.3	2.95	0.24
Oriented strandboard	13.3	2.39	0.18
Southern pine plywood	12.3	2.21	0.18
Particleboard	11.8	2.12	0.18
Polyethylene	40.3	1.9–2.2	0.05
Polystyrene	27.9	1.8	0.06
Nylon	29.8	1.5	0.05
Polyvinylchloride	9.3–11.3	2.3–2.7	0.12–0.29

Table 5.10 Carbon monoxide CO and smoke yields of selected wood products and plastics in well-ventilated fires

Material	Carbon monoxide yield, Y_{CO} (g/g)	Smoke yield, Y_s (g/g)
Hardwoods	0.004	0.015
Softwoods	0.004–0.005	0.015
Wood board and panel products	0.002–0.015	
Polyethylene	0.024	0.060
Polystyrene	0.060	0.164
Nylon	0.038	0.075
Polyvinylchloride	0.063	0.173

SFPE Handbook, 2016

recently published by Rinta-Paavola and Hostikka (2022). An important finding of this study is that a multiple reaction model, i.e., one reaction for each of the principal components of wood, did not appear to improve the accuracy of the pyrolysis rate predictions compared to the single-reaction model, i.e., wood modelled as a homogeneous material with a single set of kinetic parameters.

5.3.4 Production rate of smoke and toxic products of combustion

The primary toxic gas that is generated in the combustion of wood is carbon monoxide. However, compared to most plastics, the carbon monoxide and smoke (or soot) yields of wood are very low under well-ventilated conditions, as is evident from the yield data taken from table A.40 in the SFPE Handbook (2016) and reproduced in Table 5.10. Moreover, because Δh_g of these plastics is comparable to or lower than that of wood, see Table 5.9, this implies that the generation rate of CO and soot in well-ventilated fires is significantly higher for these plastics than for wood products under the same thermal exposure conditions. However, the CO yield increases dramatically to about 0.2 g/g when the air supply is below stoichiometric (equivalence ratio of 1.5 and higher), although under those conditions the CO yield is largely independent of the fuel (see Chapter 16 in the *SFPE Handbook*).

5.4 METHODS FOR IMPROVING THE REACTION TO FIRE PERFORMANCE OF WOOD PRODUCTS

5.4.1 Fire-retardant treatments, including surface coatings

Fire-retardant treatments of wood products, e.g. by chemical modification, may considerably improve the reaction to fire properties, even to the extent that the highest fire classifications for combustible products can be reached, i.e., Group 1 in Australia and New Zealand, Euroclass B in Europe or Class A in the U.S. for wall linings and ceiling materials (see Sections 5.2.1, 5.2.2 and 5.2.4) This allows wider use of visible wood, both as interior wall and ceiling linings and as exterior claddings, e.g. in façades.

It is relatively easy to obtain an improved reaction to fire performance of wood products. Most existing fire retardants are effective in reducing different reaction to fire parameters of wood such as ignitability, heat release and flame spread. However, high retention levels have to be used compared to ordinary preservation treatments used to protect wood against biological decay, often in a range of approximately 5–15%, depending on the type and amount of flame retardant. Common types of fire retardants contain nitrogen, phosphorous and/or boron, including combinations of those.

There are three main processes to treat wood with flame retardants:

1) Full-cell treatment via vacuum-pressure impregnation in a pressure chamber mainly with aqueous solutions or dispersions of the flame retardant as usually done for preservative treatment. This process is predominantly applied for timber but is also possible for veneer-based products (plywood, laminated veneer lumber). Treatment of the last-mentioned products can involve individual treatment of the veneers prior to gluing or treatment of the whole panel.
2) Surface treatment by dipping, spraying, brush or roll application. Compared to the full-cell treatment, the penetration depth of the flame retardant is about 1 mm or less. The formulations applied may be intumescent coatings or non-film-forming substances similar to those used in full-cell treatment.
3) Addition during the production process. Flame retardants may be sprayed onto particles, fibres or strands before, after or together with the adhesive and subsequently pressed to wood-based panels. This may result in significant strength loss compared to panels without flame retardants added, particularly when these are acidic. Surface properties may be inferior with respect to coating or application of laminates.

The aim of flame retardants used for wood is to delay the ignition and to reduce the heat released during combustion. The various flame retardants may be classified into five types, on the basis of their underlying mechanism:

1) Changing the pathway of pyrolysis
2) Coating formation on the wood surface
3) Slowing down ignition and burning by changing the thermal properties of wood
4) Reducing combustion by diluting pyrolysis gases
5) Reducing combustion by free radical trapping in the flame

The most effective flame retardants for wood reduce fuel production by increasing char production and lowering the amount of combustible gases. Therefore, most of the flame retardants used for wood fall under mechanism 1. The majority of flame retardants, however, operate by several of these mechanisms.

Fire retardants may influence the reaction to fire properties, but for the fully developed fire, the effect is minor (Nussbaum, 1988). One exception is intumescent paints that may delay the time for the start of charring and thus increase the fire resistance of wooden structures. In any case, fire retardants cannot make wood non-combustible, even though that is what the U.S. building code acceptance of FRT wood in non-combustible buildings would seem to imply.

It has been observed that fire-retardant treated (FRT) wood products, mainly but not exclusively plywood, used as roof sheathing lose their strength during service conditions. Several incidents have occurred. Extensive studies have been performed mainly in the U.S., and the main phenomena seem to have been explained. High temperatures in the roof structures have initiated a decay process in the wood caused by some types of fire retardants. New standards to predict the behaviour have been developed. A review of more than ten years of research has been published (Winandy, 2001). The mechanical strength is important for several applications of FRT wood products in the U.S., while in Europe it seems to be less important, since FRT wood is mainly used for non-structural purposes. In most cases, other properties, e.g. durability against weathering, are considered to be far more essential.

5.4.2 Durability of reaction to fire performance

The durability of the fire-retardant treatment is an important consideration. There are two mechanisms by which the long-term durability of treated wood products may be adversely affected. First, a high moisture content increases the risk of migration of flame retardant chemicals within the wood and salt crystallisation on the surface. The second mechanism results in decreased fire performance and involves the loss of flame retardant chemicals by leaching or other mechanisms. The latter is a major concern for exterior applications and is the main challenge in the development of new FRT wood products (Östman and Tsantaridis, 2017).

A European standard EN 16755 has been developed to determine the "Durability of Reaction-to-Fire performance" (DRF) classes has been developed. The system is summarised in Table 5.11. It consists of a control system for the durability properties of FRT wood-based products and suitable test procedures. The European system is based on ASTM test methods from North America (Wood Handbook, 2010).

In Japan, an accelerated weathering (JIS A 1326) is being used. Some results with natural exposure for up to three years have been published (Yoshioka et al., 2021), showing similar results as the European study.

Table 5.11 Requirements for DRF classes of FRT wood products according to EN 16755

DRF class	Intended use	Fire class, initial	Performance requirements for different end uses		
			Hygroscopic properties	Fire performance after weather exposure	
INT 1	Interior, dry applications	Relevant fire class	–	–	
INT 2	Interior, humid applications	Relevant fire class	Limited moisture content Minimum visible salt	–	
EXT	Exterior applications	Relevant fire class	Limited moisture content Minimum visible salt	Fire performance is maintained	

The relevant initial fire class shall be verified according to EN 13501-1 or IMO (International Maritime Organisation) classification systems. Persistence of reaction to fire performance after weather exposure shall be verified according to (ISO 5660-1) or the European system (EN 13501-1).

FRT wood products fulfilling both the reaction to fire requirements and the durability of reaction to fire performance are available worldwide.

REFERENCES

Abu-Zaid, M. (1988) *Effect of Water on Ignition of Cellulosic Materials.* Ph.D. Thesis, East Lansing, Michigan State University.

Anderson, J., Boström, L., Jansson, R. (2017) *Fire Safety of Facades, SP Rapport 2017:37, ISSN 0284–5172.*

Anderson, J., Boström, L., Chiva, R., Guillaume, E., Colwell, S., Hofmann, A., Tóth. P. (2021) European approach to assess the fire performance of façades. *Fire and Materials*, 45, p.5. https://doi.org/10.1002/fam.2878.

AS 1530.8.1 (n.d.) *Methods for Fire Tests on Building Materials, Components and Structures: Tests on Elements of Construction for Buildings Exposed to Simulated Bushfire Attack: Radiant Heat and Small Flaming Sources.* Standards Australia, Sydney, NSW.

AS 3959 (n.d.) *Construction of Buildings in Bushfire-Prone Areas*, Standards Australia, Sydney, NSW.

AS ISO 9239.1 (n.d.) *Reaction to Fire Tests for Floor Coverings, Part 1: Determination of the Burning Behaviour using a Radiant Heat Source,* Standards Australia, Sydney, NSW.

AS/NZS 1530.3 (n.d.) *Methods for Fire Tests on Building Materials, Components and Structures: Simultaneous Determination of Ignitability, Flame Propagation, Heat Release and Smoke Release*, Standards Australia, Sydney, NSW.

AS/NZS 3837 (n.d.) *Method of Test for Heat and Smoke Release Rates for Materials and Products Using an Oxygen Consumption Calorimeter*, Standards Australia, Sydney, NSW.

ASTM D2898 (n.d.) *Standard Practice for Accelerated Weathering of Fire-Retardant-Treated Wood for Fire Testing*, ASTM International, West Conshohocken, PA. DOI: 10.1520/D2898-10R17.

ASTM E84 (n.d.) *Standard Test Method for Surface Burning Characteristics of Building Materials*, ASTM International, West Conshohocken, PA. DOI: 10.1520/E0084-21A.

ASTM E108 (n.d.) *Standard Test Methods for Fire Tests of Roof Coverings*, ASTM International, West Conshohocken, PA. DOI: 10.1520/E0108-20A.

ASTM E119 (n.d.) *Standard Test Methods for Fire Tests of Building Construction and Materials*, ASTM International, West Conshohocken, PA. DOI: 10.1520/E0119-20.

ASTM E648 (n.d.) *Standard Test Method for Critical Radiant Flux of Floor Covering Systems Using a Radiant Heat Energy Source*, ASTM International, West Conshohocken, PA. DOI: 10.1520/E0648-19AE01.

ASTM E1321 (n.d.) *Standard Test Method for Determining Material Ignition and Flame Spread Properties*, ASTM International, West Conshohocken, PA. DOI: 10.1520/E1321-18.

ASTM E2768 (n.d.) *Standard Test Method for Extended Duration Surface Burning Characteristics of Building Materials (30 Min Tunnel Test)*, ASTM International, West Conshohocken, PA. DOI: 10.1520/E2768-11R18.

AWC (2019) Design for code acceptance: Flame spread performance of wood products used for interior finish, USA. https://awc.org/codes-standards/publications/dca1.

Babrauskas, V. (2002) Ignition of wood: A review of the state of the art. *Journal of Fire Protection Engineering*, 12, pp. 163–188. DOI: 10.1177/10423910260620482.

Boström, L. et al. (2018) *Development of a European Approach to Access the Fire Performance of Facades, Final Report*. Project: SI2.743702-30-CE-0830933/00-14.

CAN/ULC-S102 (n.d.) *Standard Method of Test for Surface Burning Characteristics of Building Materials and Assemblies*, Underwriters Laboratories of Canada, Toronto, Canada.

CAN/ULC-S102.2 (n.d.) *Standard Method of Test for Surface Burning Characteristics of Flooring, Floor Coverings and Miscellaneous Materials and Assemblies*, Underwriters Laboratories of Canada, Toronto, Canada.

CAN/ULC-S107 (n.d.) *Standard Methods of Fire Tests of Roof Coverings*, Underwriters Laboratories of Canada, Toronto, Canada.

CAN/ULC-S126 (n.d.) *Standard Method of Test for Fire Spread Under Roof-Deck Assemblies*, Underwriters Laboratories of Canada, Toronto, Canada.

CAN/ULC-S134 (n.d.) *Standard Method of Fire Test of Exterior Wall Assemblies*, Underwriters Laboratories of Canada, Toronto, Canada.

CEN/TS 1187 (n.d.) *Test Methods for External Fire Exposure to Roofs*. CEN European Committee for Standardization, Brussels.

CWC (2015) *Canadian Wood Council, Fire-Retardant-treated Wood*. https://cwc.ca/wp-content/uploads/2015/04/FRTW.pdf.

CWC (2020) *Canadian Wood Council*. https://cwc.ca/wp-content/uploads/2020/09/Fact-Sheets-Surface-Flammability-and-Flame-spread-Ratings.pdf.

deRis, J. (1969) Spread of a laminar diffusion flame, in Twelfth Symposium (International) on Combustion, Poitiers, France, pp. 241–252.

Dietenberger, M. (2004) Ignitability of materials in transitional heating regimes, in *Wood and Fire Safety*, 5th International Fire Safety Conference, Štrebské Pleso, Slovakia, April 18–22, pp. 31–29.

Dong, Y. et al. (2015) Flammability and physical–mechanical properties assessment of wood treated with furfuryl alcohol and nano-SiO2. *Journal of Wood and Wood Products*, 73, pp. 457–464.

EN 13501-1 (n.d.) *Fire Classification of Construction Products and Building Elements: Part 1: Classification Using Data from Reaction to Fire Tests*, CEN European Committee for Standardization, Brussels.

EN 13501-5 (n.d.) *Fire Classification of Construction Products and Building Elements: Part 5: Classification Using Data from External Fire Exposure to Roofs Tests*. CEN European Committee for Standardization, Brussels.

EN 13823 (n.d.) *Reaction to Fire Tests for Building Products: Building Products Excluding Floorings Exposed to the Thermal Attack by a Single Burning Item*. CEN European Committee for Standardization, Brussels.

EN 16755 (n.d.) *Durability of Reaction to Fire Performance. Classes of Fire-retardant Treated Wood Products in Interior and Exterior End Use Applications*. CEN European Committee for Standardization, Brussels.

EN ISO 9239-1 (n.d.) *Reaction to Fire Tests for Floorings: Part 1 Determination of the Burning Behaviour Using a Radiant Heat Source*, ISO International Organization for Standardization, Geneva.

EN ISO 11925-2 (n.d.) *Reaction to Fire Tests: Ignitability of Products Subjected to Direct Impingement of Flame: Part 2: Single-flame Source Test*. ISO International Organization for Standardization, Geneva.

Gérardin, P. (2016) New alternatives for wood preservation based on thermal and chemical modification of wood- a review. *Annals of Forest Science*, 73, pp. 559–570.

Hakkarainen, T., Hayashi, Y. (2001) Comparison of Japanese and European Reaction to Fire Classification Systems. *Fire Science & Technology*, 21, pp. 19–42. DOI: https://doi.org/10.3210/fst.21.19.

ISO 5660-1 (n.d.) *Reaction-to-Fire Tests: Heat Release, Smoke Production and Mass Loss Rate: Part 1 Heat Release Rate (Cone Calorimeter Method) and Smoke Production Rate (Dynamic Measurement)*. ISO International Organization for Standardization, Geneva.

ISO 9705-1 (n.d.) *Reaction to Fire Tests: Room Corner Test for Wall and Ceiling Lining Products: Part 1: Test Method for a Small Room Configuration*. ISO International Organization for Standardization, Geneva.

ISO 12468-1 (n.d.) *External Exposure of Roofs to Fire: Part 1 Test Method*. ISO International Organization for Standardization, Geneva.

ISO/TS 17431 (n.d.) *Fire Tests: Reduced-scale Model Box Test*. International Organization for Standardization, Geneva.

Janssens, M. (1991a) Measuring rate of heat release by oxygen consumption. *Fire Technology*, 27, pp. 234–249. DOI: 10.1007/BF01038449.

Janssens, M. (1991b) Rate of heat release of wood products. *Fire Safety Journal*, 17, pp. 217–238. DOI: 10.1016/0379-7112(91)90003-H.

Janssens, M. (1991c) *Thermophysical Properties of Wood and their Role in Enclosure Fire Growth*. The University of Ghent, Ghent, Belgium.

Janssens, M. (1993) Cone calorimeter measurements of the heat of gasification of wood, in 6th Interflam Conference, Oxford, UK, pp. 549–558.

Janssens, M. (2013) Analysis of cone calorimeter ignition data: A reappraisal, in Interflam 2013, 13th International Conference, London, UK, pp. 743–755.

Janssens, M., Carpenter, K., Huczek, J., Mehrafza, M., Sauceda, A. (2006) Performance of construction products in reaction-to-fire tests, in 17th Annual BCC Conference on Flame Retardancy, Stamford, CT.

JIS A 1310 (2013) *Test Method for Fire Propagation over Building Façades.* Japanese Standards Association, Tokyo, Japan.

JIS A 1326 (n.d.) *Test Method for Accelerated Weathering for Fire-retardant-treated Wood Products for Façades.* Japanese Standards Association, Tokyo, Japan.

Kokkala, M., Thomas, P., Karlsson, B. (1993) Rate of heat release and ignitability indices for surface linings. *Fire and Materials*, 17, pp. 209–216. DOI: 10.1002/fam.810170503.

Kokkala, M., Baroudi, B., Parker, W. (1997) Upward flame spread on wooden surface products: Experiments and numerical modelling, *Fire Safety Science*, 5, pp. 309–320. DOI: 10.3801/IAFSS.FSS.5-309.

Lyon, R. (2004) Plastics and rubber, in *Handbook of Building Materials for Fire Protection*, Chapter 3, McGraw-Hill, New York.

McGrattan, K., Hostikka, H., Floyd, J., McDermott, R., Vanella, M. (2021) *Fire Dynamics Simulator User's Guide.* 6th Edition. National Institute of Standards and Technology, Gaithersburg, MD, NIST Special Publication 1019. DOI: 10.6028/NIST.SP.1019.

Moghtaderi, B. (2006) The state-of-the-art in pyrolysis modelling of lignocellulosic solid fuels, *Fire and Materials*, 30, pp. 1–34. DOI: 10.1002/fam.891.

Morozovs, A., Bukšāns E. (2009) Fire performance characteristics of acetylated ash (Fraxinus excelsior) wood. *Wood Material Science & Engineering*, 4(1–2), pp. 76–79.

NBCC (2015) *National Building Code of Canada.* 15th Edition. Canadian Commission on Building and Fire Codes, National Research Council Canada, Ottawa, Canada.

NCC (2019) *National Construction Code Series*, Volumes 1–3. Australian Building Codes Board, Canberra, ACT. Available from https://ncc.abcb.gov.au/.

NFPA 268 (n.d.) *Standard Test Method for Determining Ignitability of Exterior Wall Assemblies Using a Radiant Heat Energy Source.* National Fire Protection Association, Quincy, MA.

NFPA 285 (n.d.) *Standard Fire Test Method for Evaluation of Fire Propagation Characteristics of Exterior Wall Assemblies Containing Combustible Components*, National Fire Protection Association, Quincy, MA.

NFPA 286 (n.d.) *Standard Methods of Fire Test for Evaluation Contribution of Wall and Ceiling Interior Finish to Room Fire Growth.* National Fire Protection Association, Quincy, MA.

Nussbaum, R. (1988) Effect of low concentration fire retardant impregnations on wood charring rate and char field. *Journal of Fire Sciences*, 6, pp. 290–307.

NZ Legislation. https://www.legislation.govt.nz/regulation/public/1992/0150/latest/DLM162576.html.

Östman, B., Mikkola, E. (2010) European classes for the reaction to fire performance of wood-based panels. *Fire and Materials* 34, 315–330.

Östman, B., Mikkola, E. (Ed.) (2018) *Guidance on Fire Safety of Bio-Based Facades.* Dissemination document N230-07 – COST Action FP 1404 I. ETH Zurich.

Östman, B., Tsantaridis, L. (2017) Durability of the reaction to fire performance of fire-retardant-treated wood products in exterior applications: a 10-year report. *International Wood Products Journal*, 8(2), pp. 94–100. DOI: 10.1080/20426445.2017.1330229.

Östman, B., Mikkola, E., Stein, R., Frangi, A., König, J., Dhima, D., Hakkarainen, T., Bregulla, J. (2010) *Fire Safety in Timber Buildings: Technical Guideline for Europe, SP Report 2010:19*. SP Trätek, Stockholm, Sweden.

Peacock, R., Reneke, P., Forney, G. (2021) *CFAST–Consolidated Model of Fire Growth and Smoke Transport (Version 7), Volume 2: User's Guide*. National Institute of Standards and Technology, Gaithersburg, MD, NIST Technical Note 1889v2. DOI: 10.6028/NIST.TN.1889v2.

Petterson, R. (1984) The chemical composition of wood, Chapter 2 in Rowell M., Ed., *The Chemistry of Solid Wood, Advances in Chemistry Series 207*. American Chemical Society, Washington, DC. Available from https://www.fpl.fs.fed.us/documnts/pdf1984/pette84a.pdf.

Quintiere, J. (1981) An approach for modeling wall fire spread in a room. *Fire Safety Journal*, 3, pp. 201–214. DOI: 10.1016/0379-7112(81)90044-8.

Rinta-Paavola, A., Hostikka, S. (2022) A model for the pyrolysis of two nordic structural timbers. *Fire and Materials*, 46, pp. 55–68. DOI: 10.1002/fam.2947.

RIR 21419–05 (2018) *Assessment of the Critical Radiant Flux (CRF) Performance of Solid Timber (Minimum Thickness 12 mm) and Plywood (Minimum Thickness 15 mm) when Tested in Accordance with AS/ISO 9239.1-2003, Fire Test Report*. Forest and Wood Products Australia.

RIR 30930800 RIR2.0 (2019) *An Assessment of Jarrah and Karri Enclosed Decks if Tested in Accordance with AS1530.8.1–2007 Section 20 for BAL 29 Exposure, Fire Test Report*. Forest and Wood Products Australia.

RIR 41117.9 (2019) *Fire Hazard Properties of Timber Floor, Wall and Ceiling Linings in Accordance with the Requirements of NCC 2019 Volume One, Fire Test Report*. Forest and Wood Products Australia.

RIR 45980.10 (2018) *An Assessment of Solid Timber Wall and Ceiling Linings in Accordance with AS 5637.1:2015, Fire Test Report*. Forest and Wood Products Australia.

RIR 45981.10 (2019) *Fire Hazard Properties of Plywood for Use as Internal Wall and Ceiling Linings in Accordance with AS 5637.1:2015, Fire Test Report*. Forest and Wood Products Australia.

RIR 45982.13 (2019) *Fire Hazard Properties of Timber Veneers on Standard MDF and Particleboard Substrates in Accordance with AS5637.1:2015, Fire Test Report*, Forest and Wood Products Australia.

Roberts, A. (1970) A review of kinetic data for the pyrolysis of wood and related substances. *Combustion and Flame*, 14, pp. 261–272. DOI: 10.1016/S0010-2180(70)80037-2.

SFPE Handbook (2016) *The SFPE Handbook of Fire Protection Engineering*, 5th Edition. Springer, New York. DOI: 10.1007/978-1-4939-2565-0.

Simms, D. (1960) Ignition of cellulosic materials by radiation. *Combustion & Flame*, 4, pp. 293–300. DOI: 10.1016/S0010-2180(60)80042-9.

Sundström, B. (2007) *The Development of a European Fire Classification System for Building Products: Test Methods and Mathematical Modelling*. Department of Fire Safety Engineering Lund University, Sweden.

Tran, H. (1992) Chapter 11: Wood materials. Part B. Experimental data on wood materials, in Heat Release in Fires, V. Babrauskas and S. Grayson, Eds. Elsevier Applied Science, New York, pp. 357–372. Available from https:// www.fpl.fs.fed.us/documnts/pdf1992/tran92a.pdf.

Tran, H., Janssens, M. (1991) Wall and corner fire tests on selected wood products. *Journal of Fire Sciences*, 9, pp. 106–124. DOI: 10.1177/073490419100900202.

USDA Forest Products Laboratory. https://www.fpl.fs.fed.us/products/products/ cone/introduction.php.

Wade, C., Baker, G., Frank, K., Harrison, R., Spearpoint, M. (2016) *B-RISK 2016 User Guide and Technical Manual*, Study Report SR364, Building Research Association of New Zealand, Porirua.

White, N., Delichatsios, M. (2014) *Fire Hazards of Exterior Wall Assemblies Containing Combustible Components*, Document Number: EP142293, Fire Protection Research Foundation, Quincy, MA.

White, R., Dietenberger, M., Tran, H., Grexa, O., Richardson, L., Sumathipala, K., Janssens, M. (1999) Comparison of test protocols for the standard room/corner test. *Fire and Materials*, 23, pp. 139–146. DOI: 10.1002/ (SICI)1099-1018(199905/06).

Winandy, J.E. (2001) Thermal degradation of fire-retardant-treated wood: Predicting residual service life. *Forest Products Journal*, 51(2), pp. 47–54.

Wood Handbook (2010) *Wood as an Engineering Material, General Technical Report FPL-GTR-190*, Department of Agriculture, Forest Service, Forest Products Laboratory, Madison, WI. Available from https://www.fpl.fs.fed.us /documnts/fplgtr/fpl_gtr190.pdf.

WoodSolutions (2021) https://woodsolutions.com.au.

Yoshioka, H., Nishio, Y., Kanematsu, M., Noguchi, T., Hyakawa, T., Zhao, X. (2021) Fire behaviour of fire-retardant treated wooden facades: Comparison of deterioration caused by accelerated weathering and natural exposure in Japan. 12th Asia-Oceania Symposium on Fire Science Technology (AOSFST 2021), University of Queensland, Australia.

Chapter 6

Fire-separating assemblies

Norman Werther, Christian Dagenais,
Alar Just and Colleen Wade

CONTENTS

DOI: 10.1201/9781003190318-6

SCOPE

This chapter describes the important role of fire-separating assemblies for passive fire protection in any type of building. Fire-separating assemblies provide essential compartmentation, which limits fire spread, contributing to both life safety and property protection. It gives design recommendations for providing fire resistance to timber- and wood-based separating assemblies, including walls, floors and roofs.

6.1 GENERAL

In addition to maintaining the load-carrying capacity of the structure during a fire, the concept of compartmentation is one of the most effective passive measures for providing fire protection for life safety and property protection. Without firefighting or automatic fire suppression, the concept of compartmentation is the only way of preventing a fire from spreading beyond its room of origin. This concept has become an essential requirement in both prescriptive- and performance-based building codes all over the world.

The main objective of applying fire-resistance-rated separating assemblies is to limit the probability that fire or smoke will spread from the compartment of fire origin to other compartments at the same or other storeys in a building, or to neighbouring buildings, within a defined time. By an optimum arrangement of separating assemblies, the development and spread of fire is slowed down, property damage is reduced, fire exposure to multiple sites is limited, safety of occupants is improved, and firefighting and rescue operations become more effective (Figure 6.1).

6.2 BASIC REQUIREMENTS FOR FIRE-SEPARATING ASSEMBLIES

Walls, floors and roofs acting as separating assemblies can be designed as load-bearing or non-load-bearing elements. Separating assemblies are only exposed to fire from one side at a time, even if they are designed for separate exposure in both directions. This is important for non-symmetrical assemblies where the fire resistance may not be the same whether fire occurs from one side or the other, but the assembly must be designed for exposure from both sides separately.

Where fire compartmentation is required, the elements forming the boundaries of the fire compartment must be designed and constructed in such a way that they maintain their separating function during the relevant fire exposure. The relevant time of fire exposure is normally expressed in terms of fire resistance, a specified time of exposure to the standard temperature–time curve (see Chapter 2). The time is usually specified by prescriptive building regulations.

Figure 6.1 Compartmentation between buildings by a separating firewall (Photo johnivison.com).

Most fire codes around the world specify the required fire resistance R/E/I separately for three functions of structural adequacy or load-bearing (R), integrity (E), and insulation (I), in that order, as shown in Figure 6.2. Load-bearing assemblies are presented in Chapter 7.

In addition to these criteria, some counties, mainly in Europe, require that firewalls shall withstand mechanical action (criterion M) to maintain compartmentation. Mechanical action M is the ability of the element to withstand impact, representing the case where the structural failure of another component in a fire causes an impact on the element concerned (EN 13501-2). The element is subject to impact of a predefined force shortly after the time for the desired R, E and/or I classification period. The element shall resist the impact without prejudice to the R, E, and/or I performance to have the classification supplemented by M.

The integrity criterion (E) is satisfied when no sustained flaming, or hot gases sufficient to ignite a cotton pad, occurs on the unexposed side and no cracks or openings in excess of certain dimensions open up (which could allow passage of flames or hot gases).

The insulation criterion (I) is satisfied where the average temperature rise, as measured by standard thermocouples placed on the non-exposed surface,

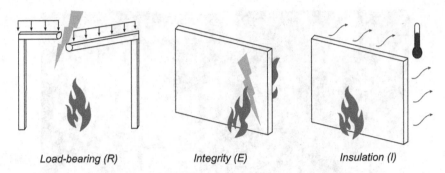

Load-bearing (R)   Integrity (E)   Insulation (I)

Figure 6.2 Performance criteria for fire resistance (TU Munich). The criteria apply to both horizontal and vertical assemblies.

is limited to 140°C, and the maximum temperature rise at any point (or thermocouple) on that surface does not exceed 180°C, when exposed to the standard fire. This prevents the ignition of objects in the neighbouring compartment due to excessive radiation emitted by that hot surface. In general, there is no risk of fire spread due to thermal radiation when criterion I (insulation) is satisfied.

With respect to non-standard design fires, which include the decay phase, Eurocode 5 recommends a different limit to the unexposed surface temperature increase of an average of 200°C and a maximum of 240°C (EN 1995-1-2, 2004).

In Europe, the performance criteria are expressed together with a time value e.g. EI 30, EI 60, EI 90 etc. In some countries, such as Canada and United States, the fire resistance of a separating assembly is determined as the time when the first performance criterion fails, without specifically stating what failure mode limited the assembly's fire resistance. In New Zealand and Australia, the three criteria are expressed together as 30/30/30, or 60/60/60.

In most countries, the spread of smoke is assumed to be fulfilled implicitly when the EI criterion is satisfied, but an explicit evaluation is required in a few building codes within specific testing standards or for some assemblies. For example, the German standard DIN 4102-2 requires the visual observation and classification of the emergence of smoke on the unexposed side in addition to the evaluation of integrity and insulation behaviour. An assessment by technical measurements of smoke leakage is done for smoke control doors on the basis of EN 13501-2.

6.3 ENCAPSULATION

Some timber elements can be used as separating assemblies with no applied fire protection. Additional protective layers are required in some cases, depending on the type of assembly and the time of protection required. Protection, which is sufficient to prevent the charring of the underlying wood, is called encapsulation, as discussed in Chapter 2. Encapsulation may be provided for several reasons:

- To increase the EI criteria for fire resistance of separating assemblies
- To exclude or reduce charring, which will decrease the load-bearing capacity (R) of structural elements
- To reduce the additional fuel load in the fire compartment due to charring

Different criteria are used for encapsulation in different countries. In Europe, K classes are defined in EN 13501-2 and encapsulation for structural fire protection is given in prEN 1995-1-2, 2021. Tests for encapsulation in Canada are described in CAN/ULC-S146. Even if the individual classification criteria differ between test and design standard, the common objective is to protect structural timber elements against charring (see Chapter 7) and contribute to the overall fire resistance of separating assemblies. The European K classes do not guarantee compartmentation, due to the different performance criteria, as shown in Figure 6.3, with temperature rise measurements recorded at different locations and times.

The European system of K classes for the fire-protection performance of coverings is defined in EN 13501-2 based on full-scale furnace testing in horizontal orientation according to EN 14135, as shown in Figure 6.4. Besides the temperature criterion behind the protective lining after different time intervals (10, 30 and 60 minutes), no collapse, burning on the substrate or falling parts are allowed. An encapsulation with K classification is required by building regulations in several European countries. Two types of K classes are defined, depending on the substrate behind the protective material. Class K_1 includes substrates with a density less than 300 kg/m^3,

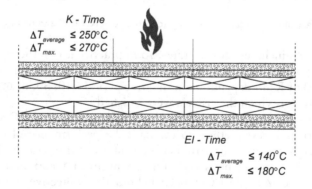

Figure 6.3 Difference between the surface temperature rise performance criteria for insulation I and encapsulation K (TU Munich).

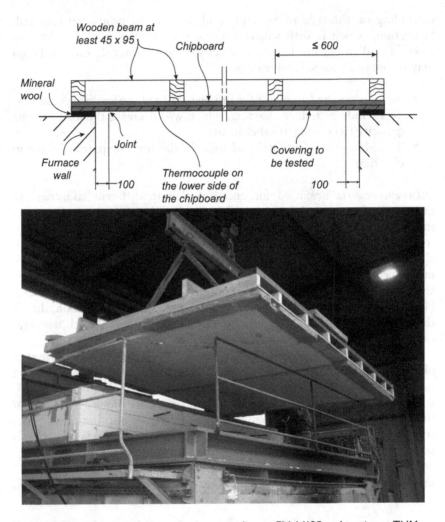

Figure 6.4 Test of encapsulation covering according to EN 14135 and testing at TUM.

while class K_2 includes all substrates, so in practice it is sufficient to verify K_2 classes for the protection of wood. The K_2 classes would then be given as $K_2 10$, $K_2 30$ or $K_2 60$ depending on the time periods (10, 30, or 60 minutes, respectively) for which the criteria were satisfied.

Similar to the European approach, in Canada, the CAN/ULC-S146 standard describes a test method to evaluate the performance of encapsulation materials for structural timber elements. Contrary to the European approach, the assessment is only for the mean and maximum increase in temperature ($\Delta T \leq 250°C/270°C$) behind the protective lining on the wooden substrate. Falling parts of the encapsulation material or a visual observation of the burned substrate are not evaluated.

In Eurocode 5 (EN 1995-1-2) the value t_{ch} (see Chapter 7) is defined as the start time of charring behind the protective lining for the calculation of the fire resistance of protected timber structures. A limiting temperature of 300°C is used for the onset of charring. The criterion of 300°C is higher than the stricter temperature criteria in EN 13501-2 and CAN/ULC-S146. This is due to the fact that in the European test standard EN 14135, there is a time delay between the end of the fire exposure and the visual observation of the wooden substrate regarding burned areas at joints or at fasteners, and thus a further thermal exposure occurs. For this purpose, the temperature criterion at the surface of the wood substrate was set below the typical 300°C criterion. An overview of the different assessment criteria used in current standards is given in Table 6.1.

However further investigation showed that starting of charring around fasteners behind the protective lining did not initiate any self-propagating smouldering in the timber substructure before reaching the critical surface temperature ($\Delta T \leq 250°C / 270°C$) according to EN 13501-2 (Mögele, 2010). The requirement regarding charring around fasteners is therefore expected to be deleted in future revisions of EN 13501-2.

The most common products used for encapsulation of timber members are gypsum plasterboards or gypsum fibreboards. Wood-based products may also be used for encapsulation (OJ, 2014), but they are usually more sensitive to dimensional changes in the original timber member. Additional wood surfaces added to the interior surfaces of a fire compartment for encapsulation may also increase the fuel load (see Chapter 3). The minimum thickness for achieving encapsulation may vary slightly, depending on

Table 6.1 Failure criteria to assess the fire protection performance of covering materials

Criterion	Fire protection system acc. to EN 1995-1-2	Encapsulation material "K" acc. to EN 13501-2	Encapsulation material acc. to CAN/ULC S146
Limitation of temperature behind the protective covering	Temperature limit for t_{ch} (start time of charring) is 300°C	No exceedance of the initial temperature by - 250°C (average) - 270°C (maximum)	No exceedance of the initial temperature by - 250°C (average) - 270°C (maximum)
Exclusion of burned or charred material	Only at the surface (joints are considered separately, fasteners are not taken into account).	Also in the area of fasteners and joints (assessed by visual observation after the test)	Not explicitly assessed
Fall-off/collapse of the protective covering	Time at which the protective covering falls off is given by t_f	Fall off or collapse (even of parts) is not permitted	Not explicitly assessed

Table 6.2 Products fulfilling K classes according EN 13501-2

Product	Minimum thickness (mm)		
	$K_2 10$	$K_2 30$	$K_2 60$
Gypsum plasterboard Type F Gypsum fibreboard	10	18 or 2×12.5	2 × 18
Particleboard, 600 kg/m³	12	25[a]	–
Plywood, 450 kg/m³	12	24[a]	–
OSB 600 kg/m³	10	30[a]	–
Solid wood panel, 450 kg/m³	13	26[a]	52*
Solid wood panelling and cladding, 450 kg/m³	15	27[a]	2 × 27*

[a]Tongue and groove required. Fixing devices shall fulfil certain requirements (OJ, 2014)

the test standard, product type and the mounting conditions and means of fixing. Examples of products fulfilling a K class according to EN 13501-2 are given in Table 6.2.

6.4 DESIGN METHODS FOR SEPARATING ASSEMBLIES

6.4.1 Methods for determining the fire resistance of separating assemblies

The fire resistance of timber elements can be assessed by standard fire tests based on EN 13501-2, ASTM E119, CAN/ULC S101, AS 1530.4 or ISO 834-1 or can be calculated by standard methods such as those in EN 1995-1-2, the National Building Code of Canada (NBCC), the International Building Code (IBC) or AS/NZS 1720.4.

While for many timber elements in new assemblies, product-specific fire tests are still widely used for the verification of the separating function, empirical and analytical design models are becoming more and more common. For instance, in the Eurocodes (prEN 1995-1-2, 2021), NBCC and IBC, three different levels are available for assessing the fire resistance of separating assemblies (see Table 6.3):

- Tabulated design data
- Simplified calculation methods
- Advanced calculation methods

Product-neutral tabulated data allow for ease of use in the design process, but at the same time they only consider a limited number of assemblies, usually on the conservative side. On the other hand, simplified and advanced calculation methods offer a wider range of applications and optimised results, but they require an increased effort in the calculation process.

Table 6.3 Characteristics of available design and calculation methods

	Tabulated design data	Simplified calculation methods	Advanced calculation methods
Field of application	Limited	intermediate	large
Accuracy	Conservative	intermediate	accurate
Complexity	Simple	intermediate	high

6.4.2 Classification based on fire testing

The experimental determination of fire resistance is, despite improvements in analytical assessment methods, still an essential tool in the assessment of the separating function for timber elements. Fire resistance tests are used especially for new products and assemblies, even if this is accompanied by a high cost and destruction of the test specimen. Furthermore, the experimental determination of fire resistance is beneficial for specific designs and optimisations in large-scale projects, for instance, to reduce the thickness of protective linings, which would typically be required using simplified or conservative design methods.

Full-scale furnace fire tests are often used as the basis for the classification of individual fire-separating elements like walls and floors. Fire exposure from only one side is considered, whereby the performance criteria from Section 6.2 must be met. Full-scale compartment fire tests are occasionally used, more often for research than for routine testing.

In order to verify the separating function of walls, floors, roofs or doors, various testing standards have been further developed in recent years, which allow for the assessment of timber assemblies. General basics and requirements for fire tests are specified, for instance, for Europe in EN 1363-1, for the US in ASTM E119, for Canada in CAN/ULC S101 or for Australia and New Zealand in AS 1530.4. Further similar test standards exist and are linked to the international standard ISO 834-1. A similar time–temperature curve is used in all these fire tests (Chapter 3). The common approach in all full-scale fire test standards is that the specimen shall reflect a size and execution typical in construction practice, limited by the size of available furnaces. For separating assemblies, a fire-exposed area of at least 3 × 3 m for walls and 3 × 4 m for floors is typically used (Figures 6.5 and 6.6), but greater dimensions may be required based on the applicable test standard. The assembly should include joints between the protective linings, realistic stud spacing and void cavities, as well as typical cut-outs for sockets and downlights, if relevant. Cut-outs and other penetrations through assemblies are often covered by separate fire resistance tests.

Even if the heat transfer within an assembly can be measured in small-scale or intermediate-scale fire resistance tests with the standard temperature–time curve, effects such as the limited deflection, modified heat

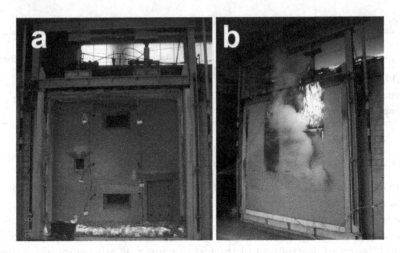

Figure 6.5 Furnace fire test for separating light timber frame wall specimen (TU Munich): (a) inside of wall furnace (after testing); (b) integrity failure (flame-through).

Figure 6.6 Furnace fire test for separating mass timber floor specimen: (a) unexposed surface of CLT floor specimen (TU Munich); (b) inside of floor furnace during testing (TU Munich); (c) integrity failure of plywood spline between mass timber floor panels (Fire TS Lab, New Zealand).

transfer coefficients, different shrinkage of materials, limited falling off of linings or void cavity insulation may not be the same as in a full-scale test. This is why full-scale tests are usually required.

When a wall assembly is to be designed as a "firewall", as defined in certain building codes, the assembly may need to be subjected to additional requirements. As such, EN 1366-2 requires that firewalls be subjected to the impact of a predefined horizontal force applied immediately after the time of the desired classification period (see Section 6.2). Even if firewalls are typically required to be constructed from non-combustible materials, the performance requirement to withstand the impact of a horizontal force at the end of the fire exposure can also be fulfilled by light timber frame or mass timber assemblies when designed accordingly. The US standard ASTM E119 requires the use of a hose stream test at the end of the fire-exposure time.

As an outcome of standardised fire tests, test reports are issued by certified or accredited laboratories, describing the setup, the fire exposure, the loading conditions and the obtained results such as visual observations and failure mode. These reports are usually the basis for issuing official classification documents or certificates according to national requirements.

6.4.3 Tabulated design data

The concept of tabulated design methods or generic tabulated fire resistance ratings exists in most countries worldwide and represents local experience and building tradition based on full-scale tests carried out in accordance with recognised fire-testing standards over many years. With respect to separating timber assemblies, the fire resistance rating is usually given for standard fire exposure only considering the criteria of integrity (E) and insulation (I). The load-bearing function (R) is also included if necessary.

Tabulated fire resistance ratings are often available for typical construction products, like solid timber, LVL, CLT, gypsum plasterboard, wood-based panels or insulation, etc., with no reference to individual manufacturers. This allows an easy application and simplified proof within a deemed-to-satisfy concept. The advantage of those tabulated listings is that they can be applied to commonly available materials, represented via national or international product standards in any country.

Compared to product-specific fire resistance tests, tabulated designs usually are very conservative, due to their liberal and simplified definition of materials and layers in an assembly such as the minimal thickness of individual layers, their dimensions, density and fixing or assembly. In addition, the application is often limited only to a specific load level or size of wall and floor assembly, unless noted otherwise in applicable building codes.

Despite these limitations, generic tabulated data for separating timber assemblies are still an essential part in determining the fire resistance for light timber frame and mass timber assemblies, as shown, for example, in Tables 6.4 and 6.5.

Several standards and guidance documents summarise the current national and international experience regarding generic fire-resistance-rated assemblies under consideration of the individual national building practices and test experience. Lists and tables of generic fire-resistance-rated wall, floor and roof assemblies can be found in tables 9.10.3.1.-A and -B of the NBCC, Section 721 of the IBC, in several National Annexes (NA) of EN 1995-1-2 (NF EN 1995-1-2/NA; ÖNORM B 1995-1-2) or other national standards (Angehm et al., 2015; DIN 4102-4). Corresponding tables typically comprise fire resistance ratings up to 120 minutes.

Due to the rapid digitalisation and harmonisation in the field of fire design, tabulated design concepts are becoming available for designers and engineers via free online platforms, like www.dataholz.eu.

Table 6.4 Deemed-to-satisfy design solution for fire resistance of load-bearing separating light timber frame wall assemblies

Panel (fire-exposed side)*		Timber stud		Panel (fire-unexposed side)*		Load (kN/m)	Fire resistance (minutes)
1st layer (surface layer) (mm)	2nd layer (mm)	Width (mm)	Depth (mm)	1st layer (surface layer) (mm)	2nd layer (mm)		
Fire-rated gypsum panel 12.5	–	≥60	≥100	Fire-rated gypsum panel 12.5	–	≤19	30
Fire-rated gypsum panel 12.5	Wood-based panel 15	≥60	≥100	Fire-rated gypsum panel 12.5	Wood-based panel 15	≤19	60
Fire-rated gypsum panel 15	Fire-rated gypsum panel 15	≥60	≥100	Fire-rated gypsum panel 15	Fire-rated gypsum panel 15	≤19	90

Stud spacing ≤ 625 mm; wall height ≤ 3000 mm; void cavity insulation ≥ 60% of stud depth, with mineral wool, wood fibre insulation or cellulose fibre insulation; *specific fixation rules apply extract from ÖNORM B 1995-1-2

Table 6.5 Deemed-to-satisfy design solution for fire resistance of separating mass timber wall assemblies

Panel (fire-exposed side)	Laminated solid timber and CLT	Panel (fire-unexposed side)	Fire resistance[a]
1st layer (mm)	Depth (mm)	1st layer (mm)	(minutes)
–	≥ 80	–	30
–	≥ 120	–	60
Fire-rated gypsum panel 12.5	≥ 50	–	30
	≥ 110		60
	≥ 150		90
Fire-rated gypsum panel 18	≥ 40	–	30
	≥ 90		60
	≥ 130		90

[a]Only separating function (insulation, integrity), separate proof of load-bearing function needed (Chapter 7). Extract from ÖNORM B 1995-1-2

Many manufacturers of specific fire-resisting products provide tabulated data for fire resistance of their assemblies. For example, in New Zealand, Winstone Wallboards (2018) provide a 100-page listing of fire resistance ratings for a large number of floor and wall assemblies using their proprietary gypsum plasterboard products on light timber frame assemblies.

6.4.4 Simplified calculation methods

In timber buildings, walls and floors are mostly built up by adding different layers to form an assembly. For the calculation of fire resistance with regard to the separating function of timber assemblies, component additive methods can be used. These methods determine the fire resistance of a layered construction by adding the contribution of each layer to obtain the fire resistance. Here, the integrity criterion is deemed to be satisfied if the insulation criterion is met. It needs to be noted that the individual contribution of each layer of material is definitely not the same as the fire resistance of that layer of material when tested individually.

Separating Function Method (Europe)

The Separating Function Method (SFM) used in Europe is capable of considering timber assemblies with an unlimited number of layers made of mass timber, glulam, CLT, LVL, wood-based boards, gypsum plasterboards, gypsum fibre boards, clay plaster, mineral wool, wood fibre and cellulose fibre cavity insulation or combinations thereof. The model was based on earlier research (Norén, 1994; Östman et al., 1994), further developed by Schleifer et al. (2007; 2009) and extended by several research projects during recent years, to improve the calculation results (Mäger et al., 2017) or to consider new relevant materials (Mäger et al., 2019; prEN 1995-1-2, 2021; Rauch et al., 2020; Winter et al., 2019). The method is able to consider the different heat transfer paths through an assembly, as shown in Figure 6.7. For light timber frame assemblies, the heat transfer path through the cavity insulation layer is prevalent and the path through the timber studs or joists can be neglected.

According to the European SFM, the total fire resistance of a timber assembly is taken as the sum of the contributions from the different layers (claddings, void or insulated cavities, mass timber elements), as shown in Equation 6.1. The layers in an assembly fulfil different functions. All fire-exposed layers have a protective function (giving protection time), while the last layer on the fire-unexposed side provides an insulation function (giving insulation time). These functions are linked to different temperature criteria, as shown in Figure 6.8.

$$t_{ins} = \sum_{i=1}^{i=n-1} t_{prot,i} + t_{ins,n} \quad [min] \tag{6.1}$$

with

$\sum_{i=1}^{i=n-1} t_{prot,i}$ Sum of the protection times $t_{prot,i}$ of the layers (in the direction of the heat flux) preceding the last layer of the assembly on the side not exposed to fire layers (according to Figure 6.9).

$t_{ins,n}$ Insulation time $t_{ins,n}$ of the last layer of the assembly on the side not exposed to fire

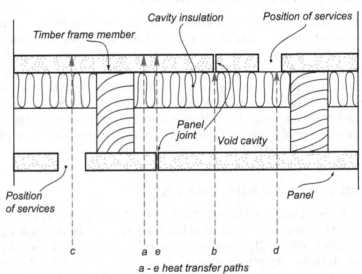

Figure 6.7 Illustration of heat transfer paths through separating multiple-layered construction (Östman et al., 2010).

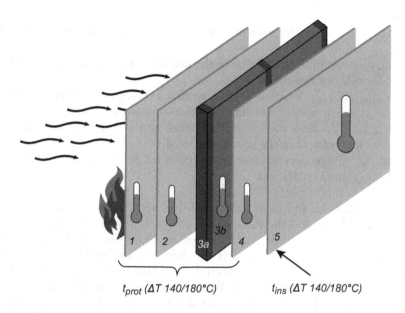

Figure 6.8 Design approach of SFM (TU Munich).

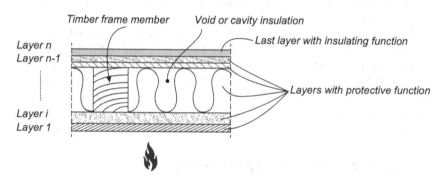

Figure 6.9 Numbering and function of the different layers (Östman et al., 2010).

Protection and insulation times of the layers can be determined according to the following Equations 6.2 and 6.3, taking into account the basic values of each layer, the coefficients for the position of the layers in the assembly, the coefficients for the joint configurations and if relevant the benefit of fire-rated claddings, which provide additional protection.

$$t_{prot,i} = \left(t_{prot,0,i} \cdot k_{pos,exp,i} \cdot k_{pos,unexp,i} + \Delta t_i \right) \cdot k_{j,i} \quad \left[min \right] \tag{6.2}$$

$$t_{ins,n} = \left(t_{ins,0,n} \cdot k_{pos,exp,n} + \Delta t_n \right) \cdot k_{j,n} \quad \left[min \right] \tag{6.3}$$

with

$t_{prot,0,i}$ Basic protection value of layer i (as shown in Figure 6.8 and Figure 6.9)

$t_{ins,0,n}$ Basic insulation value of the last layer n of the assembly on the side not exposed to fire

$k_{pos,exp,i}, k_{pos,exp,n}$ Position coefficient that takes into account the influence of layers preceding the layer considered

$k_{pos,unexp,i}$ Position coefficient that takes into account the influence of layers backing the layer considered

$\Delta t_i, \Delta t_n$ Correction time for layers protected by fire-rated claddings

$k_{j,i}, k_{j,n}$ Joint coefficient

The coefficients and basic values are dependent on the material of the investigated layer and the influence of the preceding and backing layers. These coefficients were derived by extensive finite element thermal simulations based on physical models for heat transfer through separating multi-layered constructions (see Section 6.4.5 (Benichou et al., 2001)). The material properties used for the finite element thermal simulations were calibrated and validated by fire tests using the standard temperature–time curve. The comparison between test results and the design method shows that the improved model is able to predict the fire resistance of timber assemblies safely and permits verification of the separating function of a large number of common timber assemblies.

All the protection and insulation times, position and joint coefficients of the generic materials are given in prEN 1995-1-2 (2021). Product-specific parameters are usually provided by the producers of materials. Annex G of prEN 1995-1-2, 2021 gives the procedure for determining the necessary parameters. The aforementioned methods rely on detailing rules, such as fixing of panels, oversizing of insulation or spacings, in order to avoid a premature failure such as falling off of cladding or insulation materials.

Component Additive Method (US/Canada)

The North American empirical method for calculating fire resistance of assemblies using the Component Additive Method (CAM) was developed in the 1960s (CWC, 1996). When using CAM, a designer can rapidly determine the fire resistance rating of a given wall or floor assembly in a new construction design, without the need to perform structural calculations or conduct full-scale fire resistance tests.

Appendix D-2.3 of the NBCC (2020) details the CAM for wall and floor assemblies up to 90 minutes of fire resistance. Following CAM, the fire resistance of a light timber frame assembly is taken as the sum of the assigned times of the membrane on the fire-exposed side, the framing members, and additional protective measures, such as insulation. When using this method, floor-framing elements such as timber joists, prefabricated wood I-joists, and parallel-chord wood trusses spaced at a maximum of 600 mm contribute 10 minutes to the fire resistance of a floor assembly. Timber joists are to be at least 38 × 184 mm (nominal 2″ × 8″), and studs are to be at least 38 × 89 mm (nominal 2″ × 4″). Resilient metal channels are permitted to be installed with no effect on the rating of the floor assembly. Roof assemblies consisting of timber joists spaced at a maximum of 400 mm contribute 10 minutes, while metal-plated trusses spaced at a maximum of 600 mm have an assigned contribution of 5 minutes. It is noted that these times are not the actual fire resistance afforded by the structural members but their contribution to the overall fire resistance of a light timber frame assembly.

The contribution of protective membranes can then be added to the contribution of the framing elements to determine the fire resistance. For example, the Canadian method assigns 25 minutes and 40 minutes, respectively, for one layer of 12.7 mm and 15.9 mm Type X gypsum board directly attached to timber joists or installed on resilient metal channels spaced at no more than 400 mm. Double layers of 12.7 mm Type X gypsum board provide 50 minutes and 45 minutes, when installed on resilient metal channels spaced at no than 400 mm and 600 mm, respectively. A time of 60 minutes is assigned to a double layer of 15.9 mm Type X gypsum board, installed on resilient metal channels spaced at no more than 600 mm. Additional times can also be obtained when filling the cavities with various types of insulation. Type X boards must meet the requirements of ASTM C1396 or CAN/CSA-A.82.27.

Similar provisions can be found in Section 722.6 of the IBC in the US, with minor differences for some contribution times. The US CAM also provides more options for gypsum board protection than that of the Canadian CAM.

6.4.5 Advanced calculation methods

Numerical calculation methods are the most sophisticated tools to evaluate the fire resistance of separating assemblies. These methods are not a substitute for fire resistance testing, but they are useful for assessing the fire resistance of assemblies which cannot be tested, or for developing new products or assemblies. These advanced calculation methods eliminate the costs of expensive fire testing by using validated finite element (FE) computer models or other appropriate advanced procedures to determine the thermal and structural performance of timber assemblies exposed to fire. This also includes the assessment of the separating function concerning the insulation criterion, as shown in Figure 6.10. The input data and the results of numerical calculation methods must always be verified by the results of full-scale fire resistance tests.

These advanced calculation models consider the fundamental physical nature of heat transfer to predict transient temperature distributions in

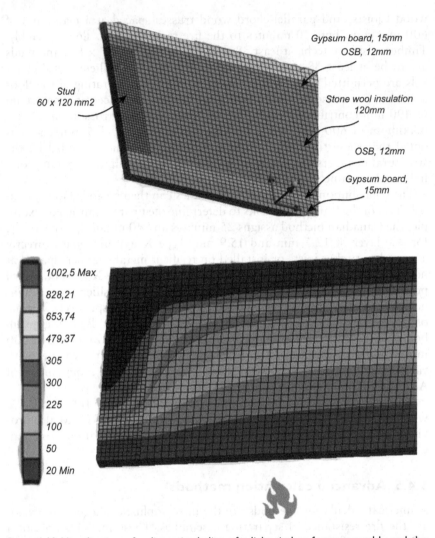

Figure 6.10 Visualisation of a discretised slice of a light timber frame assembly and the resulting temperature distribution after 75 min of standard fire exposure gained by numerical simulation (TU Munich).

assemblies and structural elements by solving complex differential equations. Such approaches may account implicitly for the complex physical and chemical phenomena so that a simple conductive heat transfer analysis for solid anisotropic materials can be carried out by applying adjusted "effective" material properties rather than using real-measured material properties. Realistic heat transfer and heat loss at the boundary surface are usually considered by a convective and radiative fraction.

Thermal simulations require the thermal conductivity, specific heat and density at elevated temperatures as input for describing the material of each individual layer. Validated temperature-dependent effective material properties proposed for standard fire exposure are available in the literature or in standards for traditional products (Benichou et al., 2001; prEN 1995-1-2, 2021; Schleifer, 2009), including timber, wood-based panels, gypsum panels, mineral wool insulation and also bio-based insulation products, clay or screed (Liblik et al., 2019; Rauch et al., 2020; Winter et al., 2019). Such values can be obtained by complex numerical approaches using fire test data as validation (Mäger et al., 2016). Effective material properties derived for a standard fire exposure cannot be used for other types of fire exposure without further validation.

The use of more realistic material properties requires the explicit consideration of highly complex phenomena within the simulation algorithms, such as the formation and oxidation of char and cracks, the evaporation and thermal transport of moisture, the constantly changing geometry for timber assemblies, including a falling off of panels, when applicable, and thermally induced deformations (Chen et al., 2020; Pečenko et al., 2014; Richter, 2019; Su et al., 2014; Werther and Matthäus, 2020). The complexity of these problems leads to an increased input effort, coupled simulations and longer calculation time. Regarding the assessment of the separating function, typically a two-dimensional model is appropriate to show the thermal influence of the individual components. In general, for the heat transfer in light timber frame assemblies, the cavity area governs the design due to a more rapid temperature formation compared to the charring of the framing members, as shown in Figure 6.10. For certain boundary conditions, this allows the application of a simple one-dimensional model for the assessment of the heat transfer conditions, as used for plane mass timber elements.

Even if several commercial software packages are available, the use of such software tools requires sufficient knowledge of the material and structural response under fire exposure, sufficient experience of the user to assess the results of the simulation, an understanding of the boundary conditions for heat transfer and structural calculations and especially, well-validated thermal and physical properties of materials used in timber assemblies (Werther et al., 2012) hence the need for verification with full-scale fire resistance testing.

Advanced calculation methods can also be used to calculate the temperature field around steel bolts or connector plates, which pass through timber wall or floor panels, especially if such details are likely to compromise the separating function of the panels (see Chapter 8).

6.5 DESIGN OF ASSEMBLIES FOR COMPARTMENTATION

6.5.1 Light timber frame walls and floors

Even if light timber frame walls and floors have been used for decades, they are subject to constant product improvement and can show excellent fire performance. A fire resistance of 30, 60, 90 or 120 minutes can be achieved by typical light timber frame assemblies.

The fire resistance of light timber frame wall and floor assemblies is assigned to the complete assembly or structure but characterised by the performance of the individual layers and components, like protective linings, framing members and cavity insulation and their arrangement in the assembly (AWC, 2018; Benichou et al., 2001; Just and Schmid, 2018; Östman et al., 1994).

For fire-separating assemblies, the contribution of the lining and cavity insulation to the fire resistance is of high importance. Cavity insulation may also increase the thermal exposure of the exposed lining but prevent heat radiation from reaching the back face. They may thus promote early fall off of the exposed lining. These products can protect the framing members from charring and prevent premature fire spread into the structure and along with the early heating up of the layers behind.

Insulation and lining materials

Typically, the fire-exposed linings and high-temperature-performance cavity insulation materials provide the most significant contribution to the fire resistance of separating assemblies. Wallpapers or air-tightening membranes are typically not considered explicitly, as such layers or further linings will not decrease the fire resistance, regardless how many layers are added and where they are located. This does not apply to thin metal-sheet lining materials used as the external cladding or within firewalls, especially if they can expand under thermal exposure. Furthermore, it can be noted that one thick layer of a lining material contributes more than several layers of the same material with the same total thickness, since individual thinner layers show an earlier falling off time (AWC, 2018; prEN 1995-1-2, 2021).

Void spaces

Void spaces or cavities in light timber frame assemblies can positively contribute to the separating function but are considered as the potential fire spread paths or areas for cavity fires and should be avoided as much as possible, especially in multi-storey timber buildings. Adding high-temperature-performance insulation in the cavities will generally increase the fire performance and the acoustic performance. At the same time, additional measures are needed to ensure the integrity of joints and junctions or at penetrations of service installations. Appropriate detailing measures are presented in Chapter 9.

Mechanical impact

Even if firewalls are often required to be from non-combustible materials, light timber frame walls can also withstand the additional mechanical impact, when designed adequately (see Section 6.2). Usually, the stud spacing is reduced and the thickness of lining increased to resist the mechanical impact without prejudicing the R, E or I criteria (P-3500/1115/07-MPA BS, 2017).

Linings

For a light timber frame wall or floor, the most important layer is the lining on the fire-exposed surface. Even if the thickness of the cavity insulation exceeds the thickness of the linings by several times, the performance of the fire-exposed lining is the most important and will provide the largest contribution to the entire fire resistance. This is shown in Figure 6.11 for a light timber frame wall with one layer of lining on each side and insulation in the cavity.

Figure 6.11a shows the temperature-failure criterion in the Separation Function Method (Section 6.4.4.1), with the contribution of each layer

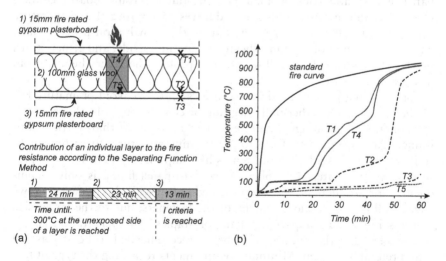

Figure 6.11 Contribution of layers to the fire resistance of the timber frame assembly. Thermocouples TI to T5 show temperatures at the indicated positions during a standard fire (TU Munich): (a) setup of timber frame wall assembly and contribution of layers to fire resistance; (b) temperature profiles in the assembly exposed to the standard fire curve.

shown separately. The 15 mm thick fire-rated gypsum plasterboard on the fire side provides a protection time of 24 minutes. The insulation provides a protection time of 23 minutes, and the last gypsum layer only contributes 13 minutes to the total fire resistance of 60 minutes, before reaching an insulation failure. Temperatures in the fire test are shown in Figure 6.11b.

Beside gypsum plasterboards, gypsum fibreboards and wood-based panels, also clay boards, cementitious boards or plastered ETICS (External Thermal Insulation Composite System), are commonly used as lining materials for timber frame assemblies.

Gypsum plasterboards

Gypsum plasterboards offer the most efficient and economic protection capacity in terms of panel thickness. This distinct protection capacity when exposed to fire results from a multiple-step dehydration reaction (calcination) in which calcium-sulphate-dihydrate ($CaSO_4 \cdot 2H_2O$) is converted to calcium-sulphate-hemihydrate ($CaSO_4 \cdot 0,5H_2O$) and further to Anhydrite ($CaSO_4$) under the release of water when heated.

Significant energy is required to evaporate the free water (2 to 3% by mass) and make the chemical changes to release the water of crystallisation (21% by mass) within the gypsum panels (Benichou et al., 2001). The most distinct result of this process is a temperature plateau of 100°C at the unexposed surface of the lining for a certain period of time, as shown in Figure 6.11. The length of the plateau depends on the panel thickness, panel density and backing material. Fire-rated gypsum boards contain glass or cellulose fibres and other additives to improve the temperature stability and reduce the shrinkage during the dehydration process. Such panels, also panels of calcium silicate, magnesium oxide, and cementitious or clay panels can delay falling off and protect the layers behind from direct fire exposure.

For all lining panels, good fixings are essential to avoid premature fall off before the panels thermally degrade or char through. In this context, horizontal ceiling linings are more prone to falling off than vertical wall linings. Panels should be fixed to the framing structure, to battens or to resilient channels with metal fasteners like staples, screws or nails, as prescribed by the material supplier. A panel-to-panel fixing is only recommended for product-specific setups, tested in full scale. Fixing with screws usually allows for a wider spacing of the fasteners, due to the larger size of the screw head compared to nails or staples. The penetration length of fasteners into the unburned timber is recommended to be at least 10 mm, preferably 20 mm. Minimum requirements regarding the type of fastener, edge distances, spacing, penetration length of fasteners or spacing of the substructure are given in several guidance documents or standards (prEN 1995-1-2, 2021) or within product literature when using proprietary systems.

Cavity insulation

Cavity insulation is an essential part of light timber frame assemblies and can provide a significant contribution to the thermal, acoustic and fire performance. Batts or loose-fill insulation, both made from mineral- or bio-based materials, like stone wool, glass wool, wood fibre or cellulose fibre insulation can be used. Flexible batts generally have a better performance than loose-fill insulation of the same type because of the higher inner cohesion and the sidewise clamping effect. Especially in horizontal floor assemblies, the contribution of the insulation to the fire resistance is low if the insulating batts fall out. This underlines the importance of high-quality fire protective lining and installation details.

Besides the insulating effect of the cavity insulation, the insulation itself can lead to a more rapid heating up of preceding layers compared to a void cavity. The accumulation of heat on the backside of the exposed linings will contribute to earlier dehydration of gypsum linings, possible falling off or faster charring of wood-based panels. Insulation materials that retain integrity when exposed to fire without melting or shrinkage, like stone wool or high-temperature-performance mineral wool can provide significant contributions to the load-bearing and separating functions of timber frame assemblies. Such materials can be identified by specific tests and are grouped to protection level 1 (PL1) (prEN 1995-1-2, 2021; Tiso, 2018).

Some insulation materials only have a limited effect on fire performance, especially those that melt (typical for glass wool starting from temperatures around 600°C), that shrink (like plastics or some bio-based insulation) or that can fall off (loose insulation products). These materials are usually grouped into protection level 2 (PL2) (prEN 1995-1-2, 2021). The performance of bio-based insulation products, like wood fibre or cellulose fibre insulation, was extensively investigated by Winter et al. (2019), showing that such products can improve the fire resistance of timber frame assemblies, when correctly installed. The contribution of the cavity insulation to fire resistance can be assessed according to the design methods given in Section 6.4.

Premature falling off of insulation batts must be prevented by using a lining, which provides sufficiently long falling off time, by adequate fixing like timber battens, resilient channels or gluing, or by tight fitting of oversize batts for walls using single-layer insulation. (prEN 1995-1-2, 2021; Sultan and Lougheed, 1997)

Framing members

To achieve sufficient load-bearing capacity of wall and floor assemblies, it is necessary to limit the charring and temperature effect on the framing members (the studs and joists) with good protective linings and fire-resistant cavity insulation. The edge of the framing in contact with the exposed lining suffers the highest thermal exposure and charring. The lateral sides have less charring, especially when protected by insulation. Protection by tight-fitting fire-resistant insulation may be needed to prevent the collapse of slender cross sections, prefabricated wood I-joists, or light trusses after the fire-exposed lining falls off. Design methods for the load-bearing function are given in Chapter 7.

6.5.2 Mass timber wall and floor panels

The wide use of mass timber panels as wall, floor or roof elements in multi-storey construction has recently become very popular, as shown in Chapter 1. Mass timber panels used as visible or lined structural assemblies have excellent fire resistance, but they provide additional fuel to the fire compartment if they are exposed to the fire. In comparison with light timber frame assemblies, mass timber panels usually have no voids or cavities and they show a uniform one-dimensional rate of charring. Mass timber elements can also maintain load-bearing and separating functions under impact loading (P-SAC02/III-635, 2019).

The fire performance of mass timber elements may be influenced by product-specific characteristics, like the layup and the dimensions of the elements or lamellas, their orientation, the existence of joints and the thermal performance of the gluelines. The charring rate of a solid wood slab exposed to the standard fire curve from one side is approximately 0.65–0.7 mm/min. However, for certain CLT elements without fire-resistant adhesive, charred lamellas have been observed to fall off the bottom of floor slabs, resulting in an increased rate of charring (Dagenais, 2016; Frangi et al., 2008; Frangi et al., 2009). For CLT wall elements, this behaviour has less impact (Dagenais, 2016; Frangi et al., 2009; Klippel et al., 2014). The influence of the glueline integrity in fire is considered via different charring models in standards or technical approvals, using a linear or stepped charring model considering the increased charring after the falling off of the protective layer, as shown in Chapter 7.

Typically, the fire resistance of mass timber assemblies is governed by their load-bearing capacity, since the good insulating behaviour of the remaining cross-section limits failure of the separating function. Mass timber elements should be tightly sealed or backed on the unexposed side to avoid convective flows through the element (Dagenais et al., 2019; Frangi, 2001). Glued mass timber elements typically show a better tightness to prevent the spread of fire and smoke than nailed elements, especially when there is no backing layer. CLT panels with edge bonded lamellas can provide excellent smoke tightness even without further lining, although many manufacturers of CLT do not glue the edges.

Design methods to assess the separating function are given in Section 6.4. Provided that there is no risk of charred layers falling off, the separating function of mass timber elements should be calculated according to Section

6.4.4.1 as one solid layer. The separating function of mass timber elements with the risk of glueline failure should be calculated, summarising the protection times of each lamella (see Section 6.4.4.1). Figure 6.12 compares a CLT panel with a glulam panel, both protected with one layer of gypsum plasterboard. For the CLT panel (Figure 6.12a), there is a risk of a charred layer falling off after the wood begins to char, if the glueline parallel to the surface is not fire-resistant. There is no similar risk for the glulam panel in Figure 6.12b.

As a simplification, the next generation of Eurocode 5 (prEN 1995-1-2, 2021) assumes that the insulation time of a mass timber assembly is 10 minutes less than the time that the charring reaches the non-fire-exposed side of the assembly (burning through). In Australia and New Zealand, AS/ NZS 1720.4 allows that where gaps have no effect, the separating function of mass timber panels can be assessed by calculating that 30 mm of residual wood remains in place after charring (30 mm: 23 mm of unaffected wood plus a 7 mm heat-affected layer).

Fire-exposed linings on mass timber elements typically show a higher contribution to fire protection than the same linings on light timber frame assemblies, due to the lower heat accumulation at the unexposed side of the lining. For gypsum plasterboard linings, typically, a 20% longer falling off time can be observed compared with the same linings on an insulated light timber frame wall (Kraudok et al., 2018; prEN 1995-1-2, 2021).

Regarding the overall fire performance of mass timber assemblies, in-plane joints between the elements must be considered in the design to prevent early failure of the separating function. Regarding the separating function of mass timber elements with a thickness of at least 75 mm and a joint width of ≤ 2 mm between neighbouring elements, only 70% of the full-protection capacity should be considered when the joints have no backing layer (prEN 1995-1-2, 2021). The effects of shrinkage and swelling must be taken into account, and measures must be designed for the maximum expected joint width. For example, glulam panels (Figure 6.12b) will tend to shrink more than CLT panels (Figure 6.12a) as they dry out, so the joints between panels will open up more. Design principles and details based on test experience for structural in-plane step joints, exterior splines, and tongue and groove joints are presented in several guidelines (Angehm et al.,

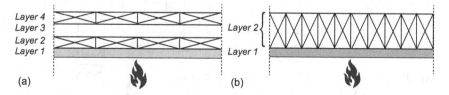

Figure 6.12 Layers considered when calculating separating function of mass timber elements: (a) with the risk of fall off of a charred layer due to glueline failure; (b) without risk of fall-off due to glueline failure (TU Munich).

2015; Dagenais et al., 2019; ÖNORM B 1995-1-2:2011-09; Werther et al., 2020). An overview of detailing for such joints is also given in Chapter 9.

6.5.3 Hollow core timber elements

Hollow core timber elements used as shear walls and floor slabs are prefabricated from sawn timber, glued wood panels, CLT, or laminated veneer lumber. The top and bottom layers are glued to the webs, and void cavities are filled with insulation, gravel or left empty for service installations. Care must be taken with all mass timber assemblies which have internal cavities.

Numerous fire tests have shown that such elements can provide a fire resistance of up to 90 minutes, depending on the design (P-SAC02/III-857, 2017; ETA 17/0941, 2018; Frangi and Fontana, 1999; O'Neill, 2013). Design methods for the separating function are given in Section 6.4. Methods to calculate the load-bearing function are given by Östman et al. (2010) and Chapter 7. Due to aesthetic requirements, the bottom skins of floors are often visible, without a protective lining.

Depending on the required fire resistance, the thickness of the bottom layer should be checked to ensure that the whole assembly has sufficient fire resistance after one-dimensional charring, as shown in Figure 6.13. Even if the bottom layer becomes non-effective due to charring, the uncharred webs may have sufficient strength for the hollow core floor to carry the fire-reduced loads as a T-beam.

In the case of perforated outer layers to improve the acoustic performance of these elements, they can be backed with absorbing insulation or wood fibre boards (Frangi and Fontana, 2004). Joints between the individual elements must be designed with sufficient fire resistance. Multiple tongue and groove joints can be used in conjunction with insulation between two adjacent webs, as shown in Figure 6.13.

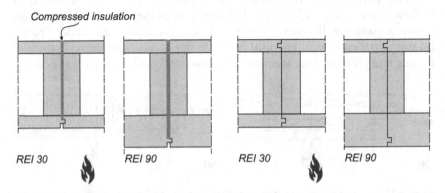

Figure 6.13 Design of outer layer and jointing details of hollow core elements. (TU Munich).

When installing service installations, penetrations and cut-outs have to be designed to exclude any fire spread within or between the elements. Some solutions include intumescent socket casings, gypsum boxes, or backing with non-combustible insulation to compensate for the cut-outs in the outer skin.

6.5.4 Timber T-beam floors

Compared to light timber frame or hollow core timber constructions, the load-bearing and separating elements of timber T-beams floors are exposed right from the start of the fire (O'Neill, 2013). This type of construction can be particularly challenging in terms of fire protection and sound insulation due to its slim and single-layer design. It is essential to ensure the insulation and integrity functions of the top flange of the T-beam. Chapter 7 provides verification methods for load-bearing capacity.

Usually, for a fire resistance of 30 minutes according to tabulated design, the top flange must have at least 50 mm of solid wood, increasing to 70 mm for 60 minutes of fire resistance (Angehm et al., 2015; DIN 4102-4) or 90 mm for 90 minutes of fire resistance.

The detailing of the joints in the top flange decking is particularly important, as convective flow of hot gases through the decking must be prevented. Multiple tongue and groove joints are often used for this purpose. All charring within the last tongue in the joint should be excluded by the joint design. An alternative is an additional layer above the load-bearing decking, as shown in Figure 6.14b. Design solutions for the consideration of fire exposure from above are described in Section 6.5.7.

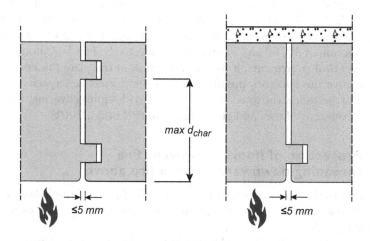

Figure 6.14 Design of outer layer and jointing details of hollow core elements. (TU Munich); (a) Double tongue and groove joint; (b) Single tongue and groove joint and additional covering.

6.5.5 Gaps for construction tolerances and shrinkage

The details in Figure 6.14 show a gap of up to 5 mm between the flanges of adjacent T-beam flooring units. Such gaps are often needed to allow for construction tolerances and for shrinkage movement as the timber floors dry out. It is essential to prevent an integrity failure occurring through such gaps. Integrity can be provided by a number of tongue and groove joints such as shown in Figure 6.14a or by providing a topping such as shown in Figure 6.14b. Larger gaps can be used if intumescent foam or compressed fire-resistant insulation is provided between the flooring units, supported by full-scale fire resistance testing wherever possible. More information is provided in Chapter 9.

6.5.6 Hybrid Timber–Concrete–Composite floors

Timber–Concrete–Composite (TCC) floors consist of a hybrid assembly of a concrete slab and timber elements, which are connected to each other with strong shear connections to provide composite structural action. Many TCC floors have excellent fire resistance.

In addition to the load-bearing capacity of TCC elements in the case of fire, which is usually characterised by the charring of the timber members and the potential heating up of the shear connectors (Frangi, 2001; Hozjan et al., 2017; Klingsch et al., 2015; Osborne, 2015; Ranger et al., 2016, O'Neill, 2013), the separating effect of the individual components and their joint configuration is essential for the fire compartmentation of TCC elements.

The top layer of the concrete usually provides sufficient protection regarding the separating function. A verification can be made using tabulated data of EN 1992-1-2 (2004) for up to 240 minutes of fire resistance. Here a floor-plate thickness of 60 mm, 80 mm and 100 mm is required for a separating function of 30, 60 and 90 minutes, respectively. For TCC floors with plane mass timber elements, a further reduction of the plate thickness and concrete covering becomes possible, due to the protective capacity of the timber. Furthermore, the top concrete layer provides joint covering, thereby avoiding convective flows and charring at joints (Frangi, 2001).

6.5.7 Protection of floors to prevent fire spreading downwards from a fire above

In addition to the fire resistance of floors exposed to fire from below, some countries also require a demonstration of fire resistance for fire exposure from the top side.

Although the fire severity to the top surface of a floor can be regarded as less severe compared to the fire exposure from the bottom, the standard fire curve is usually used when designing the fire resistance from a fire above.

This was demonstrated in a furnace test by FPInnovations (Ranger et al., 2020), where it was intended to verify whether a non-combustible encapsulation material used to protect a ceiling would perform similarly if used as protection on the top surface of a floor. The temperatures at the two surfaces (ceiling and floor) were very close, as were the heat fluxes. The installation of an assembly on the floor of the furnace worked well and suggests that the method would work well for encapsulating floors from above, without the risk or potential of falling off due to gravity when installed as ceiling protection.

Many multi-storey timber buildings have concrete screed toppings to improve acoustic performance. Even if such layers are commonly neglected when designing for fire resistance, the positive influence on the separating function may be extensive. These layers can act as an encapsulation for the floor when exposed from the top side. Tabulated data for fire resistance from below can also be used for fire exposure from above (Angehm et al., 2015; DIN 4102-4:2016-05) and the SFM can be used to assess the protection capacity of these layers (see Section 6.4.4). Dimensions and corresponding protection times for screed floor coverings are summarised in Table 6.6. In order to ensure a uniform protective capacity of the floor coverings, the joint to adjacent walls should be filled with non-combustible mineral wool, as shown in Figure 6.15.

6.5.8 Openings and penetrations in separating assemblies

The overall fire design of separating timber assemblies must consider the separating elements themselves, the joints in and between assemblies, and all penetrations for building services, in order to prevent spread of fire and smoke to other compartments or within the assembly.

Guidance for fire-safe detailing of joints and penetrations or openings for service installations is given in Chapter 9. The general concept is that joints in and between assemblies shall be designed to be tight, sealed and continuous to prevent any spread of fire or smoke.

Table 6.6 Minimum thickness of concrete screed to protect timber floors from charring for 30, 60 or 90 minutes fire exposure from above

Type of fire protection system	Minimal thickness of screed [mm] for a protection time of		
	30 min.	60 min.	90 min.
Concrete screed alone	35	60	80
Concrete screed above 15 mm impact sound insulation*	25	45	60
Concrete screed above 30 mm impact sound insulation*	20	30	45

* Impact sound insulation – density $\rho \geq 100$ kg/m³
Rauch et al., 2020

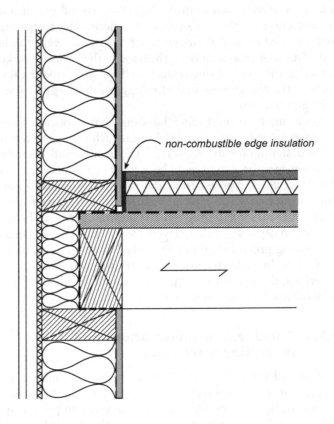

Figure 6.15 Non-combustible edge insulation.

REFERENCES

Angehm, C., Brühwiler, I., Frangi, A. and Wiederkehr, R. (2015) *Bauteile in Holz – Decken, Wände und Bekleidungen mit Feuerwiderstand. Lignum-Dokumentation Brandschutz (4.1).* Lignum, Zurich, Switzerland.

AS 1530.4 (n.d.) *Methods for Fire Tests on Building Materials, Components and Structures, Part 4: Fire-resistance Test of Elements of Construction.* Standards Australia, Sydney, NSW.

AS/NZS 1720.4 (n.d.) *Timber Structures Part 4 Fire Resistance of Timber Elements.* Standards Australia, Sydney, NSW.

ASTM C1396 (n.d.) *Standard Specification for Gypsum Board.* ASTM International, West Conshohocken, PA, USA.

ASTM E119 (n.d.) *Standard Test Methods for Fire Tests of Building Construction and Materials.* ASTM International, West Conshohocken, PA, USA.

AWC (2018) *Calculating the Fire Resistance of Wood Members and Assemblies.* Technical Report no. 10. American Wood Council, Leesburg Virginia, USA.

Benichou, N., Sultan, M.A., Mac Callum, C. and Hum, J. (2001) *Thermal Properties of Wood, Gypsum and Insulation at Elevated Temperatures.* Ottawa, Ontario, Canada.

CAN/CSA-A.82.27 (n.d.) *Gypsum Board.* CSA Group, Mississauga, ON.

CAN/ULC S101 (n.d.) *Standard Method of Fire Endurance Tests of building Construction and Material.* Underwriters Laboratories of Canada, Ontario, Canada.

CAN/ULC-S146 (n.d.) *Standard Methods of Test for the Evaluation of Encapsulation Materials and Assemblies of Materials for the Protection of Structural Timber Elements.* Underwriters Laboratories of Canada, Ontario, Canada.

CWC: Canadian Wood Council (1996) *Fire Safety Design in Buildings, A Reference for Applying the National Building Code of Canada Fire Safety Requirements in Building Design.* Ottawa, Ontario, Canada.

Chen, Z., Ni, C., Dagenais, C. and Kuan, S. (2020) A temperature-dependent plastic-damage constitutive model used for numerical simulation of wood-based materials and connections. *Journal of Structural Engineering,* 146(3), pp. 1–14.

Dagenais, C. (2016) *Fire Performance of Cross-Laminated Timber. Summary Report of North American Fire Research (Project 301010610).* FPInnovations, Quebec, Canada.

Dagenais, C., Ranger, L. and Benichou, N. (2019) *Chapter 8 Fire Performance of Cross-Laminated Timber Assemblies. Canadian CLT Handbook.* 2019 Edition. FPInnovations, Pointe-Claire, Canada.

DIN 4102-2 (n.d.) *Brandverhalten von Baustoffen und Bauteilen; Bauteile, Begriffe, Anforderungen und Prüfungen.* German Standard DIN e.V., Berlin, Germany.

DIN 4102-4 (n.d.) *Brandverhalten von Baustoffen und Bauteilen – Teil 4: Zusammenstellung und Anwendung klassifizierter Baustoffe, Bauteile und Sonderbauteile.* DIN e.V., Berlin, Germany.

EN 1363-1 (n.d.) *Fire Resistance Tests: Part 1: General Requirements.* European Standard CEN European Committee for Standardization, Brussels.

EN 1366-2 (n.d.) *Fire Resistance Tests for Service Installations – Part 2: Fire Dampers.* European Standard CEN European Committee for Standardization, Brussels.

EN 1992-1-2 (2004) *Eurocode 2 – Design of concrete structures, Part 1–2: General rules – Structural fire design.* European Standard CEN European Committee for Standardization, Brussels.

EN 1995-1-2 (2004) *Eurocode 5 – Design of timber structures, Part 1–2: General – Structural fire design.* European Standard CEN, European Committee for Standardization, Brussels.

EN 13501-2 (n.d.) *Fire Classification of Construction Products and Building Elements: Part 2: Classification Using Data from Fire Resistance Tests, Excluding Ventilation Services.* European Standard CEN European Committee for Standardization, Brussels.

EN 14135 (n.d.) *Coverings: Determination of Fire Protection Ability.* European Standard CEN European Committee for Standardization, Brussels.

ETA 17/0941 (2018) *Wood-based Composite Slab Element for Structural Purpose. European Technical Assessment.* EOTA European Organisation for Technical Assessment.

Frangi, A. (2001) *Brandverhalten von Holz-Beton-Verbunddecken.* Institute of Structural Engineering. IBK ETH Zurich, Switzerland.

Frangi, A. and Fontana, M. (1999) *Zum Brandverhalten von Holzdecken aus Hohlkastenelementen.* Institute of Structural Engineering. IBK ETH Zurich, Switzerland.

Frangi, A. and Fontana, M. (2004) *Untersuchungen zum Brandverhalten von Holzdecken aus Hohlkastenelementen.* Institute of Structural Engineering. IBK ETH Zurich, Switzerland.

Frangi, A., Fontana, M., Knobloch, M. and Bochicchio, G. (2008) Fire behaviour of cross-laminated solid timber panels. *Fire Safety Science*, 9, pp. 1279–1290.

Frangi, A., Fontana, M., Hugi, E. and Jöbstl, R. (2009) Experimental analysis of cross-laminated timber panels in fire. *Fire Safety Journal*, 44, pp. 1078–1087.

Hozjan, T., Orgin, A. and Klippel, M. (2017) *Short Review of the Research Conducted on the Fire Behaviour of Timber-concrete Composite Systems. Book of Abstracts of COST FP 1404 MC and WG Meeting Dissemination, Standardization and Implementation of Novel Improvements.* SP Report 2017:20.

IBC (2018) *International Building Code.* American National Standard Institute. Washington DC, USA.

ISO 834-1 (n.d.) *Fire-resistance Tests: Elements of Building Construction: Part 1 General requirements.* International standard ISO International Standardization Organization, Geneva.

Just, A. and Schmid, J. (eds) (2018) *Improved Fire Design Models for Timber Frame Assemblies: Guidance Document.* COST Action FP1404. ETH Zurich, Switzerland.

Klippel, M., Leyder, C., Frangi, A. and Fontana, M. (2014) Fire Tests on Loaded Cross-laminated Timber Wall and Floor Elements. *Fire Safety Science*, 11, pp. 626–639.

Klingsch, E., Klippel, M., Boccadoro, L., Frangi, A. and Fontana, M. (2015) *Fire tests on cross-laminated timber slabs and concrete-timber composite slabs.* Institute of Structural Engineering. IBK ETH Zurich, Switzerland.

Kraudok, K., Mäger, K. N. and Just, A. (2018) Fall-off times of gypsum boards. In *Book of Abstracts of the Final Conference COST FP 1404 Fire Safe Use of Bio-Based Building Products.* ETH Zürich, Switzerland.

Liblik, J., Küppers, J., Just, A., Maaten, B. and Pajusaar, S. (2019) Material properties of clay and lime plaster for structural fire design. *Fire and Materials*, 45(3), pp. 355–365.

Mäger, K. N., Brandon, D. and Just, A. (2016) Determination of the effective material properties for thermal simulations. In Proceedings of the International Network on Timber Engineering Research (INTER), Graz, Austria, Meeting 49.

Mäger, K. N., Just, A., Frangi, A. and Brandon, D. Ed. (2017) Protection by fire rated claddings in the component additive method. In Proceedings of the International Network on Timber Engineering Research (INTER), Meeting, Kyoto, Japan. Timber Scientific Publishing, vol. 50, pp. 439–451.

Mäger, K. N., Just, A., Schmid, J., Werther, N., Klippel, M., Brandon, D. and Frangi, A. (2019) Procedure for implementing new materials to the component additive method. *Fire Safety Journal*, 107, pp. 149–160.

Mögele, T. (2010) *Theoretische und experimentelle Untersuchungen zum Einfluss des Befestigungssystems auf die thermische Schutzwirkung von Brandschutzbekleidungen an Holzkonstruktionen.* Diplomarbeit, TUM, Lehrstuhl für Holzbau und Baukonstruktion.

NBCC (2020) *National Building Code of Canada.* NRC, National Research Council, Ottawa, Canada.

NF EN 1995-1-2/NA (n.d.) *Eurocode 5: conception et calcul des structures en bois – Partie 1–2: généralités – Calcul des structures au feu – Annexe nationale à la NF EN 1995-1-2:2004.*

Norén, J. (1994) *Addition Method: Calculation of Fire Resistance for Separating Wood Frame Walls (in Swedish).* Trätek – Swedish Institute for Wood Technology Research, Report I 9312070.

OJ (2014) *On the Conditions for Classification, without Testing, of Wood-based Panels under EN 13986 and Solid Wood Panelling and Cladding under EN 14915 with Regard to Their Fire Protection Ability, When Used for Wall and Ceiling Covering.* Official Journal of the European Union 5.12.2014. EC Decision of 16 July 2014. European Union, Brussels Belgium.

O'Neill, J.W. (2013). *The Fire Performance of Timber Floors in Multi-Storey Buildings.* Ph.D. thesis, University of Canterbury, New Zealand.

ÖNORM B 1995-1-2:2011–09. (n.d.) *Eurocode 5: Design of Timber Structures – Part 1–2: General: Structural Fire Design: National Specifications Concerning ÖNORM EN 1995-1-2.* National Comments and National Supplements. Austrian Standard Institute, Vienna Austria.

Osborne, L. (2015) *Fire Resistance of Long Span Composite Wood-Concrete Floor Systems (Project No. 301009649).* FPInnovations, Quebec, Canada.

Östman, B., König, J. and Norén, J. (1994) Contribution to fire resistance of timber frame assemblies by means of fire protective boards. In Proceedings of the 3rd International Fire and Materials Conference, Washington, DC.

Östman, B., Mikkola, E., Stein, R., Frangi, A., König, J., Dhima, D., Hakkarainen, T. and Bregulla, J. (2010) *Fire Safety in Timber Buildings: Technical Guideline for Europe.* SP Report 2010:19. SP Technical Research Institute of Sweden, Stockholm, Sweden.

Pečenko, R., Huč, S., Turk, G., Svensson, S. and Hozjan, T. (2014) Implementation of fully coupled heat and mass transport model to determine the behaviour of timber elements in fire. In World Conference on Timber Engineering (WCTE), Quebec City, Canada.

P-3500/1115/07-MPA BS (2017) *Classification Report Tragende raumab-schließende Wandkonstruktion mit Holzständerwerk und einer allseitigen K260 Brandschutzbekleidung REI 60 –M.* Saint-Gobain Rigips GmbH. MFPA Leipzig GmbH, Leipzig Germany.

P-SAC02/III-635 (2019) *Classification Report Tragende, raumabschließende Wandkonstruktionen in Brettsperrholzbauweisemit einer einseitigen oder zweiseitigen Bekleidung aus Gipskarton-Feuerschutzplatten oder FERMACELL Gipsfaser-Platten der Feuerwiderstandsklasse F 60-B bzw. F 90-B gemäß DIN 4102-2: 1977–09 bei einseitiger Brandbeanspruchung von der bekleideten Wandseite sowie zusatzlichen Widerstand gegen Stoßbeanspruchung gemäß DIN 4102-3: 1977–09.*

P-SAC02/III-857 (2017) *Classification Report Bauart zur Errichtung tragender, raumabschließender Deckenkonstruktionen aus Lignotrend: Brettsperrholzelementen, REI30, REI60, REI90*. Lignotrend GmbH. MFPA Leipzig GmbH, Leipzig Germany.

prEN 1995-1-2 (2021) *Final Draft Eurocode 5 – Design of Timber Structures, Part 1–2: General: Structural Fire Design*. European draft standard CEN European Committee for Standardization, Brussels.

Ranger, L., Dagenais, C. and Cuerrier-Auclair, S. (2016) *Fire-Resistance of Timber-Concrete Composite Floor Using Laminated Veneer Lumber (Project No. 301010618)*. FPInnovations, Quebec, Canada.

Ranger, L., Dagenais, C. and Benichou, N. (2020). *Encapsulation of Mass Timber Floor Surfaces (Project No. 301013624)*. FPInnovations, Quebec, Canada.

Rauch, M., Werther, N. and Winter, S. (2020) Fire design method for timber floor elements: the contribution of screed floor toppings to the fire resistance. In World Conference on Timber Engineering (WCTE), Santiago, Chile.

Richter, F. (2019) *Computational Investigation of the Timber Response to Fire*. PhD Thesis, Imperial College, London.

Schleifer, V. (2009) *Zum Verhalten von raumabschliessenden mehrschichtigen Holzbauteilen im Brandfall*. PhD Thesis No. 18156, ETH Zurich.

Schleifer, V., Frangi, A. and Fontana, M. (2007) *Experimentelle Untersuchungen zum Brandverhalten von Plattenelementen*. Institute for Structural Engineering. IBK ETH Zurich, Switzerland.

Su, J. Z., Dagenais, C., van Zeeland, I., Lougheed, G. D., Benichou, N., Berzin, R., Lafrance, P. S. and Leroux, P. (2014) Fire resistance tests of wall assemblies for use in lower storeys of mid-rise wood buildings. In World Conference on Timber Engineering (WCTE), Quebec City, Canada.

Sultan, M. A. and Lougheed, G. D. (1997) *Fire Resistance of Gypsum Board Wall Assemblies. Construction Technology Update No. 2. Institute for Research in Construction*. National Research Council of Canada. Ottawa, Canada.

Tiso, M. (2018) *The Contribution of Cavity Insulations to the Load-Bearing Capacity of Timber Frame Assemblies Exposed to Fire*. Dissertation, Tallinn University of Technology.

Werther, N. and Matthäus, C. (2020) *Wärmeenergie und Holzfeuchte als Einflussgrößen auf das Abbrandverhalten von Holz. Bautechnik*. Wiley – Ernst & Sohn, Berlin, Germany, pp. 540–548.

Werther, N., O'Neill, J. W., Spellmann, P. M., Abu, A. K., Moss, P. J., Buchanan, A. H. and Winter, S. (2012) Parametric study of modelling structural timber in fire with different software packages. In Structures in Fire: SIF'2012, Proceedings of the 7th International Conference on Structures in Fire, Zurich, Switzerland, pp. 65–74.

Werther, N., Suttner, E., Dumler, P., Kurzer, C. and Winter, S. (2020) Design principles for fire safety detailing in timber structures'. In World Conference on Timber Engineering (WCTE), Santiago, Chile.

Winstone Wallboards (2018) *Gib® Fire Rated Systems: Specification and Installation Manual*. Winstone Wallboards, New Zealand.

Winter, S., Werther, N., Hofmann, V., Kammerer, E. and Rauch, M. (2019) *Standardisierung der brandschutztechnischen Leistungsfähigkeit von Holztafelkonstruktionen mit biogenen Dämmstoffen. Band F 3101*. Fraunhofer IRB Verlag, Stuttgart, Germany.

Chapter 7

Load-bearing timber structures

Alar Just, Anthony Abu,
David Barber, Christian Dagenais,
Michael Klippel and Martin Milner

CONTENTS

DOI: 10.1201/9781003190318-7

SCOPE OF CHAPTER

This chapter gives guidance for design of load-bearing timber members exposed to a standard fire. An overview of the principles needed to predict the effect of charring and heating is presented. Simplified design models around the world are described, including design models from the second generation of Eurocode 5 (the European Charring Model and the Effective Cross-section Method). Calculation examples of timber members are also presented.

7.1 GENERAL

The design objective in the event of a fire is determined by regulatory requirements and the fire safety strategy for the building. Most fire safety strategies are for load-bearing timber structures to resist the design loads for a specified fire exposure time. In this chapter, only the standard fire exposure is considered, according to ISO 834-1 for example. More information on realistic design fires, commonly referred to as parametric fires or natural fires, is presented in Chapter 3.

Design of timber members in a standard fire situation requires an assessment of the reduction of cross-section caused by charring and the effect of heat on strength and stiffness of the residual cross-section. Charring may be influenced by protective claddings and cavity insulation. For engineered timber members, the glueline integrity in fire can also affect the charring scenario and load-bearing capacity. Any charring of structural or non-structural timber members will add to the fuel load in the fire compartment, which is discussed further in Chapter 3.

Unlike steel and concrete, thermal expansion of timber does not need to be taken into account because it is negligible. Timber members can be analysed individually without considering possible thermal actions from other timber members.

Special aspects for fire design of linear members (beams and columns), plate members (mass timber slabs and walls) and light timber frame assemblies are discussed in this chapter.

Fire resistance of structures can be assessed by fire testing or by calculations. Calculation methods should normally give conservative results compared to fire testing. The design parameters for timber and protective materials are needed for calculation methods. If these parameters are unavailable or unknown, fire testing will be the only option for verifying the fire resistance. Assessment by fire testing is described in Section 7.3 and assessment by calculation methods in Section 7.4.

Applicable fire exposures are stated in national building codes. For example, in Canada, exterior walls are to be exposed on the interior side, interior walls on either side and floors are only exposed to fire on the underside. In Europe, Australia and New Zealand, walls delimiting a fire compartment are to be designed for fire exposure from one side, walls located within a fire compartment are to be designed for fire exposure from two sides, floors and roofs are usually to be designed for fire exposure from underneath. In some countries, there are requirements to design floors for fire exposure from above (e.g. attics), see Section 6.5.7. In the UK and other countries, there may be building types and storey heights where the stability of the structure is to be maintained in the event of a fire that is not controlled by firefighters, resulting in design of the load-bearing timber structure to maintain its load-bearing function throughout the fire decay until burnout. This requires a performance-based design which is beyond the scope of this chapter. See Chapter 3.

7.2 ESTIMATION OF STRUCTURAL LOADS

There are different rules around the globe for applying the loads on timber members for design in fire situations.

The European approach is to apply the *accidental load combination* in the fire situation according to EN 1990. This load combination consists of permanent loads without extra safety factors, and live loads with reduction factors. Design live loads for the fire design are usually taken as 20–80% of characteristic load values. The reduction factors are nationally determined parameters for each country and are dependent on the load type (snow load, wind load, imposed load). For ambient design, the characteristic loads are normally increased by a safety factor to get the design loads. For permanent loads, the factor is 1.2 to 1.35 and for live loads the factor is normally 1.5. That can make the difference in design loads 1.5–5 times when comparing ambient and fire designs. Furthermore, in most of the European countries, both the wind load and snow load are not applied at the same time in a fire situation.

A similar approach as Europe is taken in *Australia and New Zealand*. The load combination for fire design does not include snow loads. The combination factor for imposed loads is 0.6 for permanent live loads and 0.4 for all other live loads. Most countries do not require consideration of lateral loads from wind or earthquake during or after a fire, but the New Zealand Building Code (2021) requires that some buildings or parts of buildings be designed to resist a lateral wind load of 0.5 kN/m² during or after fire exposure, in order to provide protection to firefighters inside or outside the building.

In the United States, the model code is the International Building Code (IBC, 2018). For timber engineering, Allowable Stress Design (ASD) is the primary means of structural assessment. Other structural materials, such as steel and concrete, use Load and Resistance Factor Design (LRFD). The National Design Specification for Wood Construction (NDS) (AWC, 2018) allows both ASD and LRFD to be used by designers and also provides conversion factors to allow engineers to swap between methods.

Under the ASD method, timber strength factors are increased for a fire exposure load case. For the LRFD method, the IBC references ASCE/SEI 7-16 "Minimum Design Loads and Associated Criteria for Buildings and Other Structures," which provides the minimum design loads for building structures. Section 2.5 "Load combinations for extraordinary events" provides a load combination for use in the fire case, with factors for reduced dead load (0.9) and live load (0.5). There is also a load case for checking the residual capacity of the structure.

In Canada, the load combinations to be used for fire design of timber structures depend on the chosen methodology. When using the traditional methodology found in Appendix D of the National Building Code of Canada (NBCC) (NRC, 2020), for glue-laminated timber beams and columns, the full factored load combination should be used to determine the load ratio applied to the timber element, e.g. 1.25 dead load+1.5 live load. When using the new fire design methodology of CSA O86:19 applicable to various timber products, the full specified load is to be used, e.g. 1.0 dead+1.0

live. According to the NBCC Structural Commentary, seismic and fire are considered rare events and the principal load factors can therefore be taken as unity. Lastly, when conducting a performance-based fire design using a time–temperature design fire other than that of standard fire, a reduced load combination for rare events can be used. In that specific scenario, a reduction of the live or snow load is allowed. As an example, the load combination for a residential building would be 1.0 dead + (0.5 live or 0.25 snow).

In some other countries, for example Japan, no reduction is allowed in the imposed loads for fire design.

7.3 ASSESSMENT OF FIRE RESISTANCE BY TESTING

Fire resistance of structures can be assessed based on fire tests. Fire tests can be performed at different scales.

Small-scale fire tests are used for research and development. It is an easy and relatively cheap way of determining some material properties for fire design. For example, cone calorimeter tests according to ISO 5660 can be used for indications of start times of charring behind fire protection materials for up to 30 or 40 minutes. There are also small-scale test methods to investigate the glueline integrity at elevated temperatures. In the United States, ASTM E1354, which is similar to ISO 5660, can be used to assess the combustion properties of materials. In Canada, there are currently no standardised small-scale test methods; however, ISO 5660 is commonly used.

Medium-scale fire test (also known as model-scale test or pilot-scale test) in furnaces should provide a fire-exposed area of at least 1 × 1 m², e.g. ISO 834-12. That scale allows assessment of the start time of charring according to EN 13381-7 and charring rates according to prEN 1995-1-2 (2021).

Medium-scale test furnace cannot generally be used to assess fall-off times of fire protection materials nor to determine the fire resistance of structural members.

Full-scale fire testing is usually performed in a furnace with minimum dimensions of 3 × 3 m² for walls and 3 × 4 m² for floors, depending on the applicable standard. In most jurisdictions, fire resistance (load-bearing and separating function) of structures, and the fall-off time of fire protection systems, must be assessed in full-scale furnaces.

In the United States, the ASTM E119 test is used and the minimum floor or wall area is required to be 9.3 m².

In Canada, CAN/ULC S101 is used, and requires that the minimum wall and partition area be at least 9.3 m², with neither dimension less than 2.75 m. Floor area is required to be at least 16.8 m², with neither dimension less than 3.66 m.

Fire tests can be loaded or unloaded. Unloaded tests are suitable to verify charring scenarios or the separating function ability of the wall or floor assembly. Load-bearing capacity can then be calculated based on the remaining cross-section of the member and conservative design methods. Loaded full-scale fire tests must be used for verification of load-bearing capacity directly. The loads that are used for assessment of the loaded fire resistance are applied as a constant load throughout the whole assessed fire duration. Testing standards in some countries require that the tested specimen be subjected to a hose stream test or other mechanical assessments at the completion of the fire test. Some other countries, e.g. Japan, require that test specimens remain loaded for some time after the end of the fire exposure (see Chapter 4).

European standards for full-scale verification testing are shown in Table 7.1. General requirements for fire-resistance tests are given in EN 1363.

In the United States and Canada, ASTM E119 and CAN/ULC S101, respectively, are used for walls, floors, beams and columns both loaded and unloaded. In Australia and New Zealand, the test standard AS 1530.4 is used.

Extrapolation from test results is often required because of the limited size of testing furnaces and the high cost of full-scale fire-resistance tests. In some countries, recognised experts are permitted to make extrapolations based on their expert opinion, but some other countries only allow such statements to be made by the laboratory which carried out the original fire test. In either case, extrapolations require good evidence and a clear understanding to ensure the extrapolation is reliable. For example, real strength of members in the loaded tests or support conditions and real buckling length of wall studs in the loaded wall tests shall be taken into account when extrapolating the results.

There are slight differences between fire-resistance test standards in different regions (Buchanan and Abu, 2017). As an example, North American standards ASTM E119 and CAN/ULC S101 require a hose stream test to be conducted on a replicate specimen exposed to fire for a period equal to one-half of that intended as the fire-resistance period, but not more than 1 hour. After fire exposure, the replicate specimen is to be immediately

Table 7.1 Standards for full-scale fire-resistance tests in Europe

Standard	Loaded fire test	Unloaded fire test
Walls	EN 1365-1	EN 1364-1
Floors	EN 1365-2	EN 1364-2
Beams	EN 1365-3	
Columns	EN 1365-4	
Balconies, walkways	EN 1365-5	
Stairs	EN 1365-6	
Protection applied to timber members	EN 13381-7	

subjected to the impact, erosion and cooling effects of a hose stream. The intent is to evaluate the residual robustness of an assembly after a given fire exposure. Moreover, North American standards typically require the elements and assemblies to be loaded to their maximum capacity so that the results can be applicable to any other structural design load ratios. Proper caution is needed when test results from a given test method are to be used for acceptance under another test method.

In the UK, fire-resistance testing can be carried out under British Standards BS 476-20. The UK accepts fire-resistance tests using the European Standards, although in the regulations this is linked to EN 13501-2. The British Standard BS 476 adopts the ISO 834 Standard fire curve but has a different approach to recording temperatures which may give increased fire-resistance ratings for some product assemblies, e.g. combination of structural timber and linings and insulation, compared to an EN standard test. The UK is undergoing rapid change in the approval and acceptance process for assemblies that have been fire-tested, so readers should check with recent updates in the UK before using fire-resistance test results.

7.4 ASSESSMENT OF FIRE RESISTANCE BY CALCULATION

Fire resistance of structures can be assessed based on calculations using the design models described below for exposure to the standard fire exposure.

There is a safety philosophy that the fire resistance of a member or assembly found by calculations should not be more than the fire resistance obtained in a full-scale test.

Fire design of timber members should take into account two phenomena:

- Reduction of cross-section by charring
- Reduction of strength or stiffness due to the elevated temperatures behind the char layer

Calculations for reduction by charring are considered slightly differently in different regions. The reduction of strength and stiffness behind the char layer can be considered by further reduction of the charred cross-section. The remaining cross-section is considered to have initial strength properties.

Another option is to reduce the average strength and stiffness properties for the whole charred cross-section. This approach ("reduced properties method") is not included in this chapter, since the method is not developed in recent decades and might give unconservative results.

The boundary conditions of a structural system may change during fire exposure, e.g. where a structural member is braced at ambient temperature and the bracing fails in the fire situation, the member must be regarded as unbraced in the structural fire design. Elements that are used for the stabilisation of the building, e.g. wood-based panels or gypsum plaster-board in wall or floor diaphragms, often lose their racking resistance in a fire situation unless they are protected from the fire. This effect on the global structural system must therefore be taken into account. In redundant structural systems, it may be advantageous to allow for premature failure if an alternative load path is possible. See also Chapter 12 on Robustness.

7.5 CHARRING OF TIMBER AND WOOD-BASED PANELS

7.5.1 Charring of unprotected timber

It is well established that timber and wood-based products tend to char at a relatively uniform rate when exposed to a standard fire. Many national and international codes give the charring rates for different wood species and timber products. Charring rates are well-known properties of wood species or wood-based products. Charring rates for standard fire exposure in the Eurocode 5 (prEN 1995-1-2, 2021) are shown in Table 7.2. Charring rates in parametric fires are covered in Chapter 11 of this guideline and in Annex A of prEN 1995-1-2 (2021).

The formation of a char layer will provide effective protection against heat flux, especially for large cross-sections behaving as thermally thick solids.

Table 7.2 Basic charring rates β_0 in Eurocode 5

Material, product	Minimum characteristic density (kg/m³)	β_0 (mm/minute)
Solid timber, glulam and CLT members		
Pine, spruce	290	0.65
LVL members made of softwood		
Pine, spruce	480	0.65
Timber members made of hardwood		
Beech	290	0.70
Ash		0.60
Oak		0.50
Wood-based panels		
Solid wood panelling and cladding	290	0.65
LVL panel	480	0.65
Particleboard, fibreboard	500	0.65
OSB	550	0.9
Plywood	400	1.0

7.5.2 Charring of protected timber

If the structure also incorporates applied protection, e.g. in the form of wood-based panels, gypsum plasterboard, stone wool batt-type insulation or other materials, the start of charring is delayed and, where the protection remains in place after the start of charring, the rate of charring is slowed down in comparison with the charring rate for initially unprotected timber elements.

Since the charring rate immediately after failure of the fire protection – i.e. after the protection has fallen off – is much greater than for initially unprotected timber (due to the combination of high temperature and absence of, or insufficient protection by, the char layer), some of the fire protection effect is lost for some time after falling off. Effective protection provided by the char layer requires a char layer thickness of about 25 mm. When the char layer has grown to that depth, the charring rate reduces to the rate for initially unprotected surfaces. A lasting protection effect is therefore only possible when a char layer thickness of 25 mm can be built up during the phase of increased charring rate immediately after failure of the fire protection.

Applied protection remaining in place provides the most effective fire protection, especially for protection materials with low thermal conductivities at high temperatures, e.g. fire-resistant gypsum plasterboards Type F (Europe) or Type X (North America), or similar proprietary boards, which exhibit longer failure times than standard types of gypsum plasterboards.

7.5.3 One-dimensional charring

Charring of timber members can be one-dimensional charring as expected for large flat surfaces, or two-dimensional charring, including the effects of cross-sectional dimensions and other effects such as corner rounding, as shown in Figure 7.1.

As a basic value, the one-dimensional charring rate β_0 is the charring rate observed for one-dimensional heat transfer under standard fire exposure of an unprotected semi-infinite timber slab without any fissures or gaps. The conditions are similar in a slab of limited thickness, as shown in Figure 7.1a, or in wide timber cross-sections remote from corner rounding effects.

The one-dimensional charring depth $d_{char,0}$ is expressed as

$$d_{char,0} = \beta_0 t \tag{7.1}$$

where t is the time of fire exposure and β_0 is the one-dimensional charring rate perpendicular to the grain for the particular wood species or wood-based product. For end charring in the direction of the grain, these charring rates are typically doubled. The one-dimensional charring rate given for

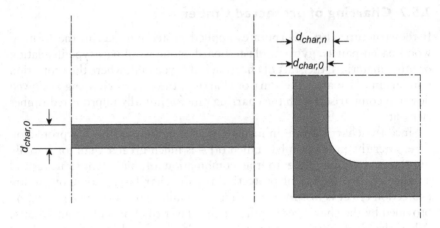

Figure 7.1 One-dimensional (*left*) and two-dimensional (*right*) charring (Östman et al., 2010).

softwoods in Table 7.2 is valid for European and North American species (0.65 mm/minute); it may also be applicable to other species, e.g. radiata pine.

The influence of density within European strength classes for softwoods (solid timber, glulam and LVL) is small and therefore neglected. A similar grouping is also implied in the United States and Canada where a fixed one-dimensional charring rate is used for structural softwoods, regardless of the density.

7.5.4 Two-dimensional charring

Near corners of, for example, rectangular cross-sections, the impinging heat flux is typically two-dimensional, resulting in a rounded shape of the residual cross-section at that location, called the corner rounding effect.

For simplicity, the residual cross-section shown in Figure 7.1b is normally replaced by an equivalent rectangular cross-section, replacing the one-dimensional charring depth and implicitly the corner rounding effect with an equivalent notional charring depth, calculated as

$$d_{\text{char},n} = \beta_n t \tag{7.2}$$

where β_n is the notional charring rate. The notional charring rate should implicitly account for the effects of fissures and corner rounding in a two-dimensional cross-section.

As an alternative to the simplification of using notional charring depths, it is possible to consider a residual cross-section with more realistic linear and rounded boundaries. The calculation of cross-sectional properties will become more complicated, but normally it is not worthwhile to consider it since the difference is negligible.

7.5.5 European Charring Model (ECM)

In Europe, the European Charring Model (ECM) is used for design. According to ECM, charring of timber members is divided into simplified linear charring phases, taking into account the presence and duration of the fire protection system.

Charring rates in the ECM

The European Charring Model consists of the following phases as shown in Figure 7.2.

For unprotected surfaces (Figure 7.2a):

- **Normal charring phase (Phase 1).** Visible exposed timber.

For protected surfaces (Figure 7.2b):

- **Encapsulated phase (Phase 0)** is the phase when no charring occurs.
- **Protected charring phase (Phase 2)** is the phase when charring occurs behind the protection while the protection is still in place.

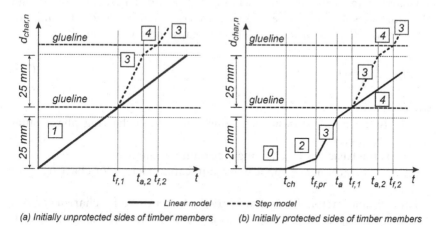

— Linear model   ----- Step model

(a) Initially unprotected sides of timber members   (b) Initially protected sides of timber members

Figure 7.2 Charring phases according to the European Charring Model (prEN 1995-1-2, 2021). (a) Initially unprotected sides of timber members. (b) Initially protected sides of timber members. Key: ①Normal charring phase (Phase 1), $d_{char,n}$ Notional charring depth, ⓪ Encapsulated phase (Phase 0), t Time, ② Protected charring phase (Phase 2), t_a Consolidation time, ③ Post-protected charring phase (Phase 3), t_{ch} Start time of charring, ④ Consolidated charring phase (Phase 4), $t_{f,pr}$ Failure time of the fire protection system.

- **Post-protected charring phase (Phase 3)** is the phase after the failure of the protection before a fully developed char layer has been formed.
- **Consolidated charring phase (Phase 4)** is the phase with a fully developed char layer.

The limits between changes of charring phases are the following times:

Start time of charring t_{ch} is the time at the beginning of fire exposure for initially unprotected timber members, or the time when the surface temperature of an initially protected timber member reaches 300°C (570°F).

Failure time of the fire protection system $t_{f,pr}$ is the time at which the collapse, fall-off or thermal degradation of the fire protection system occurs.

Consolidation time t_a is usually the time when a char layer with 25 mm depth is formed. This char layer gives sufficient protection to reduce the charring rate to that for initially unprotected timber members.

Design for *encapsulation* is intended to provide sufficient protection so that no charring will occur, hence the design objectives will have been achieved before time t_{ch} when charring begins. Timber with only *partial encapsulation* will undergo charring after time t_{ch} in one or more of phases 2, 3 or 4. See Chapter 2.

Charring rates in different charring phases are based on basic design charring rates β_0 (see Table 7.2) that are corrected by factors taking into account the effect of protection, effect of gaps and corner rounding.

$$\beta_n = \prod_{k_i} k_i \cdot \beta_0 \tag{7.3}$$

where

β_n is the notional design charring rate in one charring phase (mm/minute)

β_0 is the basic design charring rate (mm/minute)

Πk_i is the product of applicable modification factors for charring

For example, factors k_2, k_3 and k_4 are the factors used for charring phases 2, 3 and 4, respectively. Factor k_2 is given in Table 7.4. Factor k_3 is usually taken as 2.0 and factor k_4 is usually taken as 1.0. Factor k_n takes into account corner rounding for two-dimensional charring and factor k_g takes into account the effect of gaps. These factors are considered in charring rates given in Tables 7.3 and 7.5. Values for factors for charring rates can be found in Eurocode 5 (prEN 1995-1-2, 2021).

The charring rates given in Eurocode 5 are applicable for any orientation of fire-exposed surfaces and direction of fire exposure, i.e. there is

Table 7.3 Notional design charring rates β_n for linear members made of softwood, in Eurocode 5 (prEN 1995-1-2, 2021)

	Phases 1 and 4 (mm/minute)	Phase 2 (mm/minute)	Phase 3 (mm/minute)
Rectangular cross-sections			
Solid wood	0.8	$k_2 \times 0.8$	1.6
Glulam, LVL	0.7	$k_2 \times 0.7$	1.4
Circular cross-sections			
Solid wood	0.96	$k_2 \times 0.96$	1.92
Glulam, LVL	0.85	$k_2 \times 0.85$	1.69

Table 7.4 Protection factor k_2

Protection	Factor k_2
Gypsum plasterboard Gypsum fibreboard	$1 - \dfrac{h_p}{55}$
Clay plaster	$1 - \dfrac{h_p}{100}$

Table 7.5 Notional design charring rates β_n for plane members (prEN 1995-1-2, 2021)

	Phases 1 and 4 (mm/minute)	Phase 2 (mm/minute)	Phase 3 (mm/minute)
Plane members			
Gaps 0–2 mm	0.65	$k_2 \times 0.65$	1.3
Gaps 2–5 mm	0.78	$k_2 \times 0.78$	1.56
Gaps greater than 5 mm	0.78 Two-dimensional charring	$k_2 \times 0.78$	1.56 Two-dimensional charring

no distinction between vertical or horizontal surfaces. For example, for surfaces on floors with fire exposure from above, the same charring rates apply as for surfaces with fire exposure from below. For fire exposure from above, fall-off of fire-protective claddings is not relevant and need not be considered.

Effect of moisture content is normally not taken into account in the design charring rates.

The notional charring rates for linear and plane members is given in sections below. The notional charring rate β_n for wood-based panels can be calculated as follows:

$$\beta_n = k_h \cdot k_\rho \cdot \beta_0 \tag{7.4}$$

The factor for considering the effect of limited thickness k_h is given as follows:

$$k_h = \begin{cases} 1 & \text{for } h_p \geq 20\,\text{mm} \\ \sqrt{\dfrac{20}{h_p}} & \text{for } h_p < 20\,\text{mm} \end{cases}$$

where h_p is the panel thickness in mm.

The factor for density k_ρ is given as follows:

$$k_\rho = \sqrt{\dfrac{450}{\rho_k}}$$

where ρ_k is the characteristic density at 12% moisture content in kg/m³.

Effect of protection on timber member is considered by applying charring phases and representative times (Figure 7.2b) for each fire protection system – the start time of charring and the fall-off time. For some protections, the fall-off time is similar to the start time of charring and Phase 2 is missing. For example, wood-based boards or non-fire-rated gypsum plasterboards.

Charring of linear structural members

Charring of linear structural members like rectangular timber beams and columns may increase due to possible cracks, gaps and corner rounding. According to the European Charring Model, the charring rates according to Table 7.2 can be used, with some modifications.

For simplicity, the residual cross-section with rounded corners is normally replaced by an equivalent rectangular cross-section, replacing the one-dimensional charring depth and implicitly the corner rounding effect with an equivalent notional charring depth (see Figure 7.1), calculated as

$$d_{\text{char},n} = \sum \beta_{n,i} t_i \tag{7.5}$$

where i is the number of the relevant phase, t_i is the duration of the relevant phase and $\beta_{n,i}$ is the notional charring rate in the relevant charring phase.

For charring behind gypsum plasterboards or clay plaster, the protection factor k_2 can be calculated according to Table 7.4. For other materials, the factor should be determined by testing according to EN 13381-7.

In Table 7.4, h_p is taken as the thickness of the single panel or the total thickness of multiple panels of the same material.

Glueline integrity can affect the charring scenario for charring directions A and C (see Figure 7.3). When the integrity of surface gluing between lamellas is not maintained in fire, the step model of charring is applied. When the glueline integrity is maintained, the linear model is applied. For

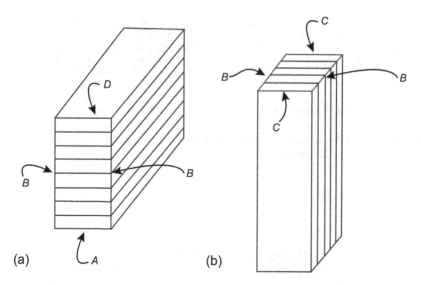

Figure 7.3 Definition of the charring direction for linear timber members (prEN 1995-1-2, 2021). (a) Horizontal member. (b) Vertical member.

directions B and D, the linear model is always applied (see Figure 7.2). See also Section 7.7 for discussion of glueline failure.

Charring of plane members

Charring of plane members like mass timber walls or floor elements is mainly dependent on the gaps between the lamellas on the fire-exposed surface that might increase the charring rate of these lamellas. According to the European Charring Model, the charring rates from Table 7.5 can be used. Gaps less than 2 mm are ignored. For gaps between 2 mm and 5 mm wide, charring is only on the fire-exposed surface, at a charring rate 20% more than for a gap-free surface. For any gaps more than 5 mm wide, two-dimensional charring occurs inside the gap, with no further reduction for corner rounding. Protection factors k_2 for protected elements are given in Table 7.4 and Figure 7.4.

When the glueline integrity of surface gluing between lamellas of the cross-laminated timber (Figure 7.4b) is maintained, and for glued-laminated slab (Figure 7.4a) the linear model of charring is applied. When the glueline integrity is not maintained, the step model should be applied. See Figure 7.2.

7.5.6 European Charring Model for light timber frame assemblies

Light timber frame assemblies (often called 2 × 4 or wood-frame construction in North America and Japan) are normally built up of the timber framing

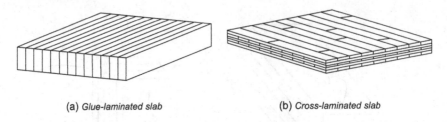

(a) *Glue-laminated slab*   (b) *Cross-laminated slab*

Figure 7.4 Examples of plane members. (a) Glued laminated slab. (b) Cross-laminated timber (Östman et al, 2010).

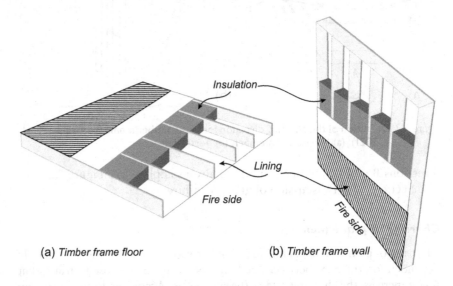

(a) *Timber frame floor*   (b) *Timber frame wall*

Figure 7.5 Examples of light timber frame constructions (Just, 2010). (a) Timber frame floor. (b) Timber frame wall.

members (floor joists or wall studs) and a protective lining attached to each side of the timber frame (the lining may be exterior cladding, sheathing or, in the case of floors, the decking or a sub-floor and additional layers). The cavities between the studs or joists may be empty, or partially or completely filled with insulation, sometimes called batts. Since the light timber frame is sensitive to fire exposure, it must be effectively protected against fire. See also Chapter 1.2.1 (Figure 7.5).

Cavity insulation

Cavity insulation can improve the fire resistance of a light timber frame assembly by protecting the lateral sides of the timber members. In the European design model, the cavity insulation is considered according to the Protection Level (PL). Level PL1 provides the best protection and the charring from the lateral sides of cross-section is prevented. Level PL2 provides some protection and the charring on the lateral sides of cross-section starts after some time from fall-off of the lining. Level PL3 is the weakest one and the charring on the fire-exposed side and lateral sides starts at the same time. See Figure 7.7 and Table 7.6.

There are two types of common mineral wool insulation that behave differently in fire – stone (rock) wool and glass wool. When the fire-protective lining is in place, the protective effect of these two mineral wool insulation types is similar. After the failure of the fire-protective lining, typical stone wool cavity insulation can resist temperatures up to 1,000°C and can provide protection to the sides of the timber members. Typical glass wool will recede after the lining falls off (at around 550°C) by a rate of 30 mm/minute (Just, 2010).

Cellulose and wood fibre insulation can provide effective protection for timber members when the shrinkage of the insulation is avoided. With these types of insulations, there might however be a risk for smouldering.

Normally, the traditional stone wool insulation is classified as PL1 and traditional glass wool is classified as PL2. Cellulose and wood fibre insulations are normally classified as PL2. Foam-based insulations are usually classified as PL3. The PL level can be assessed by a model-scale fire test according to Annex D of Eurocode 5 (prEN 1995-1-2, 2021). There are glass wool or wood-based insulation products in the market that can achieve PL1.

Light timber frame with solid wood members

In light timber frame assemblies with solid wood members, charring of the timber member is considered from the fire-exposed side, and from the lateral sides where relevant. See Figure 7.6 for the definition of sides.

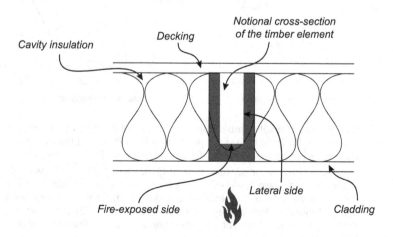

Figure 7.6 Cross-section of light timber frame assembly (Tiso, 2018).

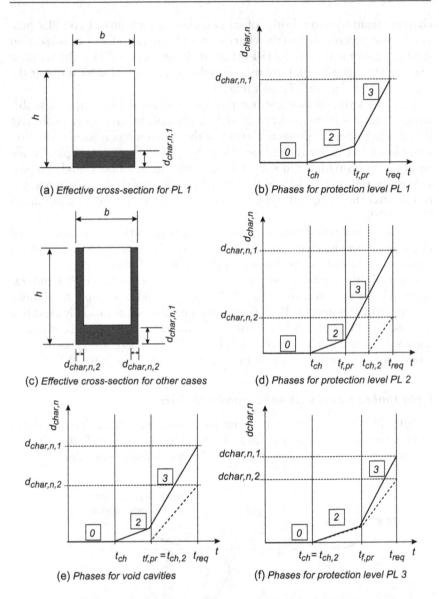

Figure 7.7 Design model for light timber frame (prEN 1995-1-2, 2021). (a) Effective cross-section for PL 1. (b) Phases for protection level PL 1. (c) Effective cross-section for other cases. (d) Phases for protection level PL 2. (e) Phases for void cavities. (f) Phases for protection level PL 3. Key: ⬚0⬚,⬚2⬚,⬚3⬚,⬚4⬚ Phase 0, Phase 2, Phase 3 and Phase 4, ——Charring for the fire-exposed side, - - -Charring for the lateral side, $d_{char,n,1}$ Notional charring depth for the fire-exposed side, $d_{char,n,2}$ Notional charring depth for the lateral side, t Time, t_{ch} Start time of charring, $t_{f,pr}$ Failure time of the protection system, $t_{ch,2}$ Start time of charring for the lateral side.

The design models given below are valid only under the assumption that the insulation remains in place. The over width of insulation batts can be used as a fixing method for walls when the thickness of the insulation is more than 120 mm. For other cases, mechanical fixing (battens, steel wires, gluing) should be used to fix the insulation in place for the post-protection phase of charring.

Charring of the timber members of light timber frame assemblies can be calculated according to charring scenarios shown in Figure 7.7. For members with PL1 cavity insulation, charring shall be considered from the fire-exposed side only. For members with PL2 and PL3 cavity insulations, charring shall be considered from the fire-exposed side (continuous line) and from the lateral sides (dashed line) (Table 7.6).

According to the European Charring Model, the charring rates calculated as given in Table 7.7 can be used.

The factor k_2 is dependent on the protective lining board (see Table 7.4). Factor k_3 is dependent on the cavity insulation and the factors $k_{s,n}$ are dependent on the cross-sectional dimensions of the members. Values for factors can be found in Eurocode 5 (prEN 1995-1-2, 2021).

Light timber frame with I-joists

The charring model for wooden I-joists in Eurocode 5 (prEN 1995-1-2, 2021) is based on the model for light timber frame assemblies with

Table 7.6 Protection level PL for typical insulation materials

Protection level PL	Insulation material	Density
PL 1	Stone (rock) wool	≥ 26 kg/m³
PL 2	Glass wool	≥ 14 kg/m³
	Wood fibre	≥ 50 kg/m³
	Cellulose fibre	≥ 50 kg/m³
PL 3	XPS	–
	PUR	–
	EPS	–

Table 7.7 Notional design charring rates for timber frame assemblies

	Phase 2 (mm/minute)	Phase 3 (mm/minute)
PL 1, PL 2, PL 3 Fire-exposed side	$k_2 k_{s,n,1}\beta_0$	$k_{3,1}k_{s,n,1}\beta_0$
PL 2 Lateral side	$k_2 k_{s,n,2}\beta_0$	$k_{3,2}k_{s,n,2}\beta_0$

rectangular cross-sections by Tiso (2018) and with I-joists by Mäger (2019, 2020). As I-joists are more sensitive to elevated temperatures compared to the rectangular timber cross-sections due to the small cross-sectional area of the flanges and thin webs, the charring calculations are more precise. That also means more complexity in the calculations. The design model takes into account different charring phases, see Figure 7.8. For void cavities, see Figure 7.7e.

The notional charring depth on the fire-exposed side of the flange may be calculated according to Annex I of prEN 1995-1-2 (2021) (Figure 7.9).

The time limit t_a should be calculated as follows:

$$t_a = 1{,}04 \cdot t_{f,\mathrm{pr}} \qquad \text{for cavity insulation PL 1} \tag{7.6}$$

$$t_a = 1{,}01 \cdot t_{f,\mathrm{pr}} \qquad \text{for cavity insulation PL 2} \tag{7.7}$$

The start time of charring for the lateral side of the flange should be calculated as follows:

$$t_{\mathrm{ch},2} = t_{\mathrm{prot,pr}} + t_{\mathrm{prot},i} \qquad \text{for cavity insulation PL 1 and PL 2} \tag{7.8}$$

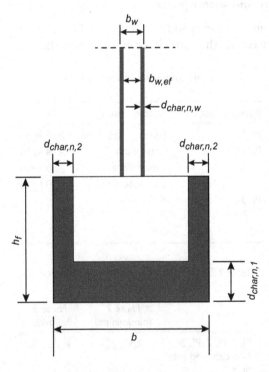

Figure 7.8 Charred cross-section of the flange and web of an I-joist with cavity insulation (prEN 1995-1-2, 2021).

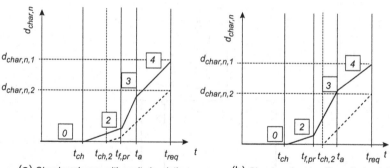

(a) Charring phases with cavity insulations qualified as PL1 and PL2 when charring on the lateral side occurs before the failure of fire protection system; $t_{ch,2} < t_{f,pr}$

(b) Charring phases with cavity insulations qualified as PL1 and PL2 when charring on the lateral side occurs after the failure of fire protection system; $t_{ch,2} > t_{f,pr}$

Figure 7.9 Design model for I-shaped timber members of light timber frame (prEN 1995-1-2, 2021). (a) Charring phases with cavity insulations PL1 and PL2 when charring on the lateral side occurs before the failure of fire protection lining; tch,2<tf,pr. (b) charring phases with cavity insulations PL1 and PL2 when charring on the lateral side occurs after the failure of fire protection lining; $t_{ch,2} >$ t_f,pr. Key: $\boxed{0},\boxed{2},\boxed{3}$ Phase 0, Phase 2 and Phase 3, — — Charring for the fire-exposed side, - - -Charring for the lateral side, h Height of the initial cross-section; b Width of the initial cross-section, $d_{char,n,1}$ Notional charring depth for the fire-exposed side, $d_{char,n,2}$ Notional charring depth for the lateral side, t Time, t_{ch} Start time of charring, $t_{f,pr}$ Failure time of the protection system, $t_{ch,2}$ Start time of charring for the lateral side.

The start time of charring for the web should be calculated as follows:

$$t_{ch,w} = t_{prot,pr} + t_{prot,i} \qquad \text{for cavity insulation PL 1 and PL 2} \tag{7.9}$$

$t_{prot,i}$ is the protection time of the layer(s) i with thickness h_i calculated according to the Separating Function Method in Eurocode 5 (see 6.4.4.1).

$t_{prot,pr}$ is the protection time of the fire protection lining according to the Separating Function Method in Eurocode 5.

The thickness h_i should be calculated as follows:

$$h_i = h_f \qquad \text{for flange, see path A-B on Figure 7.10} \tag{7.10}$$

$$h_i = h_f + 0{,}71 \cdot (b_f - b_w) \qquad \text{for web, see path A-B-C on Figure 7.10} \tag{7.11}$$

Load-bearing capacity of the I-joists should be calculated as for the normal temperature design taking into account design strength in fire and the effective cross-section.

For simplicity, the flexural capacity of the I-joist can be calculated by assessing the tension and compression capacity of the flanges only.

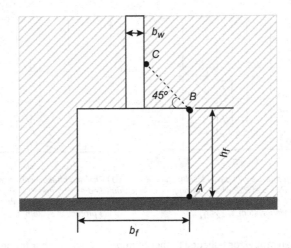

Figure 7.10 Heat paths for start of charring on the lateral side (prEN 1995-1-2, 2021).

7.5.7 Charring model in the United States

In the United States, the National Design Specification for Wood Construction (NDS) gives design procedures for charring of exposed timber members. The procedure consists of determination of non-linear char rate and the char depth. The US model was developed using imperial units and as such the parameters described below use imperial units such as inches rather than millimetres.

The non-linear char rate for this procedure can be estimated from published nominal 1-hour char rate data using the following equation:

$$\beta_t = \beta_n \quad \text{at 1 hour}$$

where

β_t is a non-linear char rate (in/hour$^{0.813}$) adjusted for exposure time t

β_n is a nominal char rate (in/hour), linear char rate based on 1 hour fire exposure

For solid wood, glulam, LVL, parallel strand lumber and cross-laminated timber, the nominal char rate is taken as $\beta_n = 1.5$ in/hour (approximately 0.64 mm/minute).

Charring depth is calculated for each exposed surface as follows (see Table 7.8):

$$a_{\text{char}} = \beta_t t^{0.813} \, (\text{in}) \tag{7.12}$$

For cross-laminated timber manufactured with laminations of equal thickness, the charring depth is calculated as follows:

Table 7.8 Char depths per US model (for $\beta_n = 1.5$ in/hour)

Required fire resistance (minutes)	Char depth a_{char}	
	(in)	(mm)
60	1.5	38.1
90	2.1	53.3
120	2.6	66.0

$$a_{char} = n_{lam} \cdot h_{lam} + \beta_t \left(t - \left(n_{lam} \cdot t_{gl} \right) \right)^{0.813} \text{(in)} \tag{7.13}$$

$$t_{gl} = \left(\frac{h_{lam}}{\beta_t} \right)^{1.23} \tag{7.14}$$

where

t_{gl} is the time for char front to reach glued interface (hour)
h_{lam} is the lamination thickness (in)

$$n_{lam} = \frac{t}{t_{gl}}$$

n_{lam} is the number of laminations charred (rounded to lowest integer)
t is the time of fire exposure (hour)

7.5.8 Charring model in Canada

A charring model was implemented in 2014 into the Canadian design standard CSA O86:19 to calculate the structural fire resistance of timber elements of large dimensions. The method is applicable for elements that are at least 70 mm in residual thickness when subjected to heating on parallel sides (i.e. presumes a thermally thick behaviour with a thermal penetration depth of 35 mm). The charring model is primarily based on the European method, but kept to a more simplistic level. The model is also only applicable to timber elements exposed to a standard fire. Table 7.9 summarises the charring rates, as provided in CSA O86:19.

Table 7.9 Charring rates for structural timber elements per CSA O86:19

Product	β_0 (mm/minute)	β_n (mm/minute)
Timbers and plank decking	0.65	0.80
Glulam	0.65	0.70
Cross-laminated timber	0.65	0.80
Structural composite lumber	0.65	0.70

In the Canadian method, a one-dimensional charring rate β_0 of 0.65 mm/minute is assigned to the softwood and engineered wood products covered in the standard, e.g. timber, glue-laminated timber, structural composite lumber and cross-laminated timber. This rate is to be used for plane elements such as wall and floor slabs, or when the effect of corner rounding is explicitly considered.

When rectangular elements are used or when the effect of corner rounding is not considered, a notional charring rate β_n is given. The notional charring rate is also to be used for cross-laminated timber (CLT) when the char layer is expected to surpass the first glueline.

For structural verification, the initial cross-section of a timber element is to be reduced by the charred layer and a zero-strength layer. The charred layer is taken as the product of the appropriate charring rate and the time, i.e. typically the fire-resistance rating. The zero-strength layer is taken as 7 mm for fire exposure greater than 20 minutes (varies linearly from 0 to 7 mm between 0 and 20 minutes).

As opposed to the European Charring Model, the Canadian model does not consider the various charring phases when fire protection membranes are used to protect the timber. When using Type X gypsum boards directly attached to the timber element, or through wood furring or resilient channels, the calculated fire-resistance time of an initially unprotected timber element can simply be increased by the following conservative times:

a) 15 minutes for one layer of 12.7 mm Type X gypsum board
b) 30 minutes for one layer of 15.9 mm Type X gypsum board
c) 60 minutes for two layers of 15.9 mm Type X gypsum boards
d) 60 minutes for two layers of 12.7 mm Type X directly attached to CLT

There is currently no charring model for light timber frames in Canada. Fire resistance of light timber frame assemblies is typically assessed by full-scale fire-resistance testing or by using the Component Additive Method, as detailed in the National Building Code of Canada. In this latter method, the fire resistance of a given assembly is determined from the sum of the various time contributions of each respective element, such as the wood joist or stud, insulation, resilient channels and Type X gypsum board protection. Work is ongoing to expand the scope of application of the charring method in CSA O86:19 for all types of wood elements and systems, including light timber frame assemblies. There is also ongoing work to revise the times assigned to Type X gypsum boards in a future edition of CSA O86 (Dagenais and Ranger, 2021).

7.5.9 Charring model in Australia and New Zealand

The charring model in Australia and New Zealand is specified in AS/NZS 1720.4. This standard gives only one charring rate which is applied to both one-dimensional or two-dimensional cross-sections, with no allowance for rounding of corners. There is no design model for light timber frames.

Because of the large number of high-density hardwoods in Australia, this standard gives the following equation for charring rate as a function of wood density, and a table of charring rates for various species, derived from the equation:

$$c = 0.4 + \left(\frac{280}{\delta}\right)^2 \qquad\qquad (7.15)$$

where

c = notional charring rate, in millimetres per minute

δ = timber density at a moisture content of 12%, in kg/m^3. For engineered wood products, this shall be based on the primary timber species not including any adhesive.

The standard specifies that for New Zealand timbers, the design density of New Zealand grown radiata pine is 550 kg/m^3, which gives the charring rate of 0.65 mm/minute as shown in Table 7.10. This charring rate applies to laminated veneer lumber (LVL) and other engineered wood products made from radiata pine, even if they have a higher density due to the manufacturing process.

Figure 7.11 compares the basic charring rate β_0 from Eurocode 5 with the density-related charring rate from AS/NZS 1720.4. The Eurocode charring rate is 0.65 mm/minute for density up to 450 kg/m^3, and 0.5 mm/minute for all higher density timber. This shows the Eurocode charring rate is not conservative for low-density timber species but conservative for high-density hardwood timber species. The basic charring rate of 0.65 mm/minute is the point that the AS/NZS 1720.4 and Eurocode 5 methods intersect.

Table 7.10 Charring rates for wood species from AS/NZS 1720.4

Timber species	Notional charring rate c (mm/minute)
Blackbutt	0.50
Cypress	0.56
Douglas fir (North America and New Zealand)	0.65
European spruce	0.65
Gum, spotted	0.46
Ironbark, grey	0.46
Ironbark, red	0.47
Jarrah	0.52
Merbau (Kwila)	0.51
Radiata pine (Australia and New Zealand)	0.65
Victorian ash and Tasmanian oak	0.59

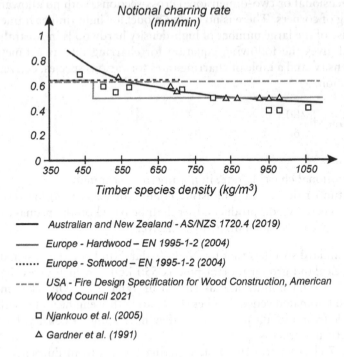

Figure 7.11 Charring rates related to wood density.

7.6 MATERIALS FOR PROTECTION OF TIMBER STRUCTURES

In the design and optimisation of protected timber members, the following points are important with respect to maximising fire resistance. These points are consistent with Harmathy's ten rules of fire endurance, illustrated in Figure 7.12:

- There is a hierarchy of contribution to fire resistance of various layers of the assembly.
- The greatest contribution to fire resistance is obtained from the layer on the fire-exposed side with respect to both insulation and failure (fall-off) of the protective cladding.
- In general, it is difficult to compensate for poor fire protection performance of the first layer by improved fire protection performance of the following layers.

Detailing is of great importance for effectiveness of fire protection. The rules for length and spacing of fasteners, and filling gaps in the joints, shall

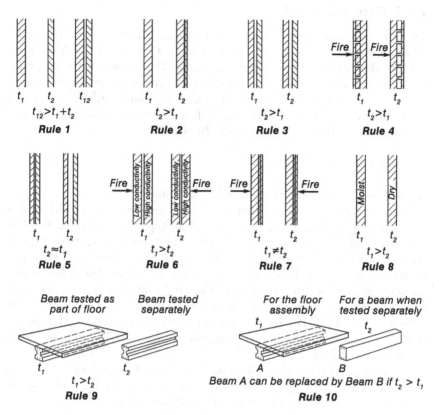

Figure 7.12 Harmathy's ten rules of fire endurance (Harmathy, 1965).

always be followed to secure the protective effect provided by fire protection lining materials.

The following guidance is given for how to consider the effect of different protection materials used for timber members.

Encapsulation criteria as K-classes according to EN 13501-2 and as tested per CAN/ULC S146 can be considered as the start time of charring t_{ch}. For more on encapsulation, see Chapters 2 and 6.

7.6.1 Wood-based protection materials

Sacrificial wood-based panels can protect structural timber members by delaying the onset of charring. Charring rates for wood-based panels are given in Table 7.2. For structural timber members protected with wood-based panels, the start time of charring is the time that the wood-based panel falls off, or when the internal surface temperature reaches 300°C.

$$t_{ch} = t_{f,pr} \tag{7.16}$$

7.6.2 Gypsum boards

There are different types of gypsum boards available around the globe. Gypsum boards consist around 20% of water that provides a delay in the start of charring until the water evaporates. That causes a certain period when temperature behind the protective board stays at approximately 100°C in case of fire. When using European gypsum boards, the start time of charring behind the gypsum board (all types) can be estimated based on thickness of the board (EN 1995-1-2, 2004).

$$t_{ch} = 2,8h_p - 14 \tag{7.17}$$

where h_p is the thickness of gypsum board in mm.

Some gypsum boards will remain in place long enough for charring to occur behind the board. Such boards include Type F according to EN 520, gypsum fibreboards according to EN 15283 or Type X meeting the requirements of ASTM C1396 or CAN/CSA-A.82.27. Several proprietary boards in Australia, New Zealand and other countries have similar or better fire-resistant properties (Buchanan and Abu, 2017). These boards will remain in place long enough to activate Phase 2 of the European Charring Model, shown in Figure 7.2. The failure time of protective linings on walls is longer than the failure time of the same protective linings on ceilings where gravity can assist the falling off.

For most other non-fire-rated boards (e.g. Type A according to EN 520), the start time of charring and failure times are assumed to be similar, so the charring behaviour jumps from Phase 1 to Phase 3 in Figure 7.2:

$$t_{ch} = t_{f,pr} \tag{7.18}$$

In Europe, evaluating more than 450 full-scale fire test reports, failure times for gypsum plasterboard linings of Type F as protection on light timber frame assemblies were collated in a database (Just et al., 2010). The tested constructions were either light timber frame assemblies, the great majority with solid timber members and some with I-joists, or in a few cases light-weight steel members. The studs or joists were placed a maximum of 600 mm on centres. Expressions based on 20% fractile values of the fire test results are given in Table 5.4 of prEN 1995-1-2, 2021. These failure times are given for light timber frame assemblies with cavity insulation, these being shorter times than for uninsulated cavities.

The failure times of gypsum plasterboards attached to large mass timber members such as glulam beams and columns or mass timber panels such as CLT may be considerably greater, especially when edge distances of screws are greater than those in light timber frame construction. In Europe, the generic failure times for gypsum linings on mass timber are taken 10% greater than the failure times on light timber frame assemblies. Table 7.11

Table 7.11 Start time of charring and protection times for gypsum plasterboards *for walls*

| Panels | Thickness of the fire protection lining (mm) | | Layers backed by | Start of charring (minutes) | Failure time (minutes) |
	Layer 1	Layer 2			
Europe (prEN 1995-1-2, 2021)					
Gypsum	12.5	–	Insulation	17	20
plasterboard	12.5	–	Panel	22	22
type A	12.5	12.5	Insulation	26	41
	12.5	12.5	Panel	36	45
Gypsum	12.5	–	Insulation	17	32
plasterboard	12.5	–	Panel	24	35
type F	15	–	Insulation	22	44
	15	–	Panel	30	48
	18	–	Insulation	29	58
	18	–	Panel	37	63
	12.5	12.5	Insulation	39	60
	12.5	12.5	Panel	49	66
	15	15	Insulation	50	82
	15	15	Panel	60	90
	18	18	Insulation	63	108
	18	18	Panel	75	119
Gypsum	12.5	12.5	Insulation	39	60
plasterboard	12.5	12.5	Panel	49	66
type F + A (type F	15	12.5	Insulation	45	71
is layer 1)	15	12.5	Panel	55	78

shows the time to start of charring and the failure time for falling off of gypsum linings on walls, from the latest draft of Eurocode 5 (prEN 1995-1-2, 2021). Table 7.12 shows the same information for ceiling linings on the underside of floors.

Since the values given in Tables 7.11 and 7.12 are conservative, especially with regard to failure times, $t_{f,pr}$, producers may wish to determine values for their products and applications to be used by designers. To determine the start time of charring, testing according to the European test Standard EN 13381-7 can be performed.

The National Research Council of Canada studied the fall-off performance of gypsum boards in standard fire tests (Roy-Poirier & Sultan, 2007; Sultan, 2010). From a review of numerous standard fire-resistance tests of lightweight assemblies protected with single and double layers of Type X gypsum boards, temperature criteria were derived to predict the time to fall-off for wall and floor assemblies. It was found that the fall-off temperature for wall assemblies was 100°C higher than that of floors with insulation in the cavity and 150°C higher than that of floors when no insulation was placed in the cavity. The higher fall-off temperature was

Table 7.12 Start time of charring and protection times for gypsum plasterboards *for floors*

Panels	Thickness of the fire protection system (mm)		Layers backed by	Start of charring (minutes)	Failure time (minutes)
	Layer 1	Layer 2			
Europe (prEN 1995-1-2, 2021)					
Gypsum	12.5	–	Insulation	17	17
plasterboard	12.5	–	Panel	19	19
type A	12.5	12.5	Insulation	26	29
	12.5	12.5	Panel	32	32
Gypsum	12.5	–	Insulation	17	25
plasterboard	12.5	–	Panel	24	28
type F	15	–	Insulation	22	28
	15	–	Panel	30	31
	18	–	Insulation	28	32
	18	–	Panel	35	35
	12.5	12.5	Insulation	39	52
	12.5	12.5	Panel	49	57
	15	15	Insulation	50	60
	15	15	Panel	60	66
	18	18	Insulation	63	69
	18	18	Panel	75	76
Gypsum	12.5	12.5	Insulation	39	52
plasterboard	12.5	12.5	Panel	49	58
type F + A (type F	15	12.5	Insulation	45	56
is layer 1)	15	12.5	Panel	55	62

Table 7.13 Summary of fall-off temperatures for light-framed assemblies (Sultan, 2010)

Assembly characteristics		Fall-off temperature		
			Double-layer assembly	
Insulation	Screw spacing (mm)	Single-layer assembly	Face layer	Base layer
No Insulation	406	460±20°C	620±50°C	430±90°C
	610	–	510±50°C	330±40°C
Insulation against gypsum board base layer	406	680±50°C	680±40°C	620±40°C
	610	–	640±40°C	480±40°C
Spray-applied insulation	406	670±40°C	–	–
	610	–	600±40 °C	380±30°C

attributed to the reduced effect of gravity on the gypsum boards, allowing them to remain in place for a longer duration when used as wall protection. Table 7.13 summarises the temperature criteria to evaluate the fall-off time of gypsum boards.

A similar review was recently performed by FPInnovations (Dagenais & Ranger 2021) to revise the times assigned to Type X gypsum boards shown in Section 7.5.8 when used to protect mass timber elements. Additional time of 30 minutes per layer was found appropriate when using single, double and triple layers of 12.7 mm (½") Type X gypsum boards. This time increases to 40 minutes per layer for single, double and triple layers of 15.9 mm (⅝") Type X gypsum boards. These times are to be added to the calculated fire resistance of unprotected mass timber elements.

7.6.3 Clay plasters

Clay and lime plaster have extensively been used in historic timber buildings to cover the walls and ceilings. In the past, plaster was the primary protection for timber structures against fire exposure. Today the combinations of timber and other ecological materials like clay plaster offer a contemporary alternative to conventional building solutions.

Clay plasters and clay boards are investigated by Liblik et al. (2020). The start time of charring behind clay plasters and clay boards can be calculated as

$$t_{ch} = 1,1h_p - 6 \tag{7.19}$$

where h_p = thickness of plaster, mm.

The equation is limited to traditional clay plaster within a density range of 1,610–1,800 kg/m³. Further, clay plaster should meet the requirements stated in product standard DIN 18947 to guarantee its mechanical strength and quality. The application technique is crucial, Standard EN 13914-2 and manufacturer's guidance should be followed.

7.6.4 Cement-based boards

No generic information is available in Europe for design models of cement-based boards, which can only be assessed by testing.

The recent changes in the United States and in Canada to timber buildings taller than six storeys triggered the need to enhance the level of fire safety in timber buildings by providing additional passive protection, such as encapsulation materials to protect the timber elements from fire exposure. In North America, the protection materials are required to be of non-combustible material using, for example, Type X gypsum board or 38 mm concrete topping placed on top of mass timber floors. However, concrete topping can slow construction timelines as it requires installation of formwork, coordination with concrete deliveries, finishing and curing. Prefabricated elements that can be installed faster, such as cement boards, are therefore an interesting alternative if they can provide the level of encapsulation required by the applicable building codes. In an attempt to evaluate the encapsulation performance of cement boards, an adapted standard fire test was conducted by Ranger et al. (2020) following the test conditions of CAN/ULC S146. A ceiling assembly was encapsulated using two layers of 15.9 mm cement boards and exposed to the standard fire of CAN/ULC S101. The double layer of cement boards achieved an encapsulation time of 39 minutes, which is insufficient for meeting the minimum 50 minutes requirement in the National Building Code of Canada.

7.6.5 Intumescent coatings

Some intumescent coatings can delay the start time of charring. Testing is required to determine the time to start of charring, and the charring rate under the intumescent coating. After the fall-off the double charring rate should be used as shown in Phase 3 of Figure 7.2. Design parameters for intumescent coatings can be assessed by testing according to EN 13381-7.

In North America, Australia and New Zealand, intumescent coatings are typically developed and used to reduce surface flammability (flame spread) and not to increase the fire resistance or delay of charring of timber elements. Similarly to Europe, the effect of intumescent coatings on the fire-resistance or delay of charring should be evaluated through relevant and appropriate testing.

It should be noted that when using coatings to protect timber elements, their long-term durability during service conditions should be considered, including water, drying, UV light, etc. More on intumescent coating is provided in Chapter 5.

7.7 EFFECT OF GLUELINE FAILURE

Adhesives are used for surface gluing and for finger joints of engineered wood products. Glued members can behave differently in fire compared to solid wood members of the same size with no gluelines (See also 2.10.5 and Chapter 3).

In Europe, for the face bonding of load-bearing timber elements, there are three requirement standards, EN 15425 for 1-component PUR adhesives, EN 301 for MUF, MF and PRF adhesives and EN 16254 for EPI adhesives. Within each adhesive group, there are large differences regarding their formulation and it is therefore impossible to generally assume that all products in an adhesive group can maintain glueline integrity in fire. A change of the mixture (e.g. thermoplastic parts and cured parts) in the adhesive product to improve certain characteristics (e.g. curing times) may counteract the fire performance. For simplicity, if non-fire-resistant adhesives are used, the charring temperature (taken at the 300°C isotherm) is typically understood to be the failure temperature of the glueline. This failure of the glueline can result in debonding and consequently fall-off of the charred lamella and faster charring of the next lamella.

Since new adhesives are being continuously developed and new engineered wood products are introduced on the market, there is a demand to assess adhesives or glued products with respect to fire performance.

In North America, adhesives used in the manufacturing of cross-laminated timber and glue-laminated timber are required to pass stringent fire tests according to ANSI/APA PRG 320 to demonstrate their integrity when the char layer approaches the glueline and to verify that the charring rate is not influenced by the glueline failure during fire exposure. For the engineered wood products described in Chapter 1, various standard test methods are mandatory for all adhesives to demonstrate their performance in elevated temperature or fire conditions, such as ASTM D2559, ASTM D7247, ASTM D7374, ASTM D7470 and CSA O177.

In Europe, a test method (glueline integrity in fire, GLIF) allows for the comparison of a cross-laminated timber product to solid timber. In the GLIF test, the performance of timber with a glueline is compared to the maximum possible mass loss of solid timber where no glueline is present.

There are also small-scale assessment methods available such as the cone heater method for finger joints loaded in tension (Mäger et al, 2021), combined cone heater and shear test methods (Sterley and Norén, 2018) or tension and shear tests at elevated temperatures according to EN 17224.

7.8 CALCULATION METHODS FOR STANDARD FIRE EXPOSURE

This section describes the calculation methods for structural fire resistance in different countries.

7.8.1 Effective cross-section method in Eurocode 5

Strength and stiffness

Strength and stiffness properties are considered differently in the fire design procedure compared to ambient design. In Europe, the design values of mechanical strength $f_{d,fi}$ and stiffness $E_{d,fi}$ properties for the fire situation are defined as follows:

$$f_{d,fi} = k_\Theta \cdot k_{fi} \cdot f_k / \gamma_{M,fi} \tag{7.20}$$

$$E_{d,fi} = k_\Theta \cdot k_{fi} \cdot E_k / \gamma_{M,fi} \tag{7.21}$$

where

$f_{d,fi}$, $E_{d,fi}$ is the fire design value of a strength or stiffness property

f_k, E_k is the characteristic value of a strength or stiffness property for normal temperature design according to EN 1995-1-1

k_Θ is the temperature-dependent reduction factor for a strength or stiffness property

k_{fi} is the modification factor for a strength or stiffness property for the fire situation

$\gamma_{M,fi}$ is the partial safety factor for the relevant mechanical material property for the fire situation

The characteristic strength of the timber member in fire is considered as the 20% fractile value instead of the 5% fractile value that is used for ambient design. This effect is taken into account with the relevant factor k_{fi} as given in Table 7.14.

For the effective cross-section method, the factor $k_\Theta = 1$. For the advanced calculations, see Figure 7.17. The partial safety factor for fire $\gamma_{M,fi} = 1$.

Effective cross-section

Timber members exposed to fire exhibit charring unless they are protected during the relevant time of fire exposure. For calculation of the resistance of timber members, the original cross-section is reduced by the effective charring depth consisting of the notional charring depth $d_{char,n}$ and the zero-strength layer depth d_0. The latter is an effective layer that compensates for the loss of strength and stiffness. The resulting cross-section is called the effective cross-section which has no reduction of strength and stiffness.

The effective charring depth is calculated as follows, shown in Figure 7.13:

$$d_{ef} = d_{char,n} + d_0 \tag{7.22}$$

Design of linear and plane timber members

Charring is calculated according to the European Charring Model and Eurocode 5.

Table 7.14 Modification factor for strength and stiffness property for the fire situation in Eurocode 5

	k_{fi}
Solid timber	1.25
Glulam and cross-laminated timber (CLT)	1.15
Wood-based panels	1.15
LVL	1.10

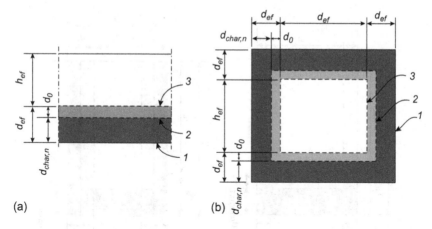

Figure 7.13 Determination of effective cross-section for timber members (prEN 1995-1-2, 2021). (a) One-dimensional charring. (b) Two-dimensional charring. Key: 1 Fire-exposed side(s) or fire-exposed perimeter, 2 Border-line of the residual cross-section, 3 Border-line of the effective cross-section, d_{ef} Effective charring depth, $d_{char,n}$ Notional charring depth, d_0 Zero-strength layer depth, b_{ef} Width of the effective cross-section, h_{ef} Height of the effective cross-section.

The values of zero-strength layer depth d_0 for the design of linear timber members are taken as follows:

- $d_0 = 10$ mm for members subjected predominantly to tension or bending
- $d_0 = 14$ mm for members subjected predominantly to compression.

The values of zero-strength layer depth d_0 for the design of plane timber members made of glulam or LVL are taken as follows:

- $d_0 = 8 + \dfrac{h}{55}$ for members with fire exposure on the tension side (mm)

- $d_0 = 9 + \dfrac{h}{20}$ for members with fire exposure on the compression side (mm)

h is the depth of the initial cross-section of the plane timber member (mm).

The values of zero-strength layer depth d_0 for the design of plane timber members made of cross-laminated timber can be taken from Eurocode 5 (prEN 1995-1-2, 2021).

Design of light timber frame floor and wall assemblies

Effective cross-section of the timber members of light timber frame assemblies should be calculated according to Figure 7.14.

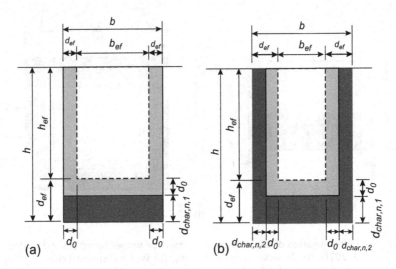

Figure 7.14 Design model for timber frame assemblies (prEN 1995-1-2, 2021). (a) Effective cross-section for PL 1. (b) Effective cross-section for other cases. Key: h Height of the initial cross-section; b Width of the initial cross-section, h_{ef} Height of the effective cross-section, b_{ef} Width of the effective cross-section, d_{ef} Effective charring depth, d_0 Zero-strength layer depth, $d_{char,n,1}$ Notional charring depth for the fire-exposed side, $d_{char,n,2}$ Notional charring depth for the lateral side.

Zero-strength layers for light timber frame assemblies depend on the protection level of cavity insulation. The zero-strength layer depths are normally in the range of 7–20 mm (Tiso, 2019). The exact zero-strength layer depths can be found in Section 7.2.4 of prEN 1995-1-2, 2021.

The cross-section of light timber floors with I-joists will be reduced by charring and the zero-strength layer that allows for the strength loss in the heated timber, see Figure 7.15. Zero-strength layer depths can be found in Annex I of prEN 1995-1-2, 2021.

Adhesives can be sensitive to elevated temperatures. In the finger joints of tension flanges, the glueline integrity can affect the load-bearing capacity. Therefore, the zero-strength layer depths are different for different classes of finger joints that take the glueline integrity into account. These can be assessed by testing according to Annex B of prEN 1995-1-2 (2021).

7.8.2 Effective cross-section method in Australia and New Zealand

The calculation method in Australia and New Zealand is specified in AS/NZS 1720.4. This is almost the same as the Eurocode method, except that the one-dimensional charring rate can be used for both linear members and

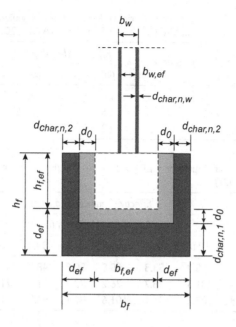

Figure 7.15 Effective cross-section of the flange and web with cavity insulation (prEN 1995-1-2, 2021).

flat panels, and the zero-strength layer thickness is kept constant at 7.0 mm. There is no increase in strength or stiffness properties for fire design.

7.8.3 Effective cross-section method in the United States

Unprotected members

For structural calculations in the United States, section properties are calculated using standard equations for area, section modulus and moment of inertia using the reduced cross-sectional dimensions. The dimensions are reduced by the effective char depth for each surface exposed to fire. A 20% increase is added to the calculated char depth (a_{char}) to consider the reduction of strength and stiffness of the heated zone and the effect of corner rounding. The effective char depth is calculated as follows:

$$a_{eff} = 1.2 \cdot a_{char} \tag{7.22}$$

For sawn lumber, glulam made of softwood, LVL, parallel strand lumber and laminated strand lumber, the char depth and effective char depth for each exposed surface are shown in Table 7.15 based on a nominal char rate $\beta_n = 1.5$ in/hour.

Table 7.15 Effective char depth for solid wood, glulam, LVL, LSL (for $\beta_n = 1.5$ in/hour) (AWC, 2021)

Required fire resistance (minutes)	Effective char depth a_{eff}	
	(in)	(mm)
60	1.8	45.7
90	2.5	63.5
120	3.2	81.3

Table 7.16 Effective char depth for CLT (with $\beta_n = 1.5$ in/hour ≈ 0.64 mm/minute) (AWC, 2021)

Required fire resistance (minutes)	Effective char depth a_{eff} (mm)								
	Lamination thickness (mm)								
	15.9	19.0	22.2	25.4	31.8	34.9	38.1	44.4	50.8
60	55.9	55.9	53.3	50.8	50.8	48.3	45.7	45.7	45.7
90	86.4	81.3	78.7	76.2	73.7	71.1	71.1	71.1	66.0
120	111.8	109.2	104.1	101.6	99.1	96.5	91.4	91.4	91.4

For CLT manufactured with laminations of equal thickness, the effective char depth for each exposed surface is shown in Table 7.16 using a nominal char rate of $\beta_n = 1.5$ in/hour (≈ 0.64 mm/minute). The US charring model for CLT accounts for glueline failure, which explains the higher char depths when compared to those of Table 7.15.

For sawn lumber, glulam made of softwood, LVL, parallel strand lumber and laminated strand lumber and cross-laminated timber, the average member strength can be approximated by multiplying reference design values by adjustment factors specified in Table 7.17. The values of these factors are given in the National Design Specification (AWC, 2018).

The strength values for bending, tension, compression and shear shall be adjusted prior to calculating the residual resistances using the equations given in the NDS.

The induced stress calculated using reduced cross-section properties determined using a_{eff} shall not exceed the member strength.

Protected members

Technical Report 10 (AWC, 2021) provides contribution times for gypsum board protecting timber elements. Similar to the Canadian method, the total fire-resistance time of a protected timber element is taken as the sum of the initially unprotected timber element and the time contribution of the gypsum board. As an example, a single layer of 12.7 mm (½ in) and 15.9 mm (⅝ in) Type X gypsum board directly attached to a timber beam or CLT can increase the fire-resistance time by 30 and 40 minutes, respectively.

Table 7.17 Adjustment factors for fire design (AWC, 2021)

			Values for allowable stress design (ASD)				
		Design stress to member strength factor	Size factor[1]	Volume factor[1]	Flat use factor[1]	Beam stability factor[2]	Column stability factor[2]
Bending strength	F_b ×	2.85	C_F	C_V	C_{fu}	C_L	
Beam buckling strength	F_{bc} ×	2.03					
Tensile strength	F_t ×	2.85	C_F				
Compressive strength	F_c ×	2.58	C_F				C_F
Column buckling strength	F_{cc} ×	2.03					

[1] Shall be based on initial cross-section.
[2] Shall be based on reduced cross-section.
For specific products, the adjustment factors may be different.

Table 7.18 Modification factor for strength property for the fire situation in CSA O86:19

Product	K_{fi}
Timbers and plank decking	1.5
Glue-laminated timber	1.35
Cross-laminated timber	
• V1–V2 stress grade	1.5
• E1–E2–E3 stress grade	1.25
Structural composite lumber	1.25

7.8.4 Effective cross-section method in Canada

In Canada, similar adjustments as in Europe are made to the strength property values. The modification factor for fire resistance, as presented in Table 7.18, is intended to convert the specified strength to mean strength values. Furthermore, the resistances are to be calculated using a short-term load duration (K_D) and a resistance factor (ϕ) of 1.0.

The calculation method in Canada is specified in CSA O86:19. This is almost the same as the Eurocode method, except that the one-dimensional charring rate can be used for both linear members and flat panels, and the zero-strength layer thickness is kept constant at 7.0 mm for exposure of at least 20 minutes (varies between 0 and 7 mm and between 0 and 20 minutes).

Similarly to the US method, the total fire-resistance time for a protected timber element is taken as the sum of the initially unprotected fire resistance and the time contribution of the gypsum board. The time contributions for Type X gypsum boards are almost the same in the Canadian and US design methods.

7.9 ADVANCED CALCULATION METHODS

Advanced calculation methods are most likely to be used with performance-based design (see Chapter 11).

For determination of the mechanical resistance of structural timber members, an advanced calculation method, e.g. using finite element modelling of fire-exposed structural timber members, comprises several steps:

1. Determination of the time–temperature curve of fire exposure
2. Determination of temperatures in the timber member, including the charring depth
3. Determination of the resistance of cross-sections using the temperature field in the timber member and the temperature-dependent reduction of strength and stiffness at each location of the cross-section
4. Determination of the structural resistance of the member (beam, column, frame, etc.)

The problem is that the data from various sources may vary considerably. Since available commercial software for heat transfer calculations does not explicitly take into account the mass transfer of water, steam and gases, these effects must be accounted for by using effective conductivity values rather than real ones (Källsner & König, 2000; König, 2006). This also applies to the formation of cracks, e.g. in the char layer or gypsum plasterboards, causing increasing heat flux which is taken into account by using increased conductivity values. For the char layer, these effects have not been considered in some sources, which give considerably lower conductivity values than EN 1995-1-2 (2004).

Since the protection provided by linings and insulation is often important for the performance of structural timber members, the software should be capable of taking into account sudden failure (fall-off) of applied protection. Examples of commercial software including this option are SAFIR, ABAQUS and ANSYS. Werther et al. (2012) examined modelling with these programs. They considered the effects that various model parameters (thermal and structural) may have on the physical interpretation of experimental data compared to the accuracy of numerical solutions. Several in-house finite element models have been developed over the years to perform a two-way coupling between heat transfer and structural analysis (Chen et al, 2020). With proper thermal properties and strength reduction factors, as well as validation and verification against test data, these models can be used to evaluate the fire performance of a broad range of timber products, assemblies and connections. See Chapter 8 for more information on fire resistance of timber connections.

For timber members, it is sufficient to assume ideal elastic–plastic behaviour for compression and purely elastic behaviour for tension, as shown in Figure 7.16. The behaviour at 20°C can be modified for other temperatures with multiplication by the temperature reduction factor k_Θ.

For advanced design, for example, using thermo-mechanical simulations and finite element analysis, the strength and stiffness can be reduced according to the effective values from Eurocode 5, presented in Figure 7.17.

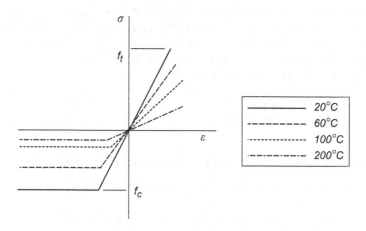

Figure 7.16 Temperature-dependent strain–stress relationships.

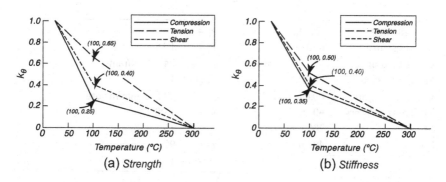

Figure 7.17 Temperature-dependent reduction factor k_Θ for strength and stiffness parallel to grain (prEN 1995-1-2, 2021). (a) Strength. (b) Stiffness. Key: — Compression, – –Tension, ….Shear, T Temperature, in °C, k_θ temperature-dependent reduction factor.

In the effective cross-section method, the temperature reduction factor k_Θ is not used. The heating effect will be replaced with a further reduction of cross-section by a fictive zero-strength layer.

The value of $\gamma_{M,fi}$ is taken as 1.0 unless the National Annex gives a different value for use in a country.

7.10 WORKED EXAMPLES

Calculations of glulam beam protected with fire-rated gypsum plasterboard 15 mm

Calculate the fire resistance of a glulam beam in an office floor. The cross-section of the beam is 200 ´ 400 mm. The glulam beam span is 8 m. Distance between the beams is 2 m. Required fire resistance is R90. The beam is exposed to fire on three sides, which are initially covered by gypsum plasterboard, Type F (or Type X) with a thickness of 15 mm.

Characteristic loads on the floor:

- Self-weight 1.0 kN/m²
- Imposed load 3.0 kN/m²

7.10.1 Effective Cross-Section method (Europe)

Load combination for fire situation according to EN 1990:

$$p_{d,fi} = 1.0g_k + 0.5q_k = 1.0 * 1.0 + 0.5 * 3.0 = 2.5\,kN/m^2$$

Linear load on the beam: $P_{d,fi} = 2.5 * 2 = 5.0$ kN/m
Maximum bending moment: $M_{max} = 5 * 8^2/8 = 40$ kNm

Strength class	GL24h	$k_\Theta =$	1	
Fire resistance	R90	$k_{fi} =$	1.15	
Cross-section	200 ´ 400 mm	$f_{c,0,k} =$	24	N/mm²
Protection system	GtF15	$\gamma_{M,fi} =$	1	
Glueline integrity	Maintained	$\beta_0 =$	0.65	mm/minute
Thickness of layers	40 mm	$k_n =$	1.08	
$b =$	200 mm	$h_p =$	15	mm
$h =$	400 mm	$t =$	90	minute
		$k_{j,i} =$	1	

Calculations

Design strength in fire $f_{d,fi} = \dfrac{k_\Theta k_{fi} f_{c,0,k}}{\gamma_{M,t}} = \dfrac{1 * 1.15 * 24}{1} = 27.6\,N/mm^2$

Protection factor for charring $k_2 = 1 - \dfrac{h_p}{55} = 1 - \dfrac{15}{55} = 0.727$

Charring in different phases:

Phase 2:	$\beta_{n,\,Phase1,4} = k_n\beta_0 = 1.08*0.65 = 0.702\,mm/minute$
Phase 3:	$\beta_{n\,Phase\,2} = k_2k_n\beta_0 = 0.727*1.08*0.65 = 0.511\,mm/minute$
Phase 4:	$\beta_{n,\,Phase\,3} = 2k_n\beta_0 = 2*1.08*0.65 = 1.404\,mm/minute$

Time limits for charring phases:

$$t_{prot,0,1} = t_{prot,1} = 30\left(\frac{h_1}{15}\right)^{1,2} = 30*\left(\frac{15}{15}\right)^{1,2} = 30\,\text{minutes}$$

$$t_{f,pr} = t_f = \left(4.6*h_p - 25\right)*1.1 = (4.6*15 - 25)*1.1 = 48.4\,\text{minutes}$$

$$t_{ch} = \min\begin{cases} t_{prot,0,1} \\ t_{f,pr} \end{cases} = 30\,\text{minutes}$$

$$t_a = \min\begin{cases} 2t_{f,pr} \\ t_{f,pr} + \dfrac{25 - \left(t_{f,pr} - t_{ch}\right)\beta_{n,\,Phase\,2}}{\beta_{n,\,Phase\,3}} \end{cases} = \min\begin{cases} 2*48.4 \\ 48.4 + \dfrac{25 - (48.4 - 30)*0.511}{1.404} \end{cases} = 66.7\,\text{minutes}$$

Total charring depths after charring phases:

Phase 2:	$d_{char,n} = \beta_{n,\,Phase\,2}*\left(t_{f,pr} - t_{ch}\right) = 0.511*(48.4 - 30) = 9.4\,mm$
Phase 3:	$d_{char,n} = \beta_{n,Phase\,3}*\left(t_a - t_{f,pr}\right) + d_{char,n-1} = 1.404*(66.7 - 48.4) + 9.4 = 35.1\,mm$
Phase 4:	$d_{char,n} = \beta_{n,\,Phase\,4}*\left(t - t_a\right) + d_{char,\,n-1} = 0.702*(90 - 66.7) + 35.1 = 51.5\,mm$

Effective cross-section:

Zero-strength layer $d_0 = 10\,mm$

Effective charring depth $d_{ef} = d_{charn} + d_0 = 51.5 + 10 = 61.5\,mm$

Effective cross-section width
$$b_{ef} = b - k_{sids}d_{\theta f} = 200 - 2*61.5 = 77\,\text{mm}$$
Effective cross-section depth
$$h_{ef} = h - k_{sids}d_{ef} = 400 - 1*61.5 = 338\,\text{mm}$$
Section modulus
$$W_y = \frac{b_{ef}h_{ef}^2}{6} = \frac{77*338^2}{6} = 1,466,131\,\text{mm}^3$$
Bending capacity for R90
$$M_{Rd} = W_y f_{d,fi} = 1,466,131*27.6*10^{-6} = 40.5\,\text{kNm}$$

40.5 kNm > 40 kNm (99% capacity)
R90 is fulfilled

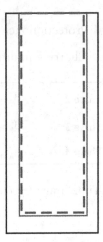

7.10.2 Effective cross-section method (Canada)

Load combination for fire situation according to CSA O86:

$$p_{d,fi} = 1.0\ g_k + 1.0\ q_k = 1.0*1.0 + 1.0*3 = 4.0\,\text{kN/m}^2$$

Linear load on the beam: $P_{d,fi} = 4.0*2 = 8$ kN/m
   Maximum bending moment: $M_{max} = 8*8^2/8 = 64$ kN m

Strength class	20f-E		$\phi =$	1	
Fire resistance	R90		$K_{fi} =$	1.35	
Cross-section	200 ´ 400 mm		$f_b =$	25.6	N/mm²
Protection system	15.9 mm Type X gypsum board		$K_D =$	1.15	
Glueline integrity	Maintained		$K_H =$	1	
Thickness of layers	40 mm		$K_T =$	1	
$b =$	200 mm		$K_L =$	1	
$h =$	400 mm		$K_{Zbg} =$	1.01	($\leq$1.3)
			$\beta_n =$	0.70	mm/minute
			$t =$	90	minute

In Canada, glulam beams required to provide fire resistance are to be man-ufactured using a special layup, as detailed in Chapter 1.
   Time contribution afforded by the 15.9 mm Type X gypsum board = 30 minutes. Therefore, the time of charring is taken as 90 − 30 = 60 minutes.

$$d_{char,n} = 0.70\,\text{mm/minute}*(90-30)\,\text{minutes} = 42\,\text{mm}$$

Effective cross-section:

Zero-strength layer
$$d_0 = 7 \text{ mm}$$
Effective charring depth
$$d_{ef} = d_{char,n} + d_0 = 42 + 7 = 49 \text{ mm}$$
Effective cross-section width
$$b_{ef} = b - 2d_{ef} = 200 - 2*49 = 102 \text{ mm}$$
Effective cross-section depth
$$d_{ef} = h - d_{ef} = 400 - 49 = 351 \text{ mm}$$
Section modulus
$$S_{ef} = b_{ef} * d_{ef}^2 / 6 = 2\ 094\ 417 \text{ mm}^3$$
Bending capacity for R90
$$M_{R,fi} = \min(M_{R,fi,1}, M_{R,fi,2}) = 82.6 \text{ kN m}$$
$$M_{R,fi,1} = f * (K_{fi}*f_b*K_D*K_H*K_{Sb}*K_T)*S_{ef}*K_x*K_z$$
$$\phantom{M_{R,fi,1}} {}_g = 84.1 \text{ kN m}$$
$$M_{R,fi,2} = f * (K_{fi}*f_b*K_D*K_H*K_{Sb}*K_T)*S_{ef}*K_x*K_L = 82.6 \text{ kN m}$$
82.6 kN m > 64 kN m (77% capacity)
R90 is fulfilled

7.10.3 Effective cross-section method (United States)

Load combination for fire situation according to NDS:

$$p_{d,fi} = 1.0\ g_k + 1.0\ q_k = 1.0 * 1.0 + 1.0 * 3 = 4.0 \text{kN/m}^2$$

Linear load on the beam: $P_{d,fi} = 4.0*2 = 8$ kN/m
   Maximum bending moment: $M_{max} = 8*8^2/8 = 64$ kN m

Strength class	20f-1.5E		$f =$	1	
Fire resistance	R90		$K_{fi} =$	1.35	
Cross-section	200 ´ 400 mm		$F_b =$	2000	psi
Protection system	15.9 mm Type X gypsum board			13.8	MPa
Glueline integrity	Maintained		$C_D =$	1	
Thickness of layers	40 mm		$C_L =$	1	
$b =$	200 mm		$C_V =$	0.91	($\leq$1.0)
$h =$	400 mm		$K =$	2.85	
			$\beta_n =$	1.50	in/hour
			$t =$	90	minute

In the United States, glulam beams required to provide a fire resistance are to be manufactured using a special layup, as detailed in Chapter 1.

Time contribution afforded by the 15.9 mm Type X gypsum board = 30 minutes. Therefore, the time of charring is taken as 90 − 30 = 60 minutes = 1 hour.

$$d_{char} = b_n * t^{0.813} = 1.5 \text{ in} = 38.1 \text{ mm}$$

Effective cross-section:

Effective charring depth
$$d_{ef} = 1.2 \ d_{char} = 1.2 * 38.1 = 45.7 \text{ mm}$$
Effective cross-section width
$$b_{ef} = b - 2d_{ef} = 200 - 2*45.7 = 108.6 \text{ mm}$$
Effective cross-section depth
$$d_{ef} = h - d_{ef} = 400 - 45.7 = 354.3 \text{ mm}$$
Section modulus
$$S_{ef} = b_{ef} * d_{ef}^2 / 6 = 2,272,065 \text{ mm}^3$$
Bending capacity for R90
$$M_{R,fi} = (F_b * C_D * C_L * K) * S_{ef} * C_V = 81.3 \text{ kN m}$$
81.3 kN m > 64 kN m (79% capacity)
R90 is fulfilled

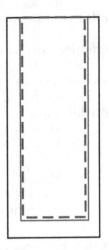

7.10.4 Summary

The calculation examples given above demonstrate that for a timber product of similar strength and stiffness, the design for fire resistance will provide similar results whether the European, Canadian or US approach is used. However, design assumptions and load combinations must be consistent with the appropriate building codes and design standards.

REFERENCES

ANSI/APA PRG 320–2018 (n.d.) *Standard for Performance-Rated Cross-Laminated Timber.* APA-The Engineered Wood Association, Tacoma, WA.

AS 1530.4 (n.d.) *Methods for Fire Tests on Building Materials, Components and Structures. Part 4: Fire-resistance Test of Elements of Construction.* Standards Australia, Sydney, NSW.

AS/NZS 1720.4 (2019) Timber structures - Part 4: Fire resistance of timber elements. Standards Australia, Sydney, NSW/ Standards New Zealand, Wellington.

ASCE/SEI 7–16 (n.d.) *Minimum Design Loads and Associated Criteria for Buildings and Other Sources.* American Society of Civil Engineers, Structural Engineering Institute, Reston, Virginia, USA.

ASTM C1396/C1396M (n.d.) *Standard Specification for Gypsum Board.* ASTM International, West Conshohocken.

ASTM D2559 (n.d.) *Standard Specification for Adhesives for Bonded Structural Wood Products for Use Under Exterior Exposure Conditions.* ASTM International, West Conshohocken.

ASTM D7247 (n.d.) *Standard Test Method for Evaluating the Shear Strength Of Adhesive Bonds In Laminated Wood Products At Elevated Temperatures.* ASTM International, West Conshohocken.

ASTM D7374 (n.d.) *Standard Practice for Evaluating Elevated Temperature Performance of Adhesives Used in End-Jointed Lumber.* ASTM International, West Conshohocken.

ASTM D7470 (n.d.) *Standard Practice for Evaluating Elevated Temperature Performance of End-Jointed Lumber Studs.* ASTM International, West Conshohocken.

ASTM E1354 (n.d.) *Standard Test Method for Heat and Visible Smoke Release Rates for Materials and Products Using An Oxygen Consumption Calorimeter.* ASTM International, West Conshohocken.

AWC (2018) *National Design Specification for Wood Construction.* Leesburg. VA: American Wood Council.

AWC (2021) *Calculating the Fire Resistance of Wood Members and Assemblies.* Technical Report No.10, Leesburg, VA: American Wood Council.

BS 476–20 (n.d.) *Fire Tests on Building Materials and Structures Part 20 Method for Determination of the Fire Resistance of Elements of Construction (General Principles).* British Standard BSI Group.

Buchanan, A.H., & Abu, A.K. (2017) *Structural Design for Fire Safety.* John Wiley & Sons, Chichester, UK.

CAN/ULC S101 (n.d.) *Fire Endurance Tests of Building Construction and Materials.* Underwriters Laboratories of Canada, Toronto, ON.

CAN/ULC S146 (n.d.) *Standard Methods of Test for the Evaluation of Encapsulation Materials and Assemblies of Materials for the Protection of Structural Timber Elements.* Toronto, ON: Underwriters Laboratories of Canada.

Chen, Z., Ni, C., Dagenais, C. & Kuan, S. (2020) Temperature-dependent plastic-damage constitutive model used for numerical simulation of wood-based materials and connections. *Journal of Structural Engineering* 146 (3), pp. 1–14.

CSA 086:14 (R2019) (n.d.) *Engineered Design in Wood.* CSA Group, Mississauga, ON, Canada.

CSA A82.27 (n.d.) *Gypsum Board.* CSA Group, Mississauga, ON, Canada.

CSA O177 (2011) *Qualification Code for Manufacturers of Structural Glued-Laminated Timber.* CSA Group, Mississauga, ON, Canada.

Dagenais, C., & Ranger, L. (2021) *Expanding Wood-Use Towards 2025: Revisiting Gypsum Board Contribution to the Fire-Resistance of Mass Timber Assemblies (Project No. 301014059).* FPInnovations, Pointe-Claire, Canada.

DIN 18947 (n.d.) *Earth Plasters: Terms and Definitions, Requirements, Test Methods (in German).* German Standard Deutsches Institut für Normung, Berlin, Germany.

EN 301 (n.d.) *Adhesives, Phenolic and Aminoplastic, for Load-bearing Timber Structures – Classification and Performance Requirements. European Standard.* CEN European Committee for Standardization, Brussels.

EN 520 (n.d.) *Gypsum Plasterboards: Definitions, Requirements and Test Methods. European Standard.* CEN European Committee for Standardization, Brussels.

EN 1363–1 (n.d.) *Fire Resistance Tests: Part 1 General Requirements. Brussels European Standard.* CEN European Committee for Standardization, Brussels.

EN 1364–2 (n.d.) *Fire Resistance for Tests for Non-loadbearing Elements: Part 2 Ceilings. European Standard.* CEN European Committee for Standardization, Brussels.

EN 1365–1 (n.d.) *Fire Resistance Tests for Loadbearing Elements: Part 1 Walls. European Standard.* CEN European Committee for Standardization, Brussels.

EN 1365–2 (n.d.) *Fire Resistance Tests for Loadbearing Elements: Part 2 Floors and Roofs. European Standard.* CEN European Committee for Standardization, Brussels.

EN 1365–3 (n.d.) *Fire Resistance Tests for Loadbearing Elements: Part 3 Beams. European Standard.* CEN European Committee for Standardization, Brussels.

EN 1365 (n.d.) *Fire Resistance Tests for Loadbearing Elements: Part 4 Columns. European Standard.* CEN European Committee for Standardization, Brussels.

EN 1365–5 (n.d.) *Fire Resistance Tests for Loadbearing Elements: Part 5 Balconies and Walkways. European Standard.* CEN European Committee for Standardization, Brussels.

EN 1365–6 (n.d.) *Fire Resistance Tests for Loadbearing Elements: Part 6 Stairs. European standard.* CEN European Committee for Standardization, Brussels.

EN 1990 (n.d.) *Eurocode: Basis of Structural Design. European Standard.* CEN European Committee for Standardization, Brussels.

EN 1995-1-2 (2004) *Eurocode 5 Design of Timber Structures. Part 1–2 General: Structural Fire Design. European Standard.* CEN European Committee for Standardization, Brussels.

EN 13381–7 (n.d.) *Test Methods for Determining the Contribution to the Fire Resistance of Structural Members. Applied Protection to Timber Members. European Standard.* CEN European Committee for Standardization, Brussels.

EN 13501–2 (n.d.) *Fire Classification of Construction Products and Building Elements: Part 2 Classification Using Data from Fire Resistance Tests, Excluding Ventilation Services. European Standard.* CEN European Committee for Standardization, Brussels.

EN 13914–2 (n.d.) *Design, Preparation and Application of External Rendering and Internal Plastering: Part 2 Internal Plastering. European Standard.* CEN European Committee for Standardization, Brussels.

EN 15283–2 (n.d.) *Gypsum Boards with Fibrous Reinforcement: Definitions, Requirements and Test Methods: Part 2 Gypsum Fibre Boards. European Standard.* CEN European Committee for Standardization, Brussels.

EN 15425 (n.d.) *Adhesives: One Component Polyurethane (PUR) for Load-bearing Timber Structures: Classification and Performance Requirements. European Standard.* CEN European Committee for Standardization, Brussels.

EN 16254 (n.d.) *Adhesives: Emulsion Polymerized Isocyanate (EPI) for Load-bearing Timber Structures: Classification and Performance Requirements. European Standard.* CEN European Committee for Standardization, Brussels.

EN 17224 (n.d.) *Determination of Compressive Shear Strength of Wood Adhesives at Elevated Temperatures. European Standard.* CEN European Committee for Standardization, Brussels.

Gardner et al. (1991) *Charring of Glue-Laminated Beams of Eight Australian-Grown Timber Species and the Effect of 13 mm Gypsum Plasterboard Protection on their Charring,* Technical Report No. 5, NSW Timber Advisory Council, Sydney, NSW, Australia.

Harmathy T.Z. (1965) Ten rules of fire endurance rating. *Fire Technology* 1(2), pp. 93–102.

IBC (2018) *International Building Code.* International Code Council, Washington, DC.

ISO 834-1 (n.d.) *Fire Resistance Tests: Elements of Building Construction: Part 1: General Requirements. International Standard.* International Organization for Standardization, Geneva, Switzerland.

ISO 834-12 (n.d.) *Fire Resistance Tests: Elements of Building Construction: Part 12 Specific Requirements for Separating Elements Evaluated on Less Than Full Scale Furnaces. International Standard.* ISO International Organization for Standardization, Geneva.

ISO 5660-1 (n.d.) *Reaction-to-Fire Tests: Heat Release, Smoke Production and Mass Loss Rate. ISO International Standard.* International Organization for Standardization, Geneva.

Just, A. (2010) *Post-protection Behaviour of Wooden Wall and Floor Structures Completely Filled with Glass Wool. Structures in Fire 2010.* DesTech Publications, East Lansing, pp. 584–592.

Just, A., Schmid, J. & König, J. (2010) *Failure Times of Gypsum Boards. Structures in Fire 2010.* DesTech Publications, East Lansing, pp. 593–601.

Källsner, B. & König, J. (2000) Thermal and mechanical properties of timber and some other materials used in light timber frame construction. In CIB W18 Meeting 33, Delft, The Netherlands.

König, J. (2006) Effective thermal actions and thermal properties of timber members in natural fires. *Fire and Materials, Volym* 30 (1), pp. 51–63.

Liblik, J., Küppers, J., Maaten, B. & Just, A. (2020) Fire protection provided by clay and lime plasters. *Wood Material Science and Engineering*, 16(2021), 290–298.

Mäger, K. N. & Just, A. (2019) *Preliminary Design Model for Wooden I-joists in Fire.* INTER, Tacoma, US. Publisher: Karlsruhe Institute of Technology. Karlsruhe, Germany.

Mäger, K. N., Just, A., Persson, T. & Wikner, A. (2020) *Fire Design Model of I-joists in Wall Assemblies.* Online, Karlsruhe Institute of Technology. Karlsruhe, Germany.

Mäger, K. N., Just, A., Sterley, M. & Olofsson, R. (2021) Influence of adhesives on fire resistance of wooden I-joists. In World Conference on Timber Engineering, Santiago, Chile, August 9–12, 2021.

New Zealand Building Code (2021) *Verification Method B1/VM1 Structure.* Ministry of Business, Innovation and Employment, Wellington.

Njankouo et al. (2005) *Fire Resistance of Timbers from Tropical Countries and Comparison of Experimental Charring Rates with Various Models.* University of Liege, Liege, Belgium.

NRC (2020) *National Building Code of Canada.* National Research Council Canada, Ottawa, Canada.

Östman, B. et al. (2010) *Fire Safety in Timber Buildings. Technical Guideline for Europe.* SP report 2010:19 SP. Technical Research Institute of Sweden, Stockholm, Sweden.

prEN 1995-1-2 (2021) *Eurocode 5. Final Draft. Design of Timber Structures. 1–2: General: Structural Fire Design. European Draft Standard.* CEN European Committee for Standardization, Brussels.

Ranger, L., Dagenais, D., & Benichou, N. (2020) *Encapsulation of Mass Timber Floor Surfaces (Project No. 301013624).* FPInnovations, Canada.

Roy-Poirier, A., & Sultan, M.A. (2007) *Approaches for Determining Gypsum Board Fall-off Temperature in Floor Assemblies Exposed to Standard Fires.* National Research Council Canada, Ottawa, ON.

Sterley, M., & Norén, J. (2018) *Fire Resistant Adhesive Bonds for Load Bearing Timber Structures. Development of a Small-scale Test Method. Smart Housing Småland.* SHS report 2018–002.

Sultan, M.A. (2010) *Comparison of Gypsum Board Fall-off in Wall and Floor Assemblies (NRCC-50843).* National Research Council Canada, Ottawa, ON.

Tiso, M. (2018) *The Contribution of Cavity Insulations to the Load-bearing Capacity of Timber Frame Assemblies Exposed to Fire.* PhD Thesis of Tallinn University of Technology. TUT Press, Tallinn.

Tiso, M., Just, A., Schmid, J., Mäger, K. N., Klippel, M., Izzi, M. & Fragiacomo, M. (2019) Evaluation of zero-strength layer depths for timber members of floor assemblies with heat resistant cavity insulations. *Fire Safety Journal.* 107, pp. 137–148. 10.1016/j.firesaf.2019.01.001.

Werther, N. et al. (2012) *Parametric Study of Modelling Structural Timber in Fire with Different Software Packages.* 7th International Conference Structures in Fire, 6–8 June 2012. Zürich, Switzerland.

Chapter 8

Timber connections

David Barber, Anthony Abu, Andrew Buchanan,
Christian Dagenais and Michael Klippel

CONTENTS

DOI: 10.1201/9781003190318-8

SCOPE OF CHAPTER

This chapter introduces structural connection typologies and provides information on potential failure modes and methods to provide fire resistance to connections exposed to a standard fire. Timber structures and their connections must be designed to have strength to resist all anticipated loads during the required fire resistance period and where required, to prevent the passage of heat and flames.

8.1 INTRODUCTION

This chapter gives guidance for design of fire resistance for connections in mass timber construction, where the timber elements are exposed to a standard fire. Connections in timber can be the weakest part of the building structure, as opposed to steel and concrete where connections are often designed to be stronger. For a connection to resist the impact of fire, a connection needs to be designed and constructed to provide the required fire resistance period, which may dictate beam and floor cross-sectional dimensions, especially for engineered wood products.

To achieve fire resistance for a load-bearing timber connection, there are three general approaches:

- For minimal levels of fire resistance where the timber is exposed to a fire, the connector is fully or partly concealed by the timber, including metallic plates, screws, bolts or dowels.
- Where a fire resistance of more than 30 minutes is required, and the timber is exposed to a fire, the connector is to be fully concealed by

the timber, so that no metallic part of the connector is exposed to the heat of the fire.

- The connection is fully encapsulated within a board system (typically non-combustible) or additional timber, so the connection and the surrounding structure are not exposed to the fire to provide the required fire resistance.

In some situations, a combination of methods may be required such as part concealment by timber and encapsulation with insulating boards, to take into account complex junctions, construction tolerances and installation needs.

Some standards such as EN 1995-1-2 (2004) provide design methods for a limited range of timber connections exposed to fire. Connection solutions stating a fire resistance need to be supported through verification by an accepted standard, or by first principles analysis, or fire test data, or advanced simulations.

This chapter primarily addresses connections in glulam members and CLT panels. Connections in sawn timber, LVL or other engineered timber members are not addressed directly, though their performance would be similar to the performance of glulam member connections in fire. It primarily focuses on connectors that are typically used in modern mass timber construction.

8.2 OVERVIEW OF BEAM-TO-COLUMN CONNECTION TYPOLOGIES

Timber connectors may be metallic, adhesive or timber. Connectors with adhesives are currently rarely used due to the difficulty of on-site construction and quality control. Connectors using only timber have a long history of use but are less popular in modern buildings. Metallic connectors are preferred for modern timber buildings as they can be custom designed to suit a particular project, and they typically fail in a ductile manner.

Where metallic connectors are exposed to the heat of a fire, they will readily conduct heat into the connected timber members and this can lead to increased localised charring and reduction in strength of the timber close to the connection, leading to a premature failure. Thus, the design of metallic connectors requires high attention to detail.

8.2.1 Timber-to-timber connections

Timber-to-timber connectors come in a wide variety of forms and includes timber dowels, timber cut-outs, intricate joinery and timber-to-timber bearing. Traditional timber buildings constructed through the 1800s and

Figure 8.1 Heavy timber beam-to-column connections. (a) With cast iron capital (image DeStefano & Chamberlain). b) Beam bearing on column connection (image David Barber).

into the early 1900s predominantly used timber-to-timber bearing connections due to their simplicity and inherent fire resistance (DeStefano, 2020) (see Figure 8.1). Timber connectors used in Japanese construction for centuries have influenced timber construction up to the modern day (Sato et al., 2000) (see Figure 8.2a).

Modern connectors bear one timber surface onto another, often seen as a beam bearing on a column (see Figure 8.2b). Another common version is a beam passing through a column and bearing directly on the column (see Figure 8.2c and d). Traditional dovetail timber-to-timber connections are also still in use (see Figure 8.3a). Many timber-to-timber connectors are inherently fire resistant due to the properties of the timber, but there are only a limited number of connectors that have been tested for their fire resistance and there is limited research on their performance in fire.

8.2.2 External metallic plates

External metallic plates, typically steel, are preferred for low-rise construction where the required structural fire resistance may be minimal (less than 15 minutes of fire resistance), or the connection deemed acceptable for use. These types of connectors are relatively easy to prefabricate and construct and can be an architectural feature for a building (see Figure 8.3b). As the steel plates and interfacing timber may have a low inherent fire resistance when exposed to fire, these types of connectors cannot be used where a building requires a substantive fire resistance rating (more than 30 minutes), unless the connector and surrounding timber is protected from fire exposure by fire-rated boards or additional timber to the timber part.

Figure 8.2 Timber-to-timber beam-to-column connections. (a) Tamedia House, Shigeru Ban Architecture. (b) Beam bearing on glulam column. (c) Two-way glulam beams bearing on glulam column. (d) Glulam beam passing through column (all images David Barber).

8.2.3 Embedded metal plates

For large timber structures, where the connectors carry significant gravity or lateral forces, a centrally embedded steel plate ("knife–plate") combined with dowels or bolts to connect the timber member to the metal plate can be used (see Figure 8.4a and b). Beam-to-column and column-to-column connections in modern timber buildings use embedded plates as they are relatively easy to design, detail and construct, given they are based on a similar design principle as for steel construction. The connectors can also achieve fire resistance due to the timber providing protection to the steel plate, provided the dowels or bolts are also protected from fire. Some steel plate connectors may be partially concealed, as shown in Figure 8.4b, where the base plate will be exposed, though fire resistance is reduced.

Figure 8.3 Beam end connections. (a) Timber-to-timber dovetail connection (image DeStefano & Chamberlain). (b) Exposed steel plate bucket-type connectors (image David Barber).

Figure 8.4 Embedded metal knife–plate connections. (a) Partially concealed knife–plate connection (image Dora Kaouki for DPR Construction). (b) Partially concealed knife–plate connection with exposed bolts (image DeStefano & Chamberlain).

8.2.4 Fully concealed connectors

Common connectors preferred by contractors and many designers are the two-part metallic connector (similar to dovetail connections used in traditional timber frame construction). These connectors are made up of two metallic parts, steel or aluminium, with each half being pre-installed to the beam or column and then connected together at the construction site (see Figure 8.5). These connectors are preferred structurally as they can resist high gravity loads and will exhibit a ductile failure mode. Column-to-column connections may also comprise a fully concealed connector.

Another type of concealed connector can be constructed through the welding of metal plates to form an internal bearing connector. These connectors may be designed specifically for a project or are proprietary to a supplier (see Figure 8.6).

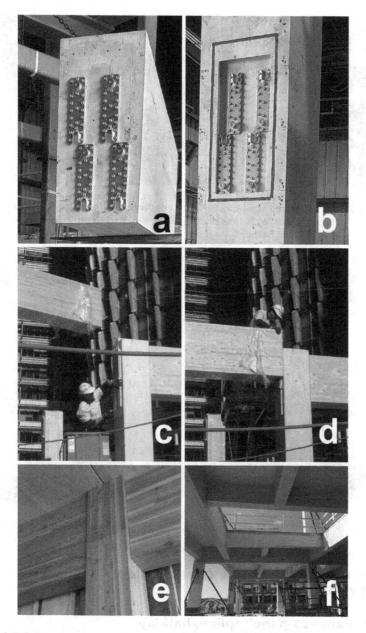

Figure 8.5 Fully concealed two-part metallic connectors: (a) installed on beam end; (b) installed on column face (images David Barber, with permission Simpson StrongTie). (c and d) Connection being installed on-site (images Arup). (e) Installed connector for glulam beam-to-column connection. (f) Multiple two-part metallic concealed connectors installed on glulam beams (images David Barber).

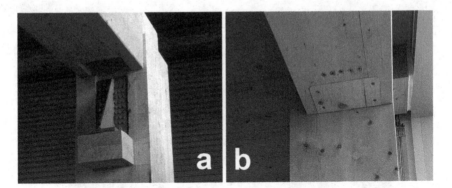

Figure 8.6 Fully concealed steel bearing plate connector: (a) beam being lowered onto the connector with timber base block pre-installed; (b) completed on-site (images David Barber).

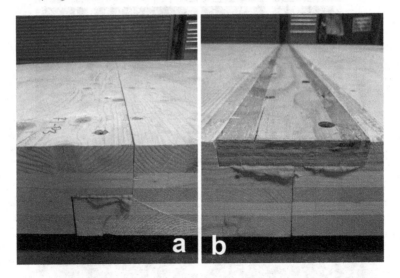

Figure 8.7 (a) CLT half-lap connection. (b) CLT single-surface spline connection (images David Barber).

8.3 MASS TIMBER PANEL CONNECTION TYPOLOGIES

8.3.1 Panel-to-panel: spline, half-lap

Mass timber panels, such as CLT, are connected by a variety of different solutions, with a half-lap and single-surface spline being the most common in construction (see Figure 8.7a and b). Each CLT manufacturer usually specifies a panel-to-panel connection for use with their panels that achieves a fire resistance proven through standardised fire testing. Some wall panel connections may not be load-bearing with fire resistance for separation only required.

8.3.2 Panel-to-panel hold-down connections

Connections are also required to CLT panels to resist lateral loading within a building. These types of connections may be steel plates exposed on the surface of the CLT, concealed within the CLT or located to the sides of the panels. Figure 8.8a shows a steel knife–plate connection between upper and lower CLT walls, and only the dowel ends are visible and exposed to fire. Figure 8.8b shows exposed hold-down brackets for CLT walls. These hold-down brackets may not need to achieve a fire resistance if the local building code does not require design for extreme lateral loads at the same time as a fire. These hold-down connectors do need to be assessed to determine if they are detrimental to the overall panel structural or separation fire resistance.

Figure 8.8 CLT wall and floor connections. (a) Concealed steel knife–plate connection between upper and lower CLT walls (image Andy Buchanan). (b) Base plate connection for CLT wall (image David Barber). (c) Steel angle ledger supporting CLT floor (image David Barber). (d) Alternative method for CLT floor supported on a steel angle ledger (image courtesy of the US Forest Service, Forest Products Laboratory).

8.3.3 CLT wall-to-floor panel connections

Where a building uses CLT for both floors and walls, there will be panel-to-panel corner connections. Where the floors bear directly on the walls (platform framing), the load transfer from the floor to the wall is by direct bearing on the top of the wall. Screws are provided to ensure connectivity between wall and floor panels. The fire resistance of the wall-to-floor connection needs to be determined based on the reduction in cross-section of both the wall and floor panels.

Where CLT panels connect into walls (balloon framing), the connection is often a timber ledger or a steel angle ledger (see Figure 8.8c and d). The ledger needs to be designed to provide a fire resistance rating to support the floor and to also prevent passage of heat and flame between floors, where the floor acts as a fire separation. A timber ledger must be designed to ensure that the uncharred timber of the residual cross-section can support the design load for the required fire resistance period. The steel angle must be protected to ensure that there is no bending failure of the bottom flange, or failure of the fasteners connecting the angle into the supporting timber wall.

8.3.4 Hybrid CLT floor to structural steel frame

A common form of construction is a "hybrid" steel frame building with CLT floors. This type of construction utilises screws to connect the top flange of the steel beam to the CLT floor (see Figure 8.9a). Where the building structure requires fire resistance, the steel beam will need to be protected. The fire resistance of the beam and floor needs to be assessed for screw resistance under heating, where the steel beam relies on the top flange for lateral buckling restraint (Barber et al., 2021). The steel section can still conduct heat into the supported CLT and weaken the screw resistance, even when protected with intumescent paint or fire-rated board.

Figure 8.9 (a) Hybrid construction with CLT floors supported on steel beam connected with regularly spaced screws. (b) Steel plate bucket connection supporting a glulam beam (images David Barber).

Where a connector fabricated from steel plate is used to connect timber (see Figure 8.9b), the connection needs to be fire-protected to ensure that the elevated temperatures of the steel plates and the fasteners into the timber do not result in structural failure. Intumescent paint may not be an option (see Section 8.7.3).

8.4 ELEVATED TEMPERATURES IN TIMBER CONNECTIONS

8.4.1 Review of fire testing results

Published research on fire testing of timber connections is primarily based on glulam, LVL or solid timber members. There are numerous fire tests or elevated temperature tests on simple tension connectors, where a knife–plate (or similar) connection is exposed to an elevated temperature and a tension force induced (Maraveas et al., 2013; Audebert et al., 2019). These types of tests are relatively easy to perform, but may not provide all information needed for building design as they are not replicating typical shear and bending forces. Audebert et al. (2014) have shown that differing tensile configurations can represent worse-case loading conditions. Fire tests on glulam connectors subject to forces that replicate actual building situations (bending and shear) are limited, due to the loading and furnace set-up required (Erchinger et al., 2010; Peng et al., 2012; Boadi, 2015; Palma et al., 2016; Palma and Frangi, 2016; Okunrounmu et al., 2020).

Fire testing has shown that the fire resistance of a connection is significantly reduced where there are metallic elements exposed to the fire, such as dowels, bolts or plates (Audebert et al., 2013; Maraveas et al., 2013; Palma et al., 2016). The fire-exposed surfaces of any metallic dowels, bolts or plates will heat up and conduct heat to the rest of the metallic components and increase the temperature of any timber that is in direct contact, reducing the strength of the timber member. Concealing the metallic components of a connector so that they are not exposed to a fire, and cannot directly transfer heat into the timber, greatly improves the resultant fire resistance. Fully concealed two-part or steel plate connections typically have the best fire performance (Audebert et al., 2019; 2020; Palma et al., 2016; Barber, 2017). Single exposed screws have been shown to have little influence on connection fire resistance (Hofman, 2016; Létourneau-Gagnon et al., 2021).

Existing published fire test information is predominately limited to smaller timber members, tested to the standard time–temperature curve, often between 30 and 60 minutes. There are few published standard fire tests taken beyond 60 minutes, using timber members that would be seen in actual multi-storey buildings (Carling, 1989; Maraveas et al., 2013; Palma et al., 2016; Barber, 2017, Brandon et al., 2019), see Figure 8.10.

Figure 8.10 Fire test of a fully concealed steel plate connector between a glulam beam and column. (a) During testing. (b) Being lifted out of the furnace after the test (Image David Barber).

8.4.2 Charring in connections

Where metallic components of the connection are exposed to the fire, charring rates at the connection will be higher than other parts of the timber member. Figure 8.11 shows temperatures through a connection at 30 minutes of standard fire exposure, with 0 mm being the outer beam edge exposed to fire and 80 mm being inside the connection. This shows that at approximately 35 mm distance from the fire face, the insulating properties of the timber result in near ambient temperature, whereas the steel dowel and bolt retain their elevated temperatures through the connection, thereby inducing higher local char rates (see also Peng et al. (2011), Maraveas et al. (2013) and Ali (2016)).

8.4.3 Influence of applied load

The thermal impact that occurs in timber connections exposed to fire change the material response of the connection components. Therefore, the load-carrying mechanisms in timber connections in fire can be different to those at ambient temperatures. For a beam connected to a column by a knife–plate or fully concealed connector, the forces induced by applied gravity loads are transferred into the column through shear. Where a timber beam bears directly on a timber column, both the beam and column undergo compressive forces.

For a beam-to-column knife–plate connector as an example, there is a rectangular volume of timber surrounding the connector that has sufficient strength and stiffness to allow the transfer of forces from the beam, through the dowels, into the steel knife–plate and then into the base plate at the column. The required member size can be determined at ambient temperatures and is influenced by the mechanical properties of the timber, the contact area of the bolts or dowels and their yield strength and embedment strength (Peng, 2010; Palma et al., 2016; Palma and Frangi, 2016; Audebert

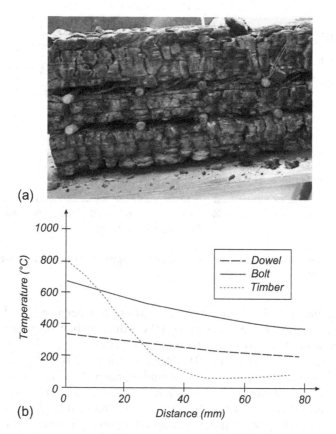

Figure 8.11 Temperatures through a connection. (a) Image of the tested connection (160 mm × 292 mm glulam with 20 mm fasteners). (b) Temperatures in the timber, a bolt and dowel after 30 minutes of standard fire exposure (Audebert et al., 2011).

et al., 2020). As the member size reduces due to cross-sectional charring, the stresses induced in the timber increase, leading to greater deflections. The failure process is further exacerbated by the temperature of the timber increasing ahead of the char layer reducing the stiffness of that timber. Any metallic components that become exposed to the increasing temperatures conduct heat into the timber, further reducing strength. Where screw fasteners are fully exposed to elevated temperatures, they will exhibit a significant reduction in strength and stiffness, changing their yield mode (when compared with ambient conditions). Therefore, the ability of the timber connection to resist the applied forces under fire conditions reduces quickly as the char front approaches the metallic connectors. This is observed in fire tests where fire resistance is influenced by the applied load and a lower applied load will improve fire resistance, over a higher loaded connection.

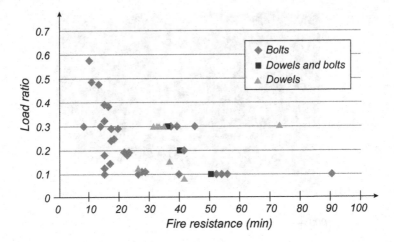

Figure 8.12 Reducing fire resistance with increasing applied load ratio (Maraveas et al., 2013).

The influence of load on connection failure has been recorded by several researchers, who have reported on the reduction in capacity under fire, with increasing load (Moss et al., 2010; Peng et al., 2011; Maraveas et al., 2013; Ali, 2016; Palma et al., 2016; Audebert et al., 2019). This is directly related to the reduction in strength of timber, regardless of connection type (see Figure 8.12). Peng et al. (2012) noted that a reduction in the ultimate applied load (load ratio) from 30% to 10% led to increased fire resistance of 7 minutes and up to 20 minutes, depending on the connection type. EN 1995-1-2 also recognises the reduction in connection capacity under fire and has a correlation that can be used to estimate this reduction. Given the link between load ratio and resultant fire resistance, any verification method for fire resistance must account for the influence of the applied load, where the connector is not kept at ambient temperature throughout the required fire duration.

8.4.4 Loss of strength behind the char layer: influence of thermal penetration depth

Accounting for the loss of strength in the timber directly behind the char layer is important for fully concealed connectors, given the elevated temperature profile directly behind the char layer in timber. For a connection to retain sufficient capacity for the duration of the fire, the connection must have adequate strength to prevent fastener pull-out or embedment failure (Frangi and Fontana, 2003; Cachim and Franssen, 2009; Schmid et al., 2014). Connector capacity is normally assessed based on ambient temperature and full-strength timber. Thus, analysis of a connector exposed to fire

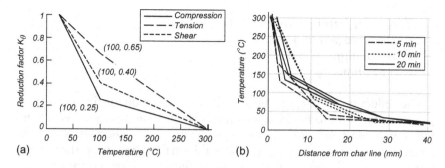

Figure 8.13 (a) Timber strength properties with increasing temperature parallel to the grain, for softwood, from EN 1995-1-2. (b) Temperature variation in timber behind the char layer with a short-duration standard fire exposure, for glulam (König and Walleij, 1999).

where the full-strength of the timber and connector capacity is assumed must be based on determining the location of ambient temperature timber. Therefore, thermal penetration depth must be accounted for.

To determine where the timber strength will reduce below a critical value requires an understanding of the temperature profile of the timber behind the char layer. The effective mechanical properties of timber exposed to fire are such that at 100°C, the tensile strength has been reduced by 35% and the compressive strength by 75% (see Figure 8.13a). It should be noted that those values are effective material properties for standard fire exposure of timber members and do not predict the material performance when exposed to constant elevated temperature. Once timber has reached 300°C, the charring process is complete, and the charred timber has lost all of its strength and stiffness.

The thermal properties of timber dictate the depth of thermal penetration and show that elevated temperatures will occur over a depth of about 35 mm ahead of the char front, as shown in Figure 8.13b. Any timber at a depth of more than 35 mm below the calculated char line can be assumed to be at ambient temperature, when exposed to a standard fire of 90-minutes duration when a glueline integrity failure can be disregarded (König and Walleij, 1999, Frangi and Fontana, 2003, Friquin, 2010, White, 2016). A method to assess thermal penetration depth as a function of time has been developed by Frangi and Fontana (2003), see Equation 8.1.

The temperature at any depth inside a timber member, when exposed to the standard fire, can be calculated as follows:

$$T(x) = 20 + 180(\beta \cdot t/x)^{\alpha} \quad \leq 300°C$$

$$\alpha(t) = 0.025t + 1.75 \tag{8.1}$$

where

$T(x)$ = temperature at depth (x) °C

β = char rate (mm/minute)

x = depth (mm)

t = time (minutes)

The reduction in strength behind the char layer due to increased temperature may need to be accounted for in connection design, especially for fully concealed connectors such as two-part metallic connectors. Where the connector and/or screw groupings have minimal timber cover, the thermal penetration depth behind the char can influence the connector capacity, given the screws are located in weakening timber.

8.4.5 Fire severity

Fire testing for timber connections has primarily focused on exposure to standard fires and therefore published assessment methodologies to determine fire resistance are also based on standard fire exposure. There are published fire experiments with timber connections exposed to physically based (natural) fires, typically CLT. There are few published fire experiments that include glulam beams and columns, and these are also not loaded (Boadi, 2015; Zelinka et al., 2018). This is an area for further research to determine if and to what extent timber connectors differ in performance when exposed to standard and physically based fires, especially under load. Where a building design uses a performance-based approach and includes the use of a physically based design fire, the calculation methodology to assess connection fire resistance will need to take account of the differences in fire severity (see Chapter 3). This is also important to consider for hybrid buildings with a steel structure that supports CLT floor panels.

8.5 DESIGN FOR FIRE RESISTANCE

The available methods for fire resistance design of connectors are limited in practice since the majority of research and engineering correlations to predict fire resistance are based on non-proprietary connectors, such as a knife–plate connector for a glulam beam-to-column connection. This is in contrast to the mass timber design and construction marketplace that prefer to use proprietary connectors, such as the two-part metallic connectors for a glulam beam-to-column connection. Proprietary connectors have significantly less published data available from fire test results and in-depth research on failure modes. Hence, guidance on achieving fire resistance is more focused on non-proprietary connectors, with guidance on proprietary connectors being more general and conservative in approach.

8.5.1 Failure modes

For the more commonly used timber connections exposed to fire, failure is typically due to the following actions:

- For a timber-bearing connection or fully concealed connection, the residual cross-section and weakened timber ahead of the char may be unable to resist the stresses induced in the timber interfacing with the fasteners and connector, resulting in failure by excessive deflections.
- For a knife–plate connection, the residual cross-section and weakened timber ahead of the char may be unable to resist the stresses induced in the timber interfacing with the bolts or dowels. Failure is exacerbated where the bolts, dowels or knife–plate are exposed to the fire, therefore increasing the heat transfer into the timber.
- In CLT floors, the reducing cross-sectional area can induce deflections that open up the panel-to-panel connection, causing integrity failure and eventually loss of structural capacity. In walls, the cross-sectional area reduction to one side leads to eccentric loading, inducing bending that in turn can lead to failure at the panel-to-panel joint.

For most connections, three separate but related assessments need to be made to determine fire resistance and failure:

1. The reduction in cross-sectional area due to charring.
2. The reduction in strength behind the char layer through thermal penetration.
3. The impact of thermal transfer from exposed metallic components of the connector into the timber member.

8.5.2 Beam-to-column bearing connections

A common form of beam-to-column connection is where the beam transfers forces from the floor into a cut-out provided at the column, or directly onto the column, placing significant compressive forces into the areas of contact. Screws are commonly used to provide stability.

Bearing connections need to be designed to account for not just the reducing cross-section due to charring, but also the thermal penetration depth behind the char, given the reduction in timber strength at temperatures below 100°C. The key issue to assess with this type of connection is the relative weakness of timber in compression both parallel and perpendicular to the grain, with small increases in temperature. These types of connections are also prone to higher deflections due to heat-induced compression at the contact zone (see Figure 8.14). A beam bearing on a column will need to consider the following:

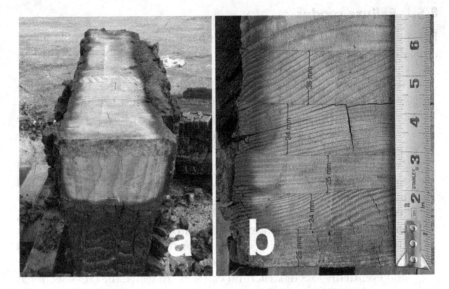

Figure 8.14 Base of a glulam beam after a standard fire test on a bearing connection. (a) The beam end and the area in contact with the bearing area of the column. (b) The reduction in glulam ply thickness at the base of the beam, where it is in contact with the column, due to heat-induced compression (images David Barber).

- The beam cross-section will reduce under fire exposure to all sides and the forces to be transferred into the bearing area need to be based on the residual cross-section. The same applies for the column bearing area, as the column will also reduce in cross-section.
- The bearing area required needs to be based on the applied load, accounting for any load reductions in the fire case (where applicable) and any factors relating to strength reduction for the timber under fire exposure.
- The bearing area needs to account for the thermal penetration depth, given that the compressive strength of timber, both perpendicular and parallel to the grain, reduces with increasing temperature. The compressive strength is very sensitive to temperature (Frangi and Fontana, 2003).
- If the beam is exposed to a longer fire, such as 90 or 120-minute exposure, the beam may displace due to compressive forces at the bearing face. As heat from the fire is conducted into the timber, the reduction in timber strength results in compression of the heat-impacted fibres. This deflection needs to be accounted for in the design, at both the beam and column bearing area, to determine if it can influence failure. A continuous floor system may lessen the impact of this deflection.

8.5.3 Beam-to-column knife–plate connectors

Steel knife–plate connectors concealed within timber can perform well, provided the connector is designed appropriately with the correct edge distances, and the bolts or dowels are also protected. The failure mode is a deformation as a result of thermal degradation of the timber in contact with the fastener. This embedment failure is due to the reduction in compressive strength of the timber and hence a reduction in shear resistance of the dowel or bolt (Erchinger et al., 2010; Audebert et al., 2011; Maraveas et al., 2013; Palma et al., 2014; Palma, 2016; Audebert et al., 2019). Yielding of the dowel or bolt can also occur in combination with the embedment failure, due to heating. Embedment failure is first seen through increased ovalisation at the dowels or bolt holes, which occurs both parallel and perpendicular to the grain (see Figure 8.15a).

The change in embedment strength with temperature has been measured by a number of researchers (Norén, 1996; Moss et al., 2010). For timber heated to 150°C, the embedment strength reduces by 40–60% in comparison to ambient (see Figure 8.15b). The embedment failure is a plastic failure and hence a desired mode and needs to be accounted for in connection design. Much of the research has been on ovenheated specimens, where the

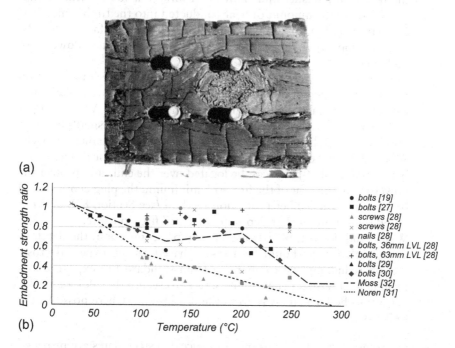

Figure 8.15 (a) Embedding failure in heated bolts (Lau 2006). (b) Reduction in embedment strength with increasing temperatures (from Norén [1996] and Moss et al. [2010]), with the differing correlations matching tests with bolts or screws (Maraveas et al., 2013).

moisture field and thermal loading duration do not match the conditions in a fire exposure. More research is needed on embedment strength in fire conditions.

For engineering design, it is important to provide the residual timber with the minimum edge, and end distances required for ambient temperature design, taking into account the depth of charring, thermal penetration depth and the elongation of the holes, as shown in Figure 8.15a. Testing has shown that cover distance to the steel knife–plate on all sides and the protection to the dowels or bolts has the most influence on fire resistance (Erchinger et al., 2010; Audebert et al., 2011; Maraveas et al., 2013; Palma et al., 2016; Audebert et al., 2019). In general, the greater the edge and end distances, and the greater the fastener spacing, the better the fire resistance (Cachim and Franssen, 2009; Khelifa et al., 2014; Owusu et al., 2019; Létourneau-Gagnon et al., 2021).

A knife–plate connector should be designed based on the following parameters:

- The use of bolts has a more negative impact on the fire resistance of the connection, compared with dowels. The primary causes are the bolt head, washer and shaft protruding outside the timber that will increase the amount of heat conducted into the timber member, compared with a dowel. Bolts can heat up twice as fast as dowels, directly impacting the structural fire resistance, hence dowels are preferred.
- The layout of dowels or bolts has been shown to have little effect on the fire resistance of a connection (see references above). The diameter of dowels or bolts does influence fire resistance. As with ambient design of timber connections, a larger number of small-diameter dowels perform better than a small number of large-diameter dowels.
- Where dowels are used, these should be recessed into the timber member so that timber plugs can be located over the ends and protect the dowels from the heat of the fire. At a minimum, the plug depth should be based on the timber protection method (see Section 8.8) and not based on minimum char depth, to standard fire exposure.
- Where additional timber is used to wrap a connection, the depth of additional timber should be assessed based on the expected cross-section reduction due to charring from standard fire exposure and include for thermal penetration. Screws to secure the additional timber are to be located away from the dowels and, where possible, the knife–plate.

Methods to assess the fire resistance of concealed steel plates are provided within EN 1995-1-2, with correlations for both dowels and bolts. The methods are applicable up to 60 minutes of standard fire resistance. Empirical correlations have also been published by Audebert et al. (2020), based on multiple years of fire testing. It should also be noted that prEN 1995-1-2 (2021) contains fire design verification methods for 90 and 120 minutes of fire resistance.

8.5.4 Charring localised to screws

To improve timber capacity, connections can be reinforced with self-drilling (self-tapping) screws. For example, screws are typically located perpendicular to the grain where a glulam beam bears onto a column to improve compressive strength. Ambient temperature testing has shown that the screw reinforcement increases both the load-carrying capacity and the ductility.

Research on screws exposed to fire has shown that charring around screw heads can penetrate 20–30 mm down the shaft and the screws have been shown to not detrimentally influence the fire resistance, provided they are well spaced apart (Hofmann et al., 2016; Petrycki and Salem, 2019; Létourneau-Gagnon et al., 2021). Of importance is that pull-out resistance can reduce where the whole length of the screw is heated. Where temperatures in the screw shaft are over 100°C, the pull-out resistance reduces by up to 50% (Hofmann et al., 2016; Létourneau-Gagnon et al., 2021). For axially loaded screws that are exposed to fire, EN 1995-1-2 provides methods to determine the reduced capacity of the screw when exposed to a standard fire. If such reinforced screws are not active in case of fire, only the additional charring in those areas should be considered for the verification of the timber member.

8.5.5 Glued-in dowels and rods

Where steel dowels or threaded rods are "glued-in," an epoxy adhesive is used to fill the gap around a dowel in an oversized hole in the wood. Using adhesives on-site can be difficult to perform and can result in inconsistent quality (Fragiacomo and Batchelar, 2012a, b). Fire resistance of glued-in dowel and rod connections has been studied by various researchers. Some epoxies can transition at temperatures near 60°C and start to lose strength (Buchanan and Barber, 1996; Harris, 2004; Di Maria et al., 2017). Other types of glue adhesives may perform differently.

Designing glued-in dowels and rods requires a detailed understanding of thermal penetration depth and behaviour of the adhesive being used at fire temperatures, due to their sensitivity to temperature. The design needs to be based on good detailing to avoid high temperatures in the steel dowels or rods where they are bolted into exposed steel plates. The dowels or rods also need to have sufficient cover distance between the exposed surface and the adhesive such that the adhesive is at ambient temperature at the rods, if no further information on the fire performance of the adhesive is available from the manufacturer.

8.6 CLT PANEL-TO-PANEL CONNECTIONS

When a mass timber wall or floor is required to provide fire resistance, the panel has to provide load-carrying capacity and a separating function. CLT panel-to-panel connections are the weak point for a CLT panel system when exposed to fire. These connections can fail through integrity under fire resistance testing (Werther et al., 2016; Dagenais, 2016; Klippel and Just, 2018). The integrity failure modes for CLT panel-to-panel half-lap and single-surface spline connections are similar.

The first mode of failure is deflection-based and occurs for both half-lap and spline connections, where deflections create gaps at the connection that induce faster charring (see Figures 8.16 and 8.17). A similar situation occurs for wall joints where the loss of cross-section induces eccentricity into the wall and deflections at the connections (typically to a lesser extent than for floors).

The second mode of failure occurs by a loss of integrity at the connection due to hot gases being able to pass through the connection. Once hot gases can pass from the fire side to the cold side, the path quickly increases in area and results in flaming on the cold side.

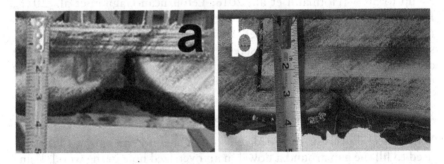

Figure 8.16 CLT floor panel connection after a standard fire test. (a) Increased charring at the spline connection, due to the deflection after a 120-minute test. (b) Increased charring at a half-lap after a 60-minute test (image David Barber).

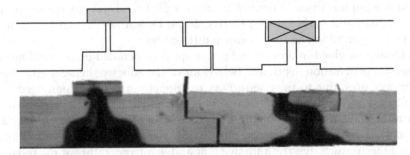

Figure 8.17 Sketch and post-fire test images of two differing timber panel connections. The joints were not airtight and rapid charring occurred (image Michael Klippel).

8.6.1 Design for fire resistance

A CLT connection must be properly detailed and sealed to prevent any convective flow through the joint, which may lead to localised increased charring in the joint. This can be achieved by fire resistance–rated caulking or sealant within the connection or construction adhesive tape to the top side of the connection. Other improvements can include a concrete topping slab or regularly screwed cladding to the top side, such as plywood or OSB (Werther et al., 2016; Klippel and Just, 2018). In some situations, the verification of fire resistance is required from the top side of a CLT floor panel, in addition to the verification from the underside, and the position and type of caulking or sealants will require further consideration.

For floor assemblies exposed to fire from underneath, the lap joint or surface spline should be positioned away from the fire-exposed side, without compromising the resistance of the lateral load resisting system. The same principle is recommended for a wall exposed to fire from one side only, with the lap joint or spline positioned away from the fire-exposed surface. For wall assemblies required to provide fire resistance from a fire occurring from either side, such as for interior walls and some exterior walls, a symmetrical joint detail should be used.

There are no methods currently available to predict or calculate the fire resistance rating of a CLT connection, based on applied load and exposure to a standard fire (load bearing, integrity and insulation). A method to determine the integrity and insulation of a CLT spline or splice connection has been developed by FPInnovations based on empirical fire testing (Dagenais, 2016). This method notes the importance of CLT coverings to the non-fire side for achieving both integrity and insulation. As the fire resistance of CLT panel-to-panel connections is strongly linked to the load applied, the fire resistance for CLT connections can only be accurately demonstrated through empirical fire testing. As cross-sectional area reduces, deflections will typically govern the connection resilience under a standard fire and determine actual fire resistance.

8.7 CONNECTIONS WITH ADDITIONAL FIRE PROTECTION

8.7.1 Protection with fire-rated board systems

Connectors can be protected from the impact of fire through protection with non-combustible or low-heat conductivity board systems, used to prevent heat transfer into the timber and connector. The most commonly used protection material is fire-rated gypsum board, though other boards or non-combustible coverings can be suitable. The board system includes the thickness of boards, support structure, fixings, spacing of fixings and any required gap sealants. The most common use is where the connector has minimal timber protection, or the connector is located on the exterior of the beam and column (see Figure 8.18).

The thickness and type of protection cannot be taken from a supplier based on protecting other structural materials, such as a steel member, or providing a fire resistance rating to a light frame wall. Fire resistance tests for other structural materials, such as a steel beam, will be based on a pass-or-fail criterion for the steel member in the range of 550-580°C and hence the board solution may not be appropriate for a timber connection given it may need to be kept below a temperature between 150°C and 300°C. Thus, when using a protective board system, the temperature behind the layers of board must be known, for the duration of standard fire exposure. Once the temperature profile for the protection is known, the connection can be designed to account for that temperature rise (Fonseca et al., 2020). Information on protection to mass timber by fire-rated gypsum board can be found within Technical Report 10 (AWC, 2021), CSA O86 (2019) and EN 1995-1-2.

Protection by boards must also be well detailed at all joints, including between the boards and more importantly, where the boards meet the timber. Most vulnerable is where the board protection stops and there is fire-exposed timber, that will get reduced in cross-section, but the fire-rated board will remain almost dimensionally unchanged. With increasing fire exposure, the gap between the board and the timber will increase as the

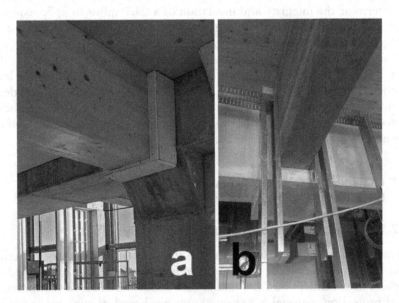

Figure 8.18 Connector encapsulation. (a) Glulam beam with externally located steel connection enclosed with first layer of encapsulation for fire protection. (b) Glulam beam connected to a steel beam protected with encapsulation (images David Barber).

timber cross-section reduces and can fail the connection through integrity. Detailing solutions to prevent this ingress of heat include the following:

- Using a high expansion intumescing seal at the board-to-timber interface
- Providing support for the protective board system to prevent an interface gap from opening up. For example, using a timber member (typically 75–150 mm deep) screwed to the glulam member and return the board protection around this small member.

8.7.2 Protection using timber

Additional timber can be used to protect connections using the inherent thermal insulative properties. Using timber to provide insulation to an already protected concealed connection or partially protected connection is a common methodology, given it is a simple application and has architectural acceptance. To date, this type of timber connection protection has had little research and testing to assist practitioners. Where additional timber is used to provide protection, detailing needs to ensure the timber will remain in place for the required fire exposure time and that any gaps are well-sealed to prevent fire ingress and ineffective protection.

To assist designers, methods to estimate the fire resistance by timber protection have been published in Technical Report 10 (AWC 2021) and EN 1995-1-2. The Technical Report 10 methodology for timber protection is based on a number of tests on timber "rim" boards, with correction factors required for single solid members accounting for the increased charring where a member chars through its full depth. The method is shown in Equation 8.2 and is valid up to 120 minutes of standard fire exposure for solid timber (sawn, glulam):

$$t_p = k_n 60 \left(d_p / 38.1 \right)^{1.23} \tag{8.2}$$

where
t_p = protection time (minutes)
d_p = thickness of protective timber member (mm)
k_n = 0.85 where one protective layer (board) is used, otherwise = 1.0

The methodology in EN 1995-1-2 (see Chapter 6) is protection of the connection by one or serval layers of timber, with assessment correlations provided. The detailing of fixings is also included in EN 1995-1-2 (see Chapter 7). These methods are valid up to 60 minutes of standard fire exposure. prEN 1995-1-2: 2021 includes similar information for up to 120 minutes.

8.7.3 Protection using intumescent paint

A common question in the structural fire design community is if external steel plates with bolts or dowels for a timber connection can be provided with intumescent paint on all metal surfaces to provide the fire resistance. Unfortunately, using intumescent paint will only provide a limited improvement in the connection's fire resistance. The weak point of the connection is the timber that is exposed to the fire, which does not have any significant improvement, from intumescent paint protection on the steel. The design issues to consider are as follows:

- Most intumescent paints do not swell and activate until they reach a temperature of 200°C and may not be fully active to protect the steel until 300–350°C. By the time the paint is protecting the steel, the timber in contact with the steel is already starting to char and is reducing in capacity.
- Even when the intumescent paint is fully expanded and active, the paint usually limits the steel temperature to about 500–550°C. At this temperature, the timber in contact with steel has charred and there is direct heat transfer into the timber through the steel dowels, bolts or plates.
- Suppliers of intumescent paint may have little knowledge of the required timber performance and temperature protection required by their products when applied to steel plates or fasteners in contact with timber.
- Failure of the intumescent-painted steel connection will occur through two processes – loss of cross-sectional area of the surrounding timber by charring; and loss of strength of timber through heating by the steel elements in direct contact with it.

Furthermore, there is very little research in this area and fire testing has shown that when a connection with external steel plates is coated with intumescent paint, there is only a small improvement in fire resistance, in the order of 10–30% when exposed to a standard fire (Frangi et al., 2009; Lau, 2006; Peng et al., 2012). Peng also noted that adhesion issues with the paint could occur during testing, with edges of the steel plates exposed.

8.7.4 Timber-to-steel connections

Careful design must occur where a timber member is connected into a steel structure, for example, a glulam beam connecting to a structural steel column or a glulam beam connecting to a steel beam. These types of connections are difficult to design for fire resistance as the fire-resistant protection for the steel structure (spray, board system or intumescent paint) has to transition to protect the connection and the timber member. The fire-resistant protection to the steel may not provide suitable temperature protection for the timber or the connection, since the steel protection is specified to keep the steel at temperatures below about 500–550°C, for the fire resistance period. Consideration of the thermal conduction from the steel into the timber and any possible temperature-induced movement in the steel must be part of the connection design.

8.8 CONNECTION DESIGN METHODS

There are only a limited number of engineering design methods to determine the fire resistance of exposed timber connections, which are empirically based on and derived from the collective fire tests completed. These approaches are relatively simplistic calculation methods and conservative. They are also specific to a connection type and must only be used with that connection type and within the applied load limits stated. If a more accurate assessment of fire resistance is required, this entails complex heat transfer modelling and may also require the determination of timber and metallic structural deformations. Thus, if undertaken, knowledge of the complexity of analysis, volume of sensitivity assessments and computing time must all be understood. A further option for project-specific connections is to undertake fire testing.

8.8.1 Char-rate methods

A "char-rate method" of assessing the fire resistance of a connection relies on calculating the residual cross-section of the original timber member, after a prescribed period of fire exposure. This is relatively simple for structural timber members (see Chapter 7) but difficult for connections, especially for three-dimensional fire exposure.

Given the range of connections that are available to the industry and the thermo-mechanical behaviour that needs to be assessed, where load, deflection, initial and changing interface gap, and number and type of connector parts needs to be included, few models have been developed, and these are limited in their engineering applicability. The most recent methods available are those from Palma and Frangi (2016) and Audebert et al. (2020), based on empirical testing of concealed or knife–plate connectors, respectively. Eurocode design methods are provided in EN 1995-1-2, with a substantive expansion of applicable methods to be included within the next edition (prEN 1995-1-2, 2021). This includes methods based on the work of both Palma and Audebert.

Other char-rate methods have been published. The US method in Technical Report 10 (AWC, 2021) is based on a minimum cover to an embedded connection of 1.14 a_{char} (where a_{char} is the required minimum char depth). It should be noted that this char depth at the connection is less than the required minimum effective char depth for the rest of a column or beam, which is required to be $1.2a_{char}$ (a_{eff}). The connection is also required to have the interface gap area protected with additional timber, with width and depth based on a_{char}. The Australian Standard AS/NZS 1720.4 and the Canadian Standard CSA O86 require a connector to be protected with timber to a depth equal to the effective depth of charring, i.e. the same char cover as the rest of the beam or column (see Figure 8.19). These published char-rate methods differ from those from Palma and Audebert and are recommended to be used conservatively, and with caution, given that the heated timber in front of the char layer is not accounted for.

8.8.2 Acceptance criteria

There are few published pass-or-fail criteria for mass timber connections and those published are conservative. Technical Report 10 (AWC, 2021) states that the temperature between the protection and the connection should be limited to an average temperature rise of less than 140°C (250°F) and a maximum temperature rise at any point of 180°C (325°F). Thus,

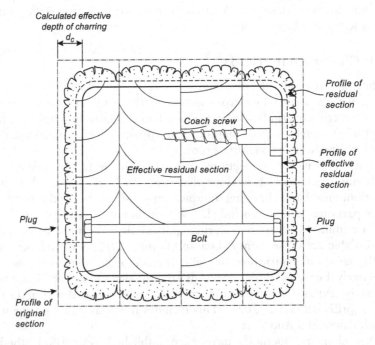

Figure 8.19 Protection of fasteners with timber depth equal to the effective depth of charring. From AS/NZS 1720.4, 2019.

for any part of a connection, whether it is concealed or partly concealed, the protection provided (being the timber surrounding the connection, or the fire-rated board system, etc.) cannot exceed an average and max temperature rise of 140°C and 180°C, respectively, for the designated fire resistance period. The Australian Standard AS/NZS 1720.4 (as referenced by the Building Code of Australia) requires protective coverings (timber, fire-rated board system, etc.) to prevent the temperature rise under the insulation from exceeding 120°C before the end of the designated fire resistance period, for metal plate connectors. Where dowels are used, these are to be protected such that temperature does not exceed 300°C.

8.8.3 Worked examples

Example 8.1 Timber protection for metallic dowels

A beam is connected to a column using a knife–plate connection with steel dowel fasteners to transfer forces from the timber beam into the knife–plate connection. The dowels are exposed at the beam surface and an additional timber member is used to protect the end of the dowels from the heat of a fire. The assessment aim is to prevent the dowel from exceeding a temperature criterion of 140°C above ambient (on average) to meet Technical Report 10 (AWC, 2021). Sixty minutes of fire resistance is required.

Using Equation 8.2:

$$t_p = k_n 60 \left(d_p / 38.1 \right)^{1.23}$$

A single layer of solid sawn timber is used as protection, so k_n is 0.85.

For 60 minutes of protection, solving for d_p, the timber protection is required to be at least 44 mm in depth (this is also equal to 1.14 a_{char}). This is 6 mm deeper than the nominal 60 minutes char depth, under Technical Report 10 (at 60 minutes $a_{char} = 38$ mm).

Example 8.2 Timber cover to two-part metallic connector, option 1

A glulam beam is connected to a glulam column by a two-part metallic connector, that is formed by two separate aluminium sides screwed to the beam end and column face, respectively. A simple method is provided to determine timber cover for the two-part connector to the beam end, such that the connector is located in ambient temperature timber (assumed to be at most 40°C for this example), when exposed to a standard fire of 90 minutes. Screw pull-out forces are based on ambient temperature timber and hence the aim is to ensure the maximum load-carrying capacity under fire with a charring rate according to EN 1995-1-2:

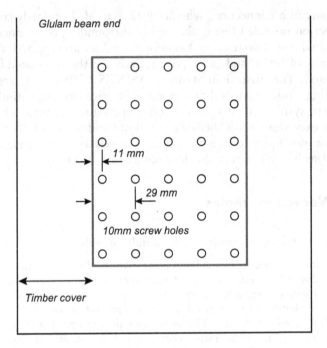

Figure 8.20 Sketch of example glulam beam end grain two-part connector.

- Connector base plate is 150 mm in width and has five rows of screws, as shown in Figure 8.20.
- The connector is more than 150 mm from the base of the beam.
- The one-dimensional char rate is $\beta_o = 0.65$ mm/minute.
- The notional char rate, including corner rounding and fissures is $\beta_n = 0.7$ mm/minute (for the glulam beam).

To account for the loss of strength and stiffness behind the char, an additional zero-strength and stiffness layer would normally be added to the char depth. As thermal penetration depth is accounted for in this approach, there is no requirement to include this additional zero-strength layer in this example.

- For the timber to be at ambient temperature so that the screws have full strength, an additional 35-mm depth is included (see Section 8.4.4).
- Thus, total cover needs to be 90 minutes ´ 0.7 mm/minute = 63 mm + 35 mm = 98 mm, which means that any steel element will have timber coverage to the fire-exposed surface and remain within ambient temperature wood.
- Total beam width required is therefore a minimum of (98 mm ´ 2) + 150 mm = 346 mm.

This calculation method represents a conservative approach based on a design assumption that the connector is at ambient temperatures for the duration of fire exposure. The beam dimensions can be reduced if more advanced methods are applied, i.e. allowing for higher temperatures at the steel elements (see Example 8.3) or thermal finite element simulations (see Section 8.9).

Example 8.3 Timber cover to two-part metallic connector, option 2

For the connector in Example 8.2, a more detailed approach can be used to further reduce the timber cover. Under fire conditions, the applied load is reduced, due to a fire case factor for the live load. Through assessing the ultimate capacity of the connector, it is determined that of the five vertical rows of screws, only the inner three rows of screws need to be in ambient temperature timber, to carry the applied load of the fire case. The outer vertical rows of screws are not needed in the fire case, but if the metallic base plate is exposed to the heat of the fire, it will transfer heat across the whole plate quickly and could lead to early failure.

Thus, two checks need to be carried out:

- Is the char front at the edge of the metallic connector, but not past?
- Are the central three rows of screws in ambient temperature timber?

The glulam beam connector is 150 mm wide and the second vertical row of screws from the centre is located at 29 mm (6 mm + 10 mm + 8 mm + 5 mm, representing edge distance, hole diameter, hole spacing, half-hole diameter) from the outer edge. The beam width is to be determined.

- At 90 minutes, the char depth is 63 mm (90 minutes ′ 0.7 mm/minute) and hence, this may be the minimum timber cover needed, to the edge of the base plate.
- The minimum timber cover distance to check is 92 mm, given the edge of the base plate to the edge of the vertical row of screw fasteners is 29 mm and the minimum depth of char is 63 mm (29 mm + 63 mm = 92 mm).

Using the method from Equation 8.1:

$$T(x) = 20 + 180(\beta \cdot t / x)\alpha$$

$$\alpha(t) = 0.025t + 1.75$$

With the charring rate $\beta_n = 0.7$ mm/minute, time = 90 minutes and $x = 92$ mm, $T = 58°C$ which is more than ambient.

Hence the timber cover distance needs to be increased.

- To keep the vertical row of screws at ambient temperature of 40°C (max), using Equation 8.1 and solving for the edge distance x, the total distance needed is 109 mm.
- With the vertical row of screws 29 mm from the edge of the connector base plate, the cover from the edge of base plate to the edge of the beam is therefore 109 mm – 29 mm = 80 mm. Total beam width is therefore (80 mm ´ 2) + 150 mm = 310 mm.
- There is no zero-strength layer added as this is included by determining where the timber reaches ambient temperature.

Compared with Example 8.2, this method results in a beam thickness that is 36 mm less in width.

8.8.4 Connection detailing

Detailing of connections is an important topic that is inadequately covered in most guides and design standards. All timber-to-timber connections require careful consideration of how the interface between members will react to fire exposure (Klippel and Just, 2018). This should include how deflections will change the interfaces, how load transfer mechanisms will change with changing cross-sectional area and how construction tolerances will affect the fire resistance. Most important will be the sealing of gaps and joints accounting for a reducing cross-section of timber (Werther et al., 2016) (see Chapter 6). Open gaps allowing flow of hot gases through an assembly must be avoided at all times. Closed gaps between timber elements have been recommended to be less than 5 mm (Aarnio and Kalliniemi, 1983); however, EN 1995-1-2 (Clause 3.4.3.1(3)) states that gaps should be less than 2 mm. It is recommended that 2 mm should be the aim for connections, also considering construction tolerances and the long-term performance of the connection.

Areas for attention are:

- A beam-to-column interface will open up as beam deflections increase with reducing cross-section. The gap between the end of the beam and column face can also be larger than expected due to on-site construction (0–10 mm should be allowed for). An intumescent seal or additional timber protection should always be designed and located to prevent ingress of hot gases and increased charring in this interface area, and also be specified to allow for a range of gap widths.
- The beam end to column face interface fire seal (intumescing fire caulk or intumescing fire tape) should be located so that it is still active in the required fire resistance period, i.e. for a 60-minute fire exposure, the fire seal should be located at least 30 mm in from the beam edge

and be a minimum of 10 mm in width, so that it can intumesce and seal gaps once a fully developed char layer has formed.

- Where fire-rated board systems are used, these need to be designed to wrap the connection and will extend past the connection to the surrounding timber. The rated board should extend a minimum of 100 mm past the edge of the connection to be protected (Lignum, 2018). Care must also be taken to design an effective connection between the fire-rated board and timber. Screws should be located away from the connection fasteners, such as dowels.
- An appropriate number and type of inspections are required during construction to ensure where a connection has been fire tested and approved for use that the finished assembly on-site is identical to that tested, including screw fixings, fire-rated sealants and dimensions.

8.8.5 Guidance documents

There are a few national guidance documents for design and detailing of timber connections. Guides available include the Swiss Lignum handbook (Lignum, 2018), US Technical Report No. 10 (AWC, 2021) and the CLT Handbook (Karacabeyli and Gagnon, 2019). The Swiss Lignum handbook provides engineers with information to plan different types of connections and also provides detailing for joints, gaps and protection, see Figure 8.21 as an example (not all details shown, see original document for all relevant information).

8.9 ADVANCED CALCULATION METHODS

For more detailed modelling, one of the most important parameters is the temperature of the components and how they vary with time. For most connection geometries, experimental temperature data or simplified temperature calculation models are typically not available. Numerical models can be used to approximate temperatures in connections and these temperatures used to inform the structural behaviour. However, a simple but comprehensive thermo-mechanical model for the design of timber connections is not currently available (Palma and Frangi, 2016; Li et al., 2020).

Advanced calculation models for fire resistance also require detailed information about the severity of the design fire (see Chapter 3).

8.9.1 Modelling of timber connections

The most commonly used numerical technique is the finite element method (or finite element models – FEM). FEM can be developed to address both heat transfer and, to a degree, the thermo-mechanical response during fire exposure. They are most commonly used for connections to model the influence

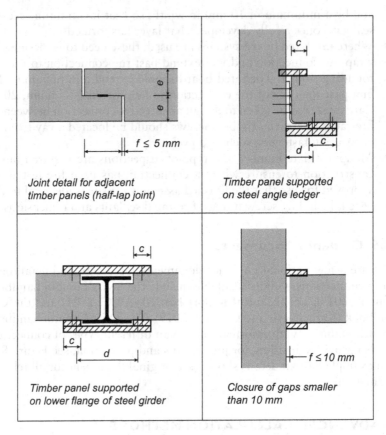

Figure 8.21 Examples of connection details given in Lignum Documentation 4.2 Timber Structures: Connections for Fire-Resistant Structures (Lignum, 2018).

of a wide range of parameters impacting temperatures within the timber that influence performance, such as timber cover depth, fastener diameter, number of fasteners, spacings, edge and end distances (Ohene, 2014). Potentially, with further validation, aspects such as intumescent sealants and gaps can also be modelled. Figure 8.22 shows a FEM model to illustrate the heat transfer into the timber due to the presence of exposed dowels (Sulon, 2020).

Performing a finite element analysis on timber connections has many challenges, including a lack of detailed validation data, imperfections in thermal contacts between connection components, accounting for moisture with increasing temperature, sensitivity to mesh size, boundary conditions and accuracy of timber material data. Also, thermal properties of timber available in the literature are derived and calibrated for standard fire exposure and their validity for non-standard fires is rarely addressed. Various software packages have been used and validated using the results of fire tests performed on timber connections, such as ABAQUS, ANSYS and

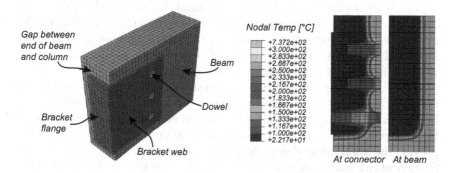

Figure 8.22 Modelled temperatures through a timber beam at a connection bracket fastened with dowels. (a) Modelling set-up. (b) Internal temperatures at two cross-sections, after 15 minutes of standard fire exposure (Sulon, 2020).

MSC-MARC (Erchinger et al., 2010; Peng, 2010, 2012; Audebert et al., 2011, 2013; Palma et al., 2016, Palma and Frangi 2019; Chen et al., 2020; Li et al., 2020). There are two approaches which are commonly adopted: thermal models with adapted load-carrying models (uncoupled models); and coupled thermo-mechanical models.

8.9.2 Uncoupled models

A less computationally intensive approach involves the development of an uncoupled heat transfer model to determine the evolution of temperatures in the connection over the duration of fire exposure. The temperatures at critical points in the connection are used to determine the reduced material properties at each time step and combined with an appropriate structural model. This type of uncoupled analysis is practical as it is relatively easy to use, especially once temperatures have been generated. Analytical models can also be derived for single fasteners to reduce the computational effort. Examples include work by Erchinger et al. (2009), Cachim and Franssen (2009), Moss et al. (2010), Peng (2010, 2012), Ohene (2014), Palma et al. (2016, 2019) and Regueira and Guaita (2018). A disadvantage of this approach is that the models do not capture the global behaviour and interactions for the various connection components.

8.9.3 Coupled thermo-mechanical models

Coupled thermo-mechanical models require the development of a heat transfer model with a coupled stress analysis. In the stress analysis, the mechanical properties are calculated based on the temperatures of the previous time increments from the heat transfer model. Examples of models using this approach include Racher et al. (2010), Audebert et al. (2011, 2013), Khelifa et al. (2014), Palma et al. (2016) and Chen et al. (2020). This approach allows the user to better understand the global behaviour of the connection and propose improvements to the geometry by evaluating the development of deformations. However, this approach requires more user input and computational time, whilst also producing load-bearing results which must be carefully interpreted and validated.

In the coupled approach, stresses will result in timber member movement and opening up at the beam-to-column or other interfaces, which need to be considered (also relevant for the uncoupled models). These methods may be non-conservative and hence require calibration against physical tests to provide the appropriate degree of accuracy (Palma et al., 2019). The verification of fire resistance of timber connections by means of finite element models should be carried out only by experienced engineers. Further, any results must be validated with appropriate fire test results.

8.10 FURTHER RESEARCH

The following areas have been identified for further research:

- Mass timber connections exposed to natural (physically based) fires to determine if and to what extent timber connectors differ in performance, compared with standard fire exposure.
- Impact on charring rate of closely spaced screws, especially in thin timber cross-sections.
- Better understanding of thermal penetration depth and timber strength reduction for longer fire durations (120 minutes).
- Assessment of fastener embedment strength under fire tests, not oven-heated tests.
- Connections between hybrid construction of steel beams and timber wall or floor panels, such as CLT or LVL.
- Effect of intumescent paint on exposed steel-to-timber connections for both standard and non-standard fires.
- Bearing area requirements and localised deflections in timber-to-timber bearing connections.

REFERENCES

Aarnio, M., Kalliniemi, P. (1983) *Fire Safety in Joints of Load Bearing Timber Construction, Technical Research Centre of Finland (VTT)*, Fire Research Report No.233.

Ali, S. (2016) *Fire Performance of Hybrid Timber Connections*, Thesis, Department of Civil and Environmental Engineering, Carleton University.

AS/NZS 1720.4 (2019) *Timber Structures: Fire Resistance for Structural Adequacy of Timber Members*, Standards Australia. Sydney, NSW.

Audebert, M., Dhima D., Taazount, M., Bouchaïr, A. (2011) Numerical investigations on the thermo-mechanical behaviour of steel to timber joints exposed to fire, *Engineering Structures* 33, pp. 3257–3268.

Audebert, M., Dhima, D., Taazount, M., Bouchaır, A. (2013) Thermo-mechanical behaviour of timber-to-timber connections exposed to fire, *Fire Safety Journal* 56, pp. 52–64.

Audebert, M., Dhima, D., Taazount, M., Bouchaïr, A. (2014) Experimental and numerical analysis of timber connections in tension perpendicular to grain in fire, *Fire Safety Journal* 63, pp. 125–137.

Audebert, M., Dhima, D., Bouchaïr, A., Frangi, A. (2019) Review of experimental data for timber connections with dowel-type fasteners under standard fire exposure, *Fire Safety Journal* 107, pp. 217–234.

Audebert, M., Dhima, D., Bouchaïr, A. (2020) Proposal for a new formula to predict the fire resistance of timber connections, *Engineering Structures* 204, 110041.

AWC (2021) *Calculating the Fire Resistance of Exposed Wood Members*, Technical Report No. 10 (TR-10), American Wood Council, Leesburg, Virginia.

Barber, D. (2017) Determination of fire resistance ratings for glulam connectors within US high rise timber buildings, *Fire Safety Journal* 91, pp. 579–585.

Barber, D., Roy-Poirier, A., Wingo, L. (2021) Modelling the fire performance of steel beam to CLT connections for hybrid construction, 12th Asia-Oceania Symposium on Fire Science and Technology, University of Queensland, Brisbane Australia.

Boadi, S. (2015) *Full Scale Tests on the Performance of Hybrid Timber Connections in Real Fires*, Thesis, Department of Civil and Environmental Engineering, Carleton University, Ottawa, Ontario.

Brandon, D., Landel, P., Ziethén, R., Albrektsson, J., Just, A. (2019) *High-Fire-Resistance Glulam Connections for Tall Timber Buildings*, Research Institutes of Sweden Report 2019:26.

Buchanan, A., Barber, D. (1996) Fire resistance of epoxied steel rods in Glulam timber, *New Zealand Timber Design Journal* 5(2), pp.12–18.

Cachim, P., Franssen, J. (2009) Numerical modelling of timber connections under fire loading using a component model, *Fire Safety Journal* 44, pp. 840–853.

Carling, O. (1989) *Fire Resistance of Joint Details in Loadbearing Timber Construction: A literature Study*, Building Research Association of New Zealand Study Report No. 18.

CSA Group (2019) *CSA O86 Engineering Design in Wood*, Canadian Standards Association, Mississauga, Ontario.

Chen, Z., Ni, C., Dagenais, C., Kuan, S. (2020) WoodST A temperature-dependent plastic-damage constitutive model used for numerical simulation of wood-based materials and connections, *Journal of Structural Engineering* 146(3), p.04019225.

Dagenais C. (2016) *Assessing the Fire Integrity Performance of Cross-Laminated Timber Floor Panel-to-Panel Joints*. Thesis, Department of Civil and Environmental Engineering, Carleton University, Ottawa, Ontario.

DeStefano, J. (2020) *Timber Structures: Tales of Engineering, Architecture, and Carpentry*, DeStefano & Chamberlain, Inc, Fairfield, Connecticut.

Di Maria, V., D'Andria, L., Muciaccia, G., Ianakiev, I. (2017) Influence of elevated temperature on glued-in steel rods for timber elements, *Construction and Building Materials* 147, pp. 457–465.

EN 1995-1-2 (2004) *Eurocode 5 Design of Timber Structures – Part 1-2: General – Structural Fire Design. European Standard.* CEN, European Committee for Standardization, Brussels.

Erchinger, C., Frangi, A., Fontana, M. (2010) Fire design of steel-to-timber dowelled connections, *Engineering Structures* 32, pp. 580–589.

Fonseca, E., Leite, P., Silva, L. (2020) Wood connections under fire conditions protected with gypsum plasterboard types A and F, *Advances in Fire Safety Engineering*, Selected Papers from the 5th Iberian-Latin-American Congress on Fire Safety, CILASCI 5, July 15-17, 2019, Porto, Portugal / edited by Paulo A. G. Piloto, João Paulo Rodrigues, Valdir Pignatta Silva. 1st edition, pp. 93–106.

Fragiacomo, M., Batchelar, M. (2012a) Timber frame moment joints with glued-in steel rods. I: design, *Journal of Structural Engineering*, ASCE 138(6), pp. 789–801.

Fragiacomo, M., Batchelar, M. (2012b) Timber frame moment joints with glued-in steel rods. II: experimental investigation of long-term performance, *Journal of Structural Engineering*, ASCE 138(6), pp. 802–811.

Frangi, A., Fontana, M. (2003) Charring rates and temperature profiles of wood sections, *Fire and Materials* 27, pp. 91–102.

Frangi, A., Erchinger, C., Fontana, M. (2009) Experimental fire analysis of steel-to-timber connections using dowels and nails, *Fire and Materials* 34(1), pp. 1–19.

Friquin, K. (2010) Charring rates of heavy timber structures for fire safety design: A study of the charring rates under various fire exposures and the influencing factors, Doctoral theses at NTNU, 2010:128, Norwegian University of Science and Technology.

Harris, S. (2004) *Fire Resistance of Epoxy-grouted Steel Rod Connections in Laminated Veneer Lumber (LVL).* Fire Engineering Research Report 04/7.

Hofmann, V., Gräfe, M., Werther, N., Winter, S. (2016) Fire resistance of primary beam–secondary beam connections in timber structures, *Journal of Structural Fire Engineering* 7(2), pp. 126–141.

Karacabeyli E., Gagnon, S. (2019) *Canadian CLT Handbook*, 2019 Edition. FPInnovations SP532-E, Pointe-Claire, Quebec.

Khelifa, M., Khennane, A., Ganaoui, M., Rogaume, Y. (2014) Analysis of the behavior of multiple dowel timber connections in fire, *Fire Safety Journal* 68, pp. 119–128.

Klippel, M., Just, A. (2018) *Guidance Document: Fire Design of CLT Including Best Practice*, COST Report N223-07, FP1404 WG2 TG1.

König, J., Walleij, L. (1999) *One-dimensional Charring of Timber Exposed to Standard and Parametric Fires in Initially Unprotected and Post-protection Situations*, Report I 9908029, Swedish Institute for Wood Technology Research, Stockholm.

Lau, P.H. (2006) *Fire Resistance of Connections in Laminated Veneer Lumber (LVL)*, Fire Engineering Research Report 06/3, University of Canterbury.

Létourneau-Gagnon, M., Dagenais, C., Blanchet, P. (2021) Fire performance of self-tapping screws in tall mass-timber buildings, *Applied Science* 11, p.3579.

Li, Z., Luo, J., He, M., He, G., Sun, Y. (2020) Analytical prediction of the fire resistance of the glulam bolted connections under coupled moment and shear, *Journal of Building Engineering* 33 101531. Doi:10.1016/j. jobe.2020.101531.

Lignum (2018) *Lignum-Dokumentation Brandschutz 4.2 Bauteile in Holz – Anschlüsse bei Bauteilen mit Feuerwiderstand (4.2 Components in Wood: Connections for Components with Fire Resistance)*, Editors Angehrn C., Brühwiler I., Wiederkehr R., Swiss Timber Industry, Zurich, Switzerland

Moss, P., Buchanan, A., Fragiacomo, M., Austruy, C. (2010) Experimental testing and analytical prediction of the behaviour of timber bolted connections subjected to fire, *Fire Technology* 46(1) pp. 129–148.

Maraveas, C., Miamis, K., Mathew, C. (2013) Performance of timber connections exposed to fire: A review, *Fire Technology*, Published online November 2013.

Norén, J. (1996) Load-bearing capacity of nailed joints exposed to fire, *Fire and Materials* 1996–20, pp. 133–143.

Ohene, A. (2014) *Modelling the Fire Performance of Hybrid Steel-Timber Connections*, Thesis, Department of Civil and Environmental Engineering, Carleton University.

Okunrounmu, O., Salem, O., Hadjisophocleous, G. (2020) Fire performance of moment-resisting concealed timber connections reinforced with self-tapping screws, 11th International Conference on Structures in Fire (SiF2020), Online, Brisbane, The University of Queensland.

Owusu, A., Salem, O., Hadjisophocleous, G. (2019) Fire performance of concealed timber connections with varying bolt patterns, 6th International Conference on Applications of Structural Fire Engineering, 13–14 June, Singapore.

Palma, P. (2016) *Fire Behaviour of Timber Connections*, Diss. ETH no. 24032, Department of Civil, Environmental and Geomatic Engineering.

Palma, P., Frangi, A. (2016) Fire design of timber connections: Assessment of current design rules and improvement proposals, International Network on Timber Engineering Research (INTER 2016), Graz, Austria, August 16–19.

Palma, P., Frangi, A. (2019) Modelling the fire resistance of steel-to-timber dowelled connections loaded perpendicularly to the grain, *Fire Safety Journal* 107(July), pp. 54–74.

Palma, P., Frangi, A., Hugi E., Cachim, P., Cruz, H. (2014) Fire resistance tests on beam-column shear connections, 8th International Conference on Structures in Fire, Shanghai, China, June 11–13.

Palma, P., Frangi, A., Hugi, E., Cachim, P., Cruz, H. (2016) Fire resistance tests on timber beam-to-column shear connections, *Journal of Structural Fire Engineering* 7 1, pp. 41–57.

Peng, L. (2010) *Performance of Heavy Timber Connections in Fire*. Thesis, Department of Civil and Environmental Engineering, Carleton University.

Peng, L., Hadjisophocleous, G., Mehaffey, J., Mohammad, M. (2011) Predicting the fire resistance of wood–steel–wood timber connections, *Fire Technology* 47, pp. 1101–1119.

Peng, L., Hadjisophocleous, G., Mehaffey, J., Mohammad, M. (2012a) Fire performance of timber connections, part 1: Fire resistance tests on bolted wood-steel-wood and steel-wood-steel connections, *Journal of Structural Fire Engineering* 3(2), pp. 107–132.

Peng, L., Hadjisophocleous, G., Mehaffey, J., Mohammad, M. (2012b) Fire performance of timber connections, part 2: Thermal and structural modelling, *Journal of Structural Fire Engineering* 3(2), pp. 133–154.

Petrycki, A., Salem, O. (2019) Structural fire performance of wood-steel-wood bolted connections with and without perpendicular-to-wood grain reinforcement, *Journal of Structural Fire Engineering Publication* date: 30 September 2019. https://doi.org/10.1108/JSFE-02-2019-0016.

prEN 1995-1-2 (2021) *Eurocode 5 Design of Timber Structures – Part 1–2: General – Structural Fire Design. European Draft Standard.* CEN, European Committee for Standardization, Brussels.

Racher, P., Laplanche K., Dhima D., Bouchaïr, A. (2010) Thermo-mechanical analysis of the fire performance of dowelled timber connection, *Engineering Structures* 32, pp. 1148–1157.

Regueira, R., Guaita, M. (2018) Numerical simulation of the fire behaviour of timber dovetail connections, *Fire Safety Journal* 96, pp. 1–12.

Sato, H., Nakahara, Y., Koichi, P. (2000) *The Complete Japanese Joinery*, Hartley and Marks Publishers.

Schmid, J., Just, A., Klippel, M., Fragiacomo, M. (2014) The reduced cross-section method for evaluation of the fire resistance of timber members: Discussion and determination of the zero-strength layer, *Fire Technology*, published online 15 July 2014.

Sulon, D. (2020) *Finite Element Modelling of Timber Connections in Fire*, Thesis, Faculty of Engineering at Stellenbosch University.

Werther N., Denzler K., Stein R., Winter S. (2016) Detailing of CLT with respect to fire resistance, Proceedings of the Joint Conference of COST Actions FP1402 and FP1404, Stockholm, Sweden.

White R. (2016) Analytical methods for determining fire resistance of wood members, Chapter 55, *Society of Fire Protection Engineers Handbook of Fire Protection Engineering*, 5th edition, Ed. Hurley, M, Springer, New York.

Zelinka, S., Hasburgh, L., Bourne, K., Tucholski, D., Ouellette, J. (2018) *Compartment Fire Testing of a Two-Story Mass Timber Building*, US Forest Service, Forest Products Laboratory, General Technical Report FPL GTR–247.

Chapter 9

Prevention of fire spread within structures

Esko Mikkola, Andrew Buchanan, Birgit Östman, Dennis Pau, Lindsay Ranger and Norman Werther

CONTENTS

DOI: 10.1201/9781003190318-9

SCOPE OF CHAPTER

This chapter describes means of preventing spread of fire and smoke between compartments in timber buildings. Much of this applies to all buildings independent of materials used, but some topics are especially relevant for timber buildings. This chapter highlights critical paths of possible spread of fire into, within and through timber structures, including solutions and detailing to prevent uncontrolled spread of fire and smoke.

9.1 INTRODUCTION

It is important that the designers of all timber buildings consider prevention of fire and smoke spread through joints in and between building elements/ assemblies, and through penetrations of building services and openings, including external walls. This chapter is complementary to Chapter 6 which describes fire-separating elements and assemblies.

More on fire exposures and fire spread on facades can be found in Chapter 5. The importance of building execution and control is covered in Chapter 13, including checking of correct installation of fire stops during the construction phase.

9.2 PREVENTING FIRE SPREAD BY DETAILED DESIGN

The correct design of joints, penetrations and opening details in buildings is critical to proper fire performance, as inappropriate details can lead to spread of fire, causing major damage. Detailing requirements and possible solutions to prevent fire spread are considered in the following sections.

9.2.1 Different types of timber constructions

Types of timber construction vary from traditional log construction to post and beam, light timber frame and mass timber construction. Today, also hybrid structures made of timber, steel and concrete are increasingly used (see Chapter 1). It is essential to understand the special features of timber structures when assessing prevention of fire spread, e.g. log construction has a large number of joints between logs, and mass timber structures typically contain few, if any, concealed void spaces. Thus, depending on the type of timber structure, differences in potential paths of fire spread may occur.

9.2.2 Typical fire spread paths and principles to prevent fire spread

In order to achieve the required level of fire safety for the entire structure, the fire behaviour of the building elements, service installations and additional safety measures have to be reviewed and assured. The evaluation criteria are interlinked, and interfaces with related requirements for fire resistance and reaction-to-fire have to be quantified.

Due to the need to connect various individual elements during the construction process, joints, gaps or voids appear which can create the potential risk that fire and smoke can spread rapidly and unnoticed. A comprehensive fire safety design must restrict the passage of flames, hot gases and smoke and as a consequence, an uncontrollable fire spread will be prevented.

A schematic illustration of typical fire spread paths for a structure is shown in Figure 9.1. The structure shown uses examples of light timber frame construction as well as mass timber construction. These paths must be taken into account within the design process to ensure an acceptable level of fire safety for buildings.

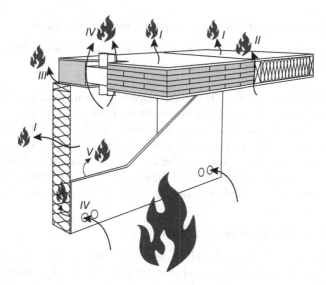

Figure 9.1 Fire spread paths in/through timber structures (Werther et al., 2020; redrawn/ with permission of TUM).

Path I: fire spread through failure of separating elements

Separating elements are typically the floors and walls of a building. Corresponding fire resistance can be derived from standardised calculation methods, taken from tabulated data or assessed within a fire test, considering the specific configuration, layout and connection of panels (see Chapter 6).

Path II: fire spread through joints

Joints between separating elements are pathways for fire spread to neighbouring compartments before the separating element fails (see Section 9.3). These joints result from the element-wise construction between elements of the same type.

Path III: fire spread through junctions

Fire spread through junctions to other building parts or intersecting elements via continuous joints before the separating element fails (see Section 9.3). These joints result from the element-wise construction between elements of a different type.

Path IV: fire spread through building services

Fire spread through penetrations or openings for building services (see Section 9.4).

Path V: fire spread through concealed construction cavities

Early and uncontrolled fire spread within the assemblies resulting in early exposure and failure of structural elements.

Early spread of fire to other fire compartments or within structural elements, as observed in real fires, can often be traced back to inadequate detailing and the lack of adequate design coordination (paths II–V). Element joints or penetrations for service installations must guarantee an equivalent fire resistance rating to the separating element (path I), which is often required to satisfy the building code clauses. Paths I–V are referred to by the subsequent sections, which further elaborate on the respective scenarios.

9.2.3 Construction tolerances

In designing joints between building elements as well as interfaces with penetrations and installations, construction tolerances need to be taken into account. This means that protective measures need to cover situations caused by actual construction tolerances, not only ideal cases with perfect fits.

Permissible tolerances vary amongst national standards and need to be checked when choosing linear joint seals or other protective materials. Typical tolerances for timber walls in national standards or guidance documents are around ±5 mm per 3 m. Also, dimensional changes on the building site caused by shrinkage when drying to a lower moisture content can increase the width of gaps which need to be considered for fire safety (if not covered by tolerance standards). Construction tolerances are considered further in Chapter 13.

9.3 FIRE SPREAD VIA SEPARATING ELEMENTS, JOINTS AND JUNCTIONS

Prevention of spread of fire and smoke within structural elements and to other fire compartments is based on proper design using fire test data and calculation methods. Applicability and limits of validity of standards, design methods and other data need to be carefully considered.

9.3.1 Fire resistance of separating elements

The fire resistance of separating timber elements can be assessed by standard fire tests or can be calculated by standard methods described in Chapter 6. If wall and floor assemblies cannot be designed by standards or calculation methods prescribed in building codes, fire testing is necessary. Even if fire test results are available for separating elements alone, joints between panels, joint seals, penetrations and other installations may have to be tested additionally, in accordance with test standards such as EN 1366 comprising several parts.

For each new timber building, specific descriptions for detailing of fire separations should be part of the drawings and specification. The following general design principles for light timber frames and mass timber assemblies should be considered to guarantee the required fire resistance rating of the separating element (paths I and II of Figure 9.1):

- Panel joints must be tightly jointed or be filled with fire-resistant material
- Joints in multilayered panels should be staggered
- All joints, penetrations and openings should be appropriately fire stopped
- Void cavities should be filled with insulation material
- Convective flow paths should be excluded or minimised
- The load-bearing function of an element supporting a separating element has to be fulfilled
- The end use conditions need to be considered, to avoid unexpected gaps or cracks due to shrinkage.

9.3.2 Fire resistance of joints between structural elements

The fire resistance of assemblies (paths I.a and I.b of Figure 9.2) can be calculated as specified in Eurocode 5 (EN 1995-1-2) or can be taken from test reports. Evaluation of the fire behaviour of the connection of a separating wall-floor structure (shown in Figure 9.2) requires that fire spread paths III.a (as in Figure 9.2) and III.b are taken into account. Besides the fire protection requirements, detailing is also influenced by structural, thermal, acoustic and air tightness performance requirements.

Based on an evaluation of a large number of fire tests (Suttner et al., 2020), principles for fire safe detailing of joints between neighbouring and intersecting elements (paths II and III.a) were derived. These principles are presented in Table 9.1 for different joint configurations and sealing methods. The scenarios include wall-to-wall, wall-to-floor and floor-to-floor element joints of light timber frames and mass timber assemblies, which are often encountered in practice. The presented details limit the spread of fire

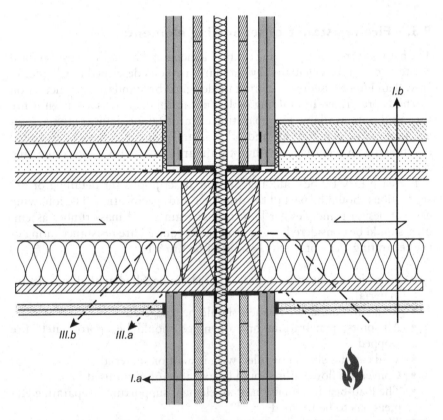

Figure 9.2 Wall–floor jointing detail with potential fire spread paths (dataholz.eu, 2020; modified with permission of dataholz).

Table 9.1 Principles for detailing element joints (Werther et al., 2020; Suttner et al., 2020) – fire spread path III.a (with permission of TUM)

No.	Joint width (s)	Detail	Fire-exposed side	Interior of the joint	Fire-unexposed side
1	$0 < s \leq 0.5$ mm	no measure	No measure required	No measure required	No measure required
2	$s \leq 2$ mm	no measure / joint filler, gypsum putty or fire rated sealant	No measure required	No measure required	Joint filler, gypsum putty or fire-rated sealant
3	$s \leq 2$ mm	no measure / sealant tape / elastomeric sealant or covering with lining	No measure required	No measure required	Sealant tape/ elastomeric sealant or covering with lining
4	$s \leq 5$ mm	joint filler, gypsum putty or fire rated sealant / insulation	Joint filler, gypsum putty or fire-rated sealant	Compressed insulation material (wood fibre, glass wool) (rate of compression 50%)	No measure required
5	$s \leq 5$ mm	covering with lining / insulation	Covering with lining	Compressed insulation material (wood fibre, glass wool) (rate of compression 50%)	No measure required
6	$s \leq 15$ mm	joint filler, gypsum putty or fire rated sealant / resilient soundproofing profile or fire rated sealant	Joint filler, gypsum putty or fire-rated sealant	Sealant/resilient soundproofing profile density $\rho \geq 200$ kg/m³/ fire-rated sealant	No measure required
7	$s \leq 15$ mm	covering with lining / resilient soundproofing profile or fire rated sealant	Covering with lining	Sealant/resilient soundproofing profile density $\rho \geq 200$ kg/m³/ fire-rated sealant	No measure required
8	$s \leq 30$ mm	stone wool	No measure required	Compressed stone wool insulation (rate of compression to 50%)	No measure required

Note: Typically, symmetrical fire stopping will be implemented for scenarios considering one-sided fire exposure from either side.

and smoke to a reasonable level and prevent the fire-stopping joints from downgrading the certified fire resistance of separating assemblies.

All details for the fire-exposed side, the fire-unexposed side and in the interior of the joint must be designed and installed separately for each compartment, with due consideration of the potential fire spread path. Normally, a fire may occur on either side of a fire-separating element, so the associated measures must be implemented for both directions of the fire spread path, which usually results in symmetrical fire-stopping details. There are scenarios where the fire exposure will be from both sides simultaneously, e.g. a load-bearing timber-framed wall supporting an intermediate floor within a fire compartment. In this case, the designer should exercise caution and consult with relevant product manufacturers, to ensure the fire-stopping system installed can still achieve adequate performance under the enhanced fire exposure.

Besides the details illustrated in Table 9.1 and when standard jointing solutions are not available, the following recommended details for in-plane element joints (path II) of mass timber and light timber frame assemblies can be considered, as shown in Figure 9.3. These recommendations are based on testing experience from a number of research, e.g. Werther et al. (2020), Exova (2017) and Dagenais (2015). The joints illustrated in

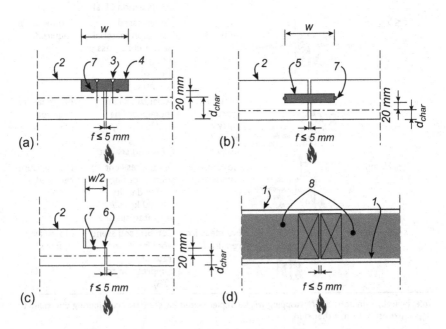

Figure 9.3 Details for fire-resistant detailing of in-plane element joints. Key: (1) surface lining; (2) mass timber element; (3) fasteners; (4) exterior spline, thickness ≥20 mm; (5) tongue–groove joint, tongue thickness ≥20 mm; (6) half-lap joint; (7) additional sealing strip; (8) light timber frame.

Figure 9.3 have gaps ≤5 mm and sealing strips to prevent any flow of hot gases through the joint. For gaps less than 5-mm wide, there will be little or no charring inside the joints, and to ensure effective stopping performance, the sealing strips are located in a thermally unexposed area of the cross section where the residual cross section covering the jointing system (exterior splines, half-lap joints or tongue–groove connections) should be ≥20 mm. Alternatively, panel coverings on the unexposed side can also mitigate the impact of convective flows.

If any gaps are greater than 5mm wide, due to construction tolerances or shrinkage of the panels before or during the fire, charring will occur inside the gap, hence the thickness of the spline or the half-lap joints must be $d_{char} + 20$ mm, unless the joint is filled with fire-resistant sealant or compressed non-combustible insulation material, as shown in Table 9.1. In order to ensure an actual gap width of ≤5mm, it is recommended a gap of ≤2mm be specified, to allow for construction tolerance and any movement on site.

Figure 9.3a and b shows a tightly fitted exterior spline and a tongue–groove joint ($w ≥ 90$ mm) between mass timber panels, with a remaining cross section of at least 20 mm below the tongue–groove or the exterior spline, for the relevant time of fire exposure. Figure 9.3c shows a half-lap joint (or step joint), overlapping ≥45 mm, with a remaining cross section below the step joint of at least 20 mm for the relevant time of fire exposure. Figure 9.3d shows a joint (≤5 mm) between two light timber-framed panels with a fire-resistant lining sheet on both sides.

Another essential aspect to achieve the required fire resistance in the jointing area of elements is to avoid displacements or gap opening due to movements between the elements, and between elements and adjacent components. Flexible sealant or additional layers such as a concrete topping or a ceiling lining may reduce the risk of an early failure.

Voids at joints between timber elements and other building elements like concrete walls or girders (path III) should be tightly filled with stone wool insulation over the entire depth, or by a backing of the joint or by flexible fire-resistant sealant.

Fire spread to other fire compartments through intersecting/flanking elements (via fire spread path III.b in Figure 9.2) also needs to be prevented in intersections of assemblies. Solid blocking with timber members or mineral wool rated to a high temperature (at least 700°C) is recommended. Typical solutions to prevent fire spread through cavities in a floor over a fire-resisting wall are shown in Figure 9.4a and b. To avoid fire spread through intersecting joints between adjacent floor elements, as shown in Figure 9.4c, the joints should be sealed with fire-resistant materials.

9.3.3 Seismic gaps

Timber buildings generally have good seismic performance because of their low mass, but seismic movements can be larger compared to other

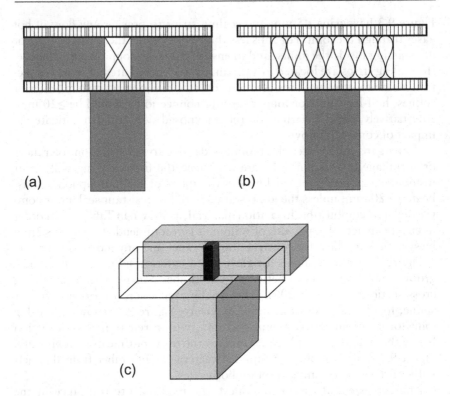

Figure 9.4 Examples of preventing spread of fire through a cavity over a fire-resisting wall. (a) Solid timber. (b) Mineral wool. (c) Fire stopping between adjacent floor elements (Werther et al., 2020; with permission of TUM).

construction materials because of the low stiffness of wood. When two timber buildings are located adjacent to each other, in a seismic zone, a seismic gap must be provided between the two buildings to allow for independent seismic movement. These seismic gaps can be large, depending on the height of the building (width up to 1% of the storey height). These seismic gaps require specific detailing, and should be fire stopped with flexible fire-stopping products, e.g. proprietary fire-rated blanket, barrier etc., which allow for adequate movement tolerance in several directions without compromising the seismic performance.

9.4 FIRE SPREAD VIA BUILDING SERVICE INSTALLATIONS AND PENETRATIONS

Modern buildings have increasingly high demands to incorporate technological equipment to achieve functionality, and sustainability such as energy efficiency. As the number of services in buildings increases, poorly sealed service installations in walls and floors can introduce a high potential for uncontrolled fire spread. This section provides a general description of penetration-sealing systems for building service installations such as electrical cables, hydraulic (heating, cooling, water and sewage) pipes and ventilation systems suitable for use in timber buildings.

9.4.1 General requirements of fire-stopping building services

The necessity of service installations and the requirements to mitigate fire spread require all building parts and construction methods (concrete, masonry, drywall or timber constructions) to have certified fire-stopping systems, e.g. fire-rated collars, fire-rated dampers/shutters, etc., for various penetrations. The fire-stopping systems must have fire resistance ratings of no less than the fire separations in which they are installed.

Service penetrations must be tested in accordance with the test methods set out in appropriate standards. In Europe, the failure criteria of penetration-sealing systems and linear joint seals according to EN 13501-2 are measured in terms of integrity (E) and insulation (I). Practical confirmation of the performance of penetration seals is provided by full-scale fire tests in accordance with EN 1366-3, and for linear joint seals in accordance with EN 1366-4. In New Zealand, the fire resistance rating of fire-stopping systems is determined by fire testing to AS 1530.4 or NZS/BS 476-21 and 22 and AS 4072.1. The tested setup should be representative of the fire-stopping systems installed in service, including the expected installation method, the type of service penetrations (e.g. metal, plastic, etc.), the gap size and the type of fire separation (e.g. concrete, timber, etc.) where the fire-stopping system is installed. Depending on jurisdiction and the support of justifiable performance-based design intents, the insulation rating of the fire-stopping system is sometimes relaxed, i.e. given the presence of sprinkler or if combustibles could remain adequately distanced from the penetration to mitigate ignition.

The fire performance of penetrating service installations is affected by the selection of fire-stopping products, the presence of additional support frame or fixing and any provisions related to joint movement to accommodate thermal contraction and expansion. The service installation structures are expected to sustain their own load but are generally not required to carry any additional live loads, as for example induced by maintenance activities. Approved fire-stopping systems for penetrations and fire dampers/shutters are widely available for concrete or drywall construction, but test data for application in mass timber or light timber frame assemblies are still limited. Because of the similarity of fire behaviour in many engineered wood products, the fire-stopping systems tested for one type of mass timber, such as CLT, LVL and glulam, may be applicable for other types of mass timber when the conditions are similar. These kinds of application rules should be included in product certifications wherever possible, and adequate consultation with product suppliers and relevant experts is required during the design process.

9.4.2 Concepts of fire protection to building services in multi-storey buildings

Penetrations of building services systems through separating elements are unavoidable. For multi-storey building design, thoughtful consideration on the location of separating elements, the runs of services and the type of fire-stopping systems could result in simple, cost-effective fire-stopping solutions for the whole building, enhancing building space utilisation. The basic concept in designing building service systems for multi-storey building is to congregate the services via a dedicated central fire-rated conduit configuration, such as in a central protected shaft, from which horizontal services runs distribute only the relevant services to specific areas and compartments. This approach greatly simplifies the design coordination process, and helps to avoid unnecessary penetrations or missed fire stopping of unplanned penetrations. The protected shaft strategy concentrates all penetrations to specific locations where a feasible, compliant fire-stopping solution can be applied easily.

Building service installation layouts should be developed in the planning phase with the services running to defined fire compartments. This can be done by the application of the design concepts of Table 9.2. All solutions must also satisfy the requirements for acoustic, moisture and thermal performance. Furthermore, the accessibility for inspection and maintenance and if necessary additional installation should be provided.

9.4.3 Types of building service installations

Penetration seals are installed in separating elements such as walls, floors, shafts and ducts, to prevent spread of fire and smoke. These systems ensure that the specified fire resistance rating of the separating elements is maintained independent of the type and size of penetration. Simultaneously, the fire-stopping systems should also ensure that there is no spread of fire in the separating elements themselves by the penetrations of building service installations.

For the classification of building service installations passing through separating elements, the type, number, size and material can be distinguished, as shown in Figure 9.5, which shows the arrangements and type of penetration seals which are available. Every type of building service installation passing through fire-separating elements has its own fire performance, so there is no single solution or product that will protect all services. The use of specific individual solutions, like sealing compounds or fire dampers adapted to the type of separating element, needs careful consideration.

Table 9.2 Concepts of fire protection for installation of building services/fire compartmentation

Concepts	Description
Service shaft/service ducts 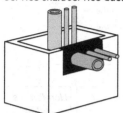	This concept is based on continuous conduits in separate fire-rated service shafts. The fire resistance rating of these shafts is the same as the separating elements. All installations that pass through the separating elements of the shafts have to be sealed with tested fire-stopping systems which maintain the fire resistance of separating elements in the area of the penetration. This solution is commonly adopted for multi-storey and tall building design
Penetration seals 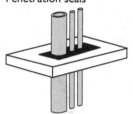	Sealing of penetrations for building service installations in separating elements (walls, floors) with approved fire-stopping systems, to maintain the fire resistance rating of the separating elements. This solution is the basic scenario commonly adopted in most low- and mid-rise buildings where the number of services penetrations is manageable
Continuous encasing 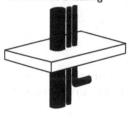	This arrangement is similar to the principle of service shafts, but with a single encasement of each cable or pipe over the entire length in accordance with the fire resistance rating of separating elements. This solution can be expensive due to complexity in design and installation, and is typically applied in special cases, e.g. a life-safety mechanical exhaust duct removing smoke from a compartment which cannot be fire stopped with conventional fire dampers, or a kitchen extract where conventional fire dampers would not work due to grease accumulation on damper blades

(Östman et al., 2010)

Designers should source adequate technical information and consult with the product manufacturers or relevant experts to ensure the fire-stopping system developed will meet the design intent.

9.4.4 Penetration fire–stopping systems for walls and floors

Approved proprietary systems are available from many manufacturers to provide suitable protection for penetrations of cable bundles, trunking, pipes and ducts through concrete, masonry, drywall and timber constructions. The type of penetration-sealing system used depends on the size, material, content and number of pipes or cables. When using approved and

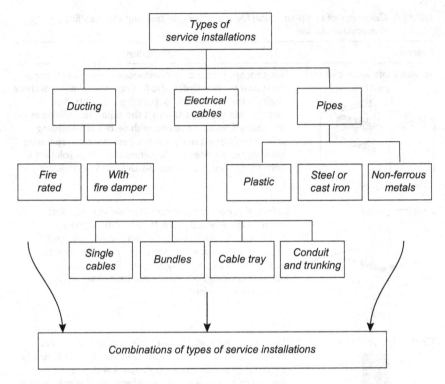

Figure 9.5 Types of service installations (Östman et al., 2010).

tested fire-stopping systems for mass timber– or light timber–frame structures, their use must comply with the specific installation instructions of
the manufacturer.

Single service penetrations are often easily treated but in a scenario where
there are multiple service penetrations going through timber separating elements, the common approach often involves forming a dedicated cut-out in
the element, which is replaced by tested, non-combustible fire stops, e.g. fire-
rated mortar, fire-rated mineral batts, etc., to contain all service penetrations. For light timber frame structures, this approach would line the area
surrounding the penetrations or openings with a non-combustible lining,
such as gypsum plasterboard or rigid mineral batts, over the entire thickness
of the separating element, as shown in Figure 9.6a. For mass timber, many
CLT manufacturers have tested a wide range of fire-resistant penetration
solutions through their CLT panels, using a variety of proprietary systems. A
summary of successful fire resistance tests of metal and plastic pipe penetrations is presented by Ranger et al. (2018). As an example, systems with intumescent materials ("heat activated sealant systems") which expand when
exposed to high temperatures can efficiently seal the gap between the sealing
system and mass timber element, as shown in Figure 9.6b.

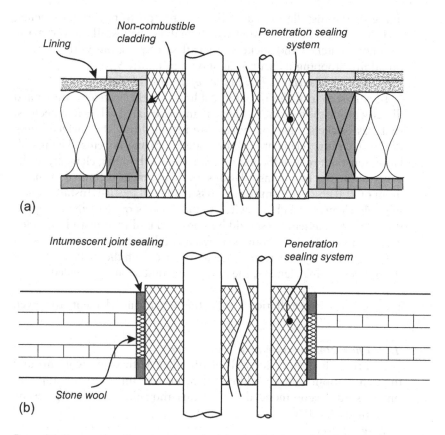

Figure 9.6 Penetration fire–stopping principles in (a) light timber frame structures and (b) mass timber structures.

9.4.5 Service installations embedded within building elements

Normally fire resistance tests of separating building elements are carried out without service installations or installed equipment. The installation of valves and accessories, electrical switches, cables and pipes being embedded in fire-separating building elements is permitted, provided that the remaining cross section of the separating element retains the required fire resistance rating. There are two principles where the embedded service installation is deemed to not affect the fire resistance of the building components:

1. *Presence of an installation cavity or installation of services outside the separating element*

   Services located in an installation cavity which runs outside of the separating wall or floor elements ensure that the fire resistance of the separating elements is not affected. In this configuration, the outer lining of the installation cavity has no fire protection requirements, and can remain non-fire rated which enables the installation of many common switches and sockets, regardless of position, without additional fire-stopping measures, as shown in Figure 9.7a.

2. *Installation of services inside the separating element*

In this scenario, the required cabling or other services are routed directly inside the separating element and associated valves (connectors, switches, sockets, junction boxes) are installed into the surface lining of the separating element with fire-rated encasing systems. The installation must not reduce the fire resistance of the building element, so the penetration size of the installations and equipment is limited: a maximum of 200 cm² is recommended (Östman et al., 2010). Installation of multiple elements, such as switches and sockets or elements on opposite sides of an assembly, should be staggered in different stud-bounded cavities, more than 150 mm away from combustible components, such as studs or beams. Where cables penetrate through the surface lining of the fire-separating element, the remaining joint should be sealed.

Examples of service installation inside the separating element are given below:

1. *Local non-combustible insulation layer*

Figure 9.7b shows the void locally filled with stone wool around the penetration, and an additional protective lining to the adjacent timber stud. Recommended dimensions and thicknesses are given by Östman et al. (2010).

2. *Gypsum putty*

Figure 9.7c shows sockets and switches encased with gypsum putty for the same protection time as that of the surface lining of the fire-separating building element, with $d \geq 30$ mm for 30-minute fire resistance and $d \geq 40$ mm for 60-minute fire resistance. This arrangement should only be used in combination with full insulation of the cavity with flexible insulating mats.

3. *Gypsum box*

Figure 9.7d shows sockets and switches encased with fire-resistant gypsum plasterboard (type F or type X) for the same protection time as that of the surface lining of the fire-separating element, with $d \geq 15$ mm. In this design, full cavity insulation is not required. Many manufacturers of gypsum plasterboard have tested and published proprietary details for protecting service penetrations.

9.4.6 Service installation within protected shafts and ducts

The use of vertical protected shafts or ducts which are fire-rated construction provides a simple, cost-effective solution for the distribution of services

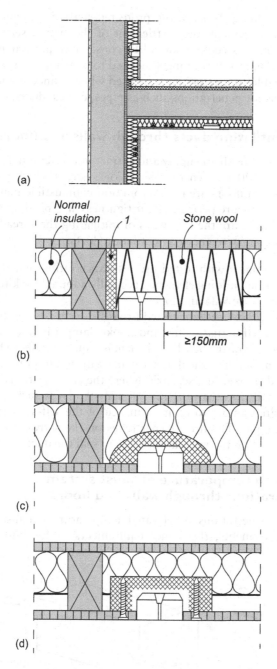

Figure 9.7 Details of fire protection with service installations: (a) installation cavity, b) mineral wool and gypsum board (I), (c) gypsum putty and (d) gypsum box (Östman et al., 2010)

throughout a building. When using protected shafts in timber structures, the differential movements and settlement of the shafts, service installations and the timber structure should be considered, particularly for constructions using platform framing subjected to wetting or drying. Flexible spacers or movable connectors must be used when connecting to walls and floors, as well as for penetrations such as pipes, cables, ducts, etc.

9.4.7 Air ventilation ducts through walls and floors

The temperature of fresh air supply and extract in ventilation ducts does not normally exceed 100°C. General ventilation systems which pass through fire-rated walls and floors are made from non-combustible materials, such as sheet metal, to ensure permanent air tightness at typical operating pressures. The following are the two ways of mitigating the spread of fire and smoke to other compartments via ventilation ducts:

- Ventilation ducts with tested fire resistance ratings equal to those of the separating building elements, typically adopted for kitchen extract ducts (see Figure 9.8a and b).
- Ventilation ducts without tested fire resistance, i.e. non-fire rated. In this case, a self-actuating fire and smoke damper in line with the location of the separating building element is required (refer to Figure 9.8c). In the event of a fire, the damper blade will shut to prevent the spread of fire and smoke. In a scenario where the cross-sectional area of the duct is small (less than 0.1 m²), design standards such as AS 1668.1 allow a fire damper to be used in lieu of the self-actuating fire and smoke damper. The fire damper relies on the activation of intumescent material on the ventilation grille to seal the duct.

9.4.8 Elevated temperature exhaust system penetrations through walls and floors

Generally, for penetrations of elevated temperature exhaust systems in fire-separating elements, the same requirements as for building service

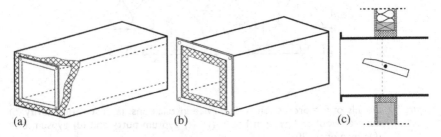

Figure 9.8 Design of fire-rated ventilation ducts (a and b) and fire and smoke damper (c) (Östman et al., 2010).

Table 9.3 Clearance distance related to exhaust temperature

Class of exhaust gas temperature (°C)	Clearance distance to combustible materials (m)
80–160	0.1
200–400	0.2
450–600	0.4

installations apply. Continuous fire-rated service shafts are applicable for this purpose (see Figure 9.8a and b).

In New Zealand, the design and construction of the associated flue systems and solid fuel, gas-burning or oil-fired appliances are in accordance with specific standards, e.g. for solid fuel, gas-burning and for oil-fired (AS/NZS 2918; NZBC Clause G11; AS 1691).

If exhaust systems penetrate separating elements in timber construction, sufficient clearance distances for uninsulated exhaust pipes must be provided or covered with insulation material to avoid direct contact. Detailed requirements are dependent on exhaust gas temperature. Distances related to exhaust gas temperature are suggested in Table 9.3 (Lignum, 2020).

Where combustible surfaces are lined directly with non-combustible encasing claddings of class $K_2 30$ or covered by non-combustible cladding with a 20-mm ventilation void, the distances in Table 9.3 can be halved (Östman et al., 2010).

Approved systems have to be used for penetrations of exhaust systems through separating building elements. These systems are designed to keep temperatures below the ignition temperature of wood. Minimum thicknesses of insulation material and cladding requirements around the exhaust systems are determined by national standards and building regulations.

9.5 FIRE SPREAD VIA BUILDING CAVITIES AND VENTILATION GAPS

9.5.1 Main principles to prevent spread of fire and smoke

Concealed cavities are often provided in walls or suspended ceiling spaces to accommodate building services. To prevent spread of fire and smoke in these cavities, the following principles should be applied, considering the suitability in each design scenario:

- Whenever possible, fill void spaces with non-shrinking material (such as mineral wool)
- Use certified fire-stopping products and check compatibility with timber structures
- Avoid possible ignition sources in cavities (e.g. connectors in electrical installations)

In cavities, e.g. for a rainscreen cladding system, the air and moisture movements are required under normal conditions, which means these cavities are either restricted or thermally activated fire stops systems are used. See more for facades in Section 9.6.

9.5.2 External and internal wall cavities and suspended ceiling spaces

Hidden voids in the construction of a building provide vertical or horizontal pathways for spread of smoke and flame to other areas, with the potential to threaten occupants remote from the location of fire origin. Hidden fires from cavities in wall, floor and roof structures may spread upwards, downwards and horizontally. Any spread of fire in concealed spaces which is hidden and difficult to access can result in delayed firefighting. Even for a sprinkler-protected building, fire spread in non-sprinklered cavities, e.g. small suspended ceiling space is typically non-sprinkler protected, can be a serious problem.

Horizontal and vertical cavities within buildings are illustrated in Figure 9.9 together with suitable fire barriers to ensure adequate fire stopping. Voids within and between structural elements in internal and external

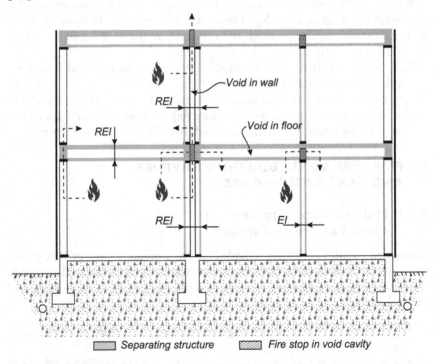

Figure 9.9 Fire stops in voids. REI: loadbearing and separating element; EI: separating element.

walls can form routes for fire spread upwards (can be also downwards) and horizontally. For preventing fire spread in facades, see Section 9.6. For volumetric modular elements, see Section 9.5.3.

9.5.3 Cavities between elements of modular construction

Three-dimensional modular elements (volumetric construction) have become popular in Northern Europe as an efficient way of producing building units in a factory under well-controlled conditions similar to car production (see also Chapter 1). They may have structures of timber or other materials.

Each modular element has surrounding structures with fire resistance of typically 30–90 minutes. Cavities between the modules must be carefully designed to prevent the spread of fire. There have been reported accidental fires involving modular construction, e.g. in Sweden and Austria. In one case, a small fire initiated on the top floor, which first spread upwards to the attic and then down into the building via an unprotected cavity. The rescue service accident investigations reported several shortcomings in fire stopping, despite the fact that the buildings were built in accordance with current regulations (Östman and Stehn, 2014). This indicates a lack of verification methods for the fire protection performance and also specific regulations to cover these cavity fires.

Different types of fire stops for cavities in modular houses have been proposed (Brandon et al., 2016; Just and Brandon, 2017; Stein, 2015). Some practical guidelines were also presented on how fire stops should be designed and used in modular constructions (see below).

Guidelines for fire stops in modular constructions

These guidelines aim to show how different types of fire stops in modular buildings could be designed:

- Fire stops must be installed in cavities between modules to prevent hidden fires from occurring and spreading between compartments.
- Fire stops must be verified by fire testing according to the European test method prEN 1364-6 or similar procedures. The test conditions need to be representative of the fire exposure of concern.
- Fire stop in cavities with combustible linings must be tested together with the combustible linings. Particleboard may be used as a standard lining for wood-based linings.

Examples of products that meet the requirements are uncompressed glass and stone wools $\geq$25 kg/m^3 and compressed mineral wools $\geq$50 kg/m^3. The fire stop should span the entire length of the gap with the section size of at least $t \times 3t$, where t is the thickness of the cavity (see Figure 9.10). They

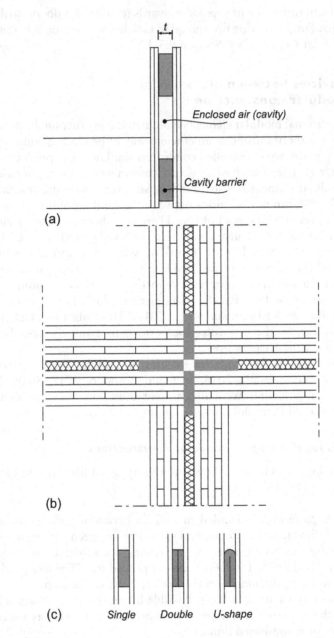

Figure 9.10 Examples of fire stops suitable for modular construction. (a) Cavity barriers at the bottom and top of a flat height. (b) Cavity barriers at the junction between four volume elements. (c) Single, double and U-shaped mineral wool barriers.

must be folded when mounted in a U-shape. Plastic foil around the fire stop must not melt with flaming droplets in the event of a fire, i.e. they must meet at least European reaction-to-fire class E.

Wood can also be used as a cavity barrier; the minimum thickness should be calculated using a one-dimensional charring rate in accordance with Eurocode 5 (EN 1995-1-2). The remaining uncharred thickness should be 25 mm after the required insulation time, for example 60 minutes. It should also be ensured that the fasteners maintain their function for the required insulation time in a standard fire. The wooden cavity barriers should be placed tightly against both opposing surfaces of the cavity and air channels across or along the wood should be avoided.

The actual design of the fire stop is crucial to guarantee fire performance and must be checked during the construction period – see further guidelines in Chapter 13.

9.6 VERTICAL FIRE SPREAD IN EXTERIOR FACADE CAVITIES

Cavities are often provided in facades or exterior wall systems. These cavities are necessary for improving moisture control and maintaining weather tightness, but they can be a serious problem for fire safety. Vertical spread of fire in external facades (exterior walls) has shown to be a serious problem, and a number of disastrous facade fires around the world have resulted in a serious loss of life and property. Vertical fire spread within exterior cavities can occur without notice and very rapidly via re-radiation and channelling effects to other spaces on multiple floors of high-rise apartments or to attic spaces through ventilation openings below the roof. Fire spread to attic spaces via a wall cavity can be prevented by providing a certified fire-stopping product or by a special fire-resistant eaves structure.

Vertical fire spread can also occur through thermal breakage of non-fire-rated windows or through open windows. This topic, which applies to buildings of all materials, is beyond the scope of this guide.

Overhanging floor slab edges, or fire stops behind the exterior facade help to restrict the vertical spread of fire within the cavity, mitigating the spread of fire over several storeys (see Figure 9.11). Fire stops in facade cavities have the following benefits:

- They can prevent stack effect in the ventilation cavity
- They deflect flames from the facade surface
- They reduce the fire intensity inside the ventilation cavity

The most important technical problem in incorporating fire stops in the ventilation cavities behind facades is that the structure must retain its

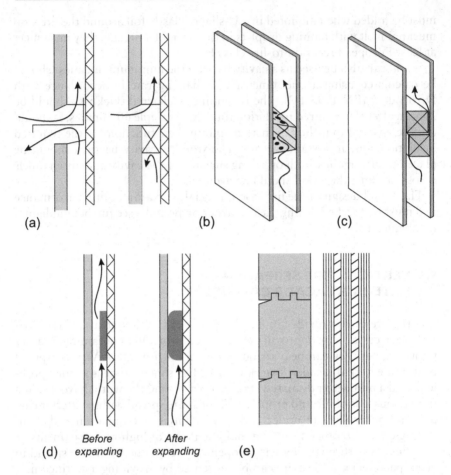

Figure 9.11 Basic principles of restricting the spread of fire in ventilation cavities of facades. (a) Fire stops with steel sheets and timber battens. (b) Perforated steel profile. (c) Airflow restriction with timber battens. (d) Self-expanding cavity barrier. (e) Two types of mass timber structures forming the outermost facade surface (Östman et al., 2010; Östman and Mikkola, 2018).

moisture protection functionality for draining and drying. Creating a functional fire sealing solution for the ventilation cavity may require a compromise between the fire protection and moisture protection requirements.

Different means of restricting spread of fire in ventilation cavities are described in Figure 9.11. Spread of fire in a ventilation cavity can be mitigated at each floor level using methods shown in Figure 9.11a by using steel sheets or timber battens, (b) a perforated steel profile, and (c) a fire stop made from two offset timber battens. In one particular steel profile fire stop (Hietaniemi et al., 2003), the diameter of the holes was 18 mm and the spacing was 140 mm so that the holes had a total area of 5% of the area of the fire stop. According to fluid dynamics calculations, this is sufficient for moisture protection, as long as no significant amounts of rainwater can flow directly into the ventilation cavity. A plain-tongued and grooved timber panelling, for example, was sufficiently rain-tight to meet the requirements. The airflow-restricting type of seal was made of timber battens of a size to leave a 7-mm wide gap, equal to about 20% of the cross-sectional area of the cavity. In practice, the gap of the timber batten fire stop cannot be made much narrower because it would encroach too much on allowances for dimensional accuracy, moisture-related dimensional changes, litter/dust and other similar matters.

In ventilation cavities of facades, one solution to prevent fire spread is the use of self-expanding (intumescing) cavity barriers, which allow ventilation at normal conditions, as shown in Figure 9.11d. Figure 9.11e shows also a mass timber structure forming the outermost facade surface. This outer structure prevents spread of fire to the inner layers of the external wall for a time which can be calculated based on the charring rate of the timber product. Thus, depending on the thickness of the timber structure, hidden facade fires can be avoided in practice, provided that all the details are designed carefully.

REFERENCES

AS 1530.4 (n.d.) *Methods for Fire Tests on Building Materials, Components and Structures Fire-resistance Structures: Tests for Elements of Construction.* Australian Standard Standards Australia, Sydney, NSW.

AS 1668.1 (n.d.) *The Use of Ventilation and Air Conditioning in Building, Part 1: Fire and Smoke Control in Buildings.* Australian Standard. Standards Australia, Sydney, NSW.

AS 1691 (n.d.) *Domestic Oil-fired Appliances: Installation.* Australian Standard. Standards Australia, Sydney, NSW.

AS 4072.1 (n.d.) *Components for the Protection of Openings in Fire-resistant Separating Elements Service Penetrations and Control Joints.* Australian Standard. Standards Australia, Sydney, NSW.

AS/NZS 2918 (n.d.) *Domestic Solid Fuel Burning Appliances: Installation.* Australian/New Zealand Standard. Standards Australia, Sydney, NSW.

Brandon, D., Just, A., Jansson McNamee, R. (2016) Behaviour of cavity barriers in modular houses: A revised test methodology. In Proceedings of Interflam, Windsor, UK.

Dagenais, C. (2015) *Fire Performance of Cross-Laminated Timber Panel-to-Panel Joints.* FPInnovations Report 301009649, Canada.

Dataholz.eu (2020) *Catalogue of Wood and Wood-based Materials, Building Materials, Components and Component Connections for Timber Construction Covering Thermal, Acoustic, Fire and Ecological Performance Levels.* Released by Accredited Testing Institutes, Holzforschung, Austria.

EN 1366 (n.d.) *Fire Resistance Tests for Service Installations: Part 1: Ducts; Part 2: Fire Dampers; Part 3: Penetration Seals; Part 4: Linear Joint Seals; Part 5:*

Service Ducts and Shafts; Part 6: Raised Access and Hollow Core Floors; Part 7: Conveyor Systems and Their Closures; Part 8: Smoke Extraction Ducts; Part 10: Smoke Control Dampers. European Standard, CEN European Committee for Standardization, Brussels.

EN 13501-2 (n.d.) Fire Classification of Construction Products and Building Elements. Part 2: Classification Using Data from Fire Resistance Tests, Excluding Ventilation Services. European Standard, CEN European Committee for Standardization, Brussels.

EN 1995-1-2 (n.d.) *(Eurocode 5) Design of Timber Structures – Part 1–2 General Rules Structural Fire Design*. European Standard, CEN European Committee for Standardization, Brussels.

Exova (2017) *Fire resistance of Various Joints in LVL Walls and Floors in Accordance with AS1530.4-2014*. EWFA Report 30073900.1. Exova Warringtonfire, Australia.

Hietaniemi, J. et al. (2003). *Ontelotilojen paloturvallisuus. Ontelopalojen leviämisen katkaiseminen (Firesafety of Cavity Spaces. Prevention of Fire Spread in Building Voids)*. VTT Tiedotteita – Research Notes 2202. Espoo, Finland.

Just, A., Brandon, D. (2017) *Fire Stops in Buildings*. SP Report 2017:10. SP Technical Research Institute of Sweden. Borås, Sweden.

Lignum (2020) *Dokumentation Brandschutz 6.1 Haustechnik: Installationen und Abschottungen*, Holzwirtschaft Schweiz, Zürich.

NZBC (1992) *Compliance Document for New Zealand Building Code. Clause G11. Gas as an Energy Source*. Department of Building and Housing, Wellington, New Zealand.

NZS/BS 476-21 (n.d.) *Fire Tests on Building Materials and Structures: Methods for Determination of the Fire Resistance of Loadbearing Elements of Construction*. British/New Zealand Standard. British Standard BSI Group.

NZS/BS 476-22 (n.d.) *Fire Tests on Building Materials and Methods for Determination of the Fire Resistance of Non-loadbearing Elements of Construction*. British/New Zealand Standard. British Standard BSI Group.

Östman, B. et al. (2010) *Fire Safety in Timber Buildings: Technical Guideline for Europe*. SP Report 2010:19. Stockholm, Sweden.

Östman, B., Mikkola, E. (eds) (2018) *Guidance on Fire Safety of Bio-Based Facades*. Dissemination document N230-07 – COST Action FP 1404 I ETH Zurich.

Östman, B., Stehn, L. (2014) *Fire in Residential Multi-storey Timber Buildings: Analyses, Recommendations and R&D Needs (in Swedish)*. SP Report 2014:07. SP Technical Research Institute of Sweden.

prEN 1364-6 (n.d.) *Fire Resistance Tests for Non-loadbearing Elements – Part 6: Cavity Barriers*. Draft European Standard, CEN European Committee for Standardization, Brussels.

Ranger, L., et al. (2018) Fire Performance of Firestops, Penetrations, and Fire Doors in Mass Timber Assemblies. In 2018 World Conference on Timber Engineering, Seoul, Korea.

Stein, R. (2015) *Evaluation of influencing factors of fire spread-mechanisms in junctions of pre-fabricated timber assemblies (in German)*. Dissertation, TUM Technical University of Munich.

Suttner, E., Werther, N., Dumler, P. (2020) *Stand der Technik zur Qualität von Bauteil- und Elementfugen in Holzbaukonstruktionen im Hinblick auf den Durchtritt von Feuer und Rauch.* Technische Universität München, Germany DOI: 10.13140/RG.2.2.20566.80964.

Werther, N. et al (2020) Design principles for fire safe detailing in timber structures. In World Conference on Timber Engineering. Santiago, Chile DOI: 10.13140/RG.2.2.29537.33121.

Chapter 10

Active fire protection by sprinklers

Birgit Östman, David Barber,
Christian Dagenais, Andrew Dunn, Kevin Frank,
Michael Klippel and Esko Mikkola

CONTENTS

DOI: 10.1201/9781003190318-10

SCOPE OF CHAPTER

A wide variety of active fire protection systems are available to fire safety practitioners. In addition to passive fire protection measures, some level of active fire protection is normally required to meet the expected minimum level of fire safety in modern buildings. Active fire protection can also be used to increase the fire safety in order to achieve a more flexible fire safety design and an acceptable level of fire safety in buildings. There are many types of active fire protection systems, but this chapter deals mainly with automatic fire sprinkler systems, since they are often used to facilitate the use of timber as structure, internal linings and external facades in large or complex buildings. Sprinklers are required in some countries for taller timber buildings, as described in Chapter 4.

10.1 GENERAL CONCEPTS OF ACTIVE FIRE PROTECTION

The main reason for using active fire protection is to manage fire impacts, protect property and provide more time for safe evacuation. It is always advantageous to include expert fire safety input at a very early stage in the design of a building in order to ensure that the building will be acceptably safe in the event of fire, and also cost-effective to design, build, operate and maintain. Several standards for active fire protection systems are available, e.g. within ISO, EN and NFPA (see below).

In contrast to passive fire protection which remains in place with no activation required, active fire protection systems become operational only when a fire occurs. Active fire protection measures include the following:

- Automatic fire detection systems – smoke, heat, flame, combustion gas, etc. to trigger alarms
- Fire alarm systems – audible or voice alarm, visual, tactile, vibrating
- Smoke management systems, including smoke fans and closing of smoke dampers in ducts

- Automatic closing of doors to provide fire compartments, refuges, etc.
- Other access control systems on doors, gates, turnstiles, etc. to aid evacuation
- Fixed fire suppression systems – water, chemical agents, inert gases

Fixed fire suppression systems control, suppress or extinguish a fire by cooling and/or wetting unburnt material, lowering the oxygen level or by chemical reaction, and thus inhibit or delay the combustion process. Water-based and gaseous fire suppression systems are most important.

Fixed fire suppression systems are typically designed and installed to control or suppress fire growth, as opposed to completely extinguishing the fire. It is often assumed that manual intervention will complete the suppression. This chapter covers mainly automatic sprinkler systems, since they may have the greatest influence on the use of timber in buildings.

10.2 DETECTION, ALARM AND SMOKE MANAGEMENT SYSTEMS

Detection and alarm systems and smoke management systems are very important active fire protection systems, but they are not discussed in detail here because their use is not specific to timber buildings.

10.3 SPRINKLER SYSTEMS

Sprinkler systems were developed by the insurance industry for property and asset protection and have been in existence for more than a century. They were subsequently found to have significant benefits for life safety as well. Sprinkler installations save lives, which is especially important in residential buildings. Some sprinkler system standards have been developed which focus on life safety objectives only for a limited range of buildings, e.g. EN 12259, EN 12845, NFPA 13R, but these sprinkler systems also provide some property protection benefits as well.

10.3.1 Objectives of sprinkler systems

Sprinkler systems have a long and successful history. The design and calculations take into account the size and construction of the building, the category of goods stored in it and the type of occupancy. Given that very few sprinkler heads may activate in the event of a fire, water damage from sprinkler systems is often minimal, although there have been cases of expensive damage from operation of one or two heads. Accidental discharge of water from sprinkler heads is a rare event, as is water leaking from sprinkler pipe work – see sprinkler reliability below.

Automatic sprinkler systems are often used to reduce risks that arise from the particular use of a building or to protect vulnerable individuals, high-value contents or to compensate for the materials used in the building's construction. They are often required in modern buildings, such as airports or storage facilities where there is limited compartmentalisation and extensive open spaces. They may also be used to protect premises that are geographically isolated. It is important that the reason for installation is understood and that, in those cases where the system has been provided as part of a fire engineering design for the building, responsible persons are made aware of and understand the interaction of the system and other building design and service features.

Sprinklers include provisions to assist the fire services to rescue occupants and minimise property damage and/or fight the fire. Both active and passive fire protection systems must be serviced and maintained in order to ensure that they will function in the event of a fire. Large fires are usually due to inbuilt fire precautions being disabled or compromised, e.g. doors left open or a delivery of equipment in combustible packaging temporarily being stored in an inappropriate location such as an atrium or stairwell.

10.3.2 Components of sprinkler systems

A sprinkler system consists essentially of a reliable water supply feeding an array of individual sprinklers mounted at defined spacing on an appropriately sized network of hydraulic pipes. Water may be supplied from one or more tanks by gravity or pumps or taken directly from the water main which will require pumps unless the mains supply can provide sufficient pressure and flow at all times.

Most sprinkler heads have an individual thermally activated element such as a fusible link or glass bulb. The thermally activated element supports a seal that holds back pressurised water in the sprinkler hydraulic network. Once the requisite thermal conditions are achieved, the thermally activated element releases the seal and allows water to flow from that element.

Glass bulbs are commonly used as a thermal element. A fluid in the glass bulb expands with the application of heat until the expanding fluid causes the bulb to fail, releasing the seal. Fusible links are another type of thermal element and are designed to melt at a particular temperature. A typical operating temperature for a sprinkler head is 68°C. However, there is thermal inertia in the mass of the sprinkler head, so an inherent delay may occur once fire gases have reached this temperature in the proximity of the sprinkler head, increasing the time for activation and discharge of water. This delay is characterised by the Response Time Index (RTI). A typical response time for a sprinkler head is a few minutes. Fast response sprinkler heads activate more quickly than standard heads. NFPA 13 provides information about sprinkler head temperature ratings, classification and colour

Figure 10.1 Typical sprinkler heads for (a) pendent, (b) concealed pendent (residential) and (c) water mist sprinklers

coding for glass bulbs. A few different types of sprinkler heads are shown in Figure 10.1.

The activation of sprinkler heads is affected by the fire plume and the resulting ceiling jet conditions. Factors that can impair sprinkler performance include fuel package shielding, obstructions of the fire plume or ceiling jet, and also fans and other ventilation equipment that may disrupt flow. Fire protection of high ceiling height areas with sprinklers can be challenging because the plume of hot gases from the fire will be weaker as it entrains cool air and cools down. It is also possible to get "sprinkler skipping" as water flow from an activated sprinkler head cools the fire plume or ceiling jet and nearby sprinkler heads. It is important to recognise that sprinkler systems are not activated by smoke or low-temperature gases, so they are not capable of activating in response to a smouldering fire.

There are several variations on the basic sprinkler design, intended to deal with unusual ambient conditions or particular risks. Many codes and standards are available to cover aspects of the design, specification, installation and maintenance of sprinkler systems, e.g. EN 12845 and NFPA 13. Installations in timber buildings may require special attention to secure correct fixings, some guidance is given by Lignum (2019). The spray pattern and the design density of sprays are important aspects of any sprinkler system design (Figure 10.2).

10.3.3 Wet-pipe and dry-pipe fire sprinkler systems

There are several different types of sprinkler systems, including wet-pipe systems, dry-pipe systems, closure systems and group release systems.

Wet-pipe systems are by far the most common due to their simplicity and reliability. In such a system, all sprinklers are connected to a pipe network filled with pressurised water. The water in turn pressurises each individual sprinkler head. Wet-pipe systems are more reliable than dry-pipe systems

Figure 10.2 Activated sprinklers.

because there is no delay for pipe-filling and there are fewer things that could go wrong. However, sprinklers will not operate in cold areas if the water inside the distribution pipes has frozen, so anti-freeze is added to sprinkler pipe water in some instances, for this reason.

Dry-pipe sprinklers have no permanent water pressure in the pipe system. Instead, the pipe network is pressurised with air. This is advantageous in cold rooms where there is a risk of freezing in the system. When a pressure drop occurs such as if a sprinkler is activated or a leak is present in the system, water fills the sprinkler pipe network. In a fire situation, this creates an additional delay for water to reach the fire compared to a wet-pipe system.

Pre-action or closure systems are dry-pipe systems with two separate release mechanisms. In addition to the sprinkler thermal element, an external fire detector must also be triggered to activate the system. This reduces the risk of water damage that can occur in the event of a malfunction of the system. At the same time, this reduces safety as two mechanisms must work.

Deluge or group release systems are activated by a separate fire detection system or are triggered manually. Water flow through a "zone" of sprinklers is initiated at the same time. The main advantage is that the system effectively prevents the spread of fire through a lot of water on a local surface.

10.3.4 Residential sprinkler systems

Residential domestic sprinklers were introduced in North America in the late 1970s as a new tool to reduce the high number of fire victims and increase safety of citizens and firefighting personnel in the United States (NCFPC, 1973). Residential sprinkler systems are simpler systems than standard sprinkler systems. They usually have the water supply directly connected to the ordinary tap water system (NFPA 13D, NFPA 13R, INSTA 900-1, EN 12259-14). The main aim of residential sprinkler systems is to control the fire and to increase the time for safe evacuation. These are often limited to low-rise residential construction. They are often cheaper than other types of sprinklers and they have a reliable water supply, since lack of tap water supply is immediately observed. However, their effectiveness is dependent on the pressure in the water supply system, they often have less robust inspection, testing and maintenance requirements, and they may not provide full coverage of all spaces in a building. Unoccupied areas like attics are usually not protected.

10.3.5 Water mist systems

Water mist systems were originally developed to suppress fires in ship engine rooms. They discharge water as much finer droplets than do wet-type sprinkler systems, and typically have a relatively low water delivery rate. Unlike wet-type sprinkler systems, water mist systems tend not to use interchangeable components, and therefore the design and installation of the complete system must be carefully controlled by a single supplier. The primary mechanism is gas-phase cooling rather than cooling of fuel packages by surface application of water.

Water mist systems from different manufacturers may operate at different pressures, they may use pumps or pressurised cylinders to provide the driving force to discharge water through heads which have activated and they may produce different droplet sizes. Application of a water mist system is based on specific testing of a particular hazard. In addition, individual components are subjected to component tests, which have been adopted from similar tests for wet-type sprinkler system components.

Water mist systems in buildings are proving to be particularly cost-effective in retrofit applications such as residential units, or to protect specific risks such as computer cabinets. They discharge water through specially developed nozzles to produce fine droplets, much smaller than traditional sprinkler systems. Triggering is typically similar to that used with wet-type sprinkler systems, and typically uses an individual glass bulb or fusible link to activate individual heads. The droplets are sufficiently small to be entrained by convective currents produced by the fire and can be drawn into and quickly cool the combustion zone or seat of the fire. They work best when used to control a growing fire in a relatively small compartment or room (as opposed to an assembly hall or warehouse), where the relatively fine water mist provides fire suppression by a combination of wetting, cooling and oxygen displacement.

Currently, the NFPA 750 and EN 14972 standards are available for design and installations, and further standardisation work is ongoing. Users of water mist fire suppression systems are very much dependent on information and data produced by the manufacturers. Consequently, the design and installation of a successful water mist system must take into account the probable type and location of fire, the fuel and the immediate environment.

A recent full-scale experiment demonstrates the effectiveness of a water mist system in an open-plan compartment with an exposed timber ceiling (Kotsovinos et al., 2022).

10.3.6 Sprinklers in earthquake areas

Sprinkler systems, like all mechanical and structural systems, are sensitive to earthquakes. In recent years, some design methods and standards have been developed (EN TS 17551) which specify requirements for earthquake protection of automatic sprinkler systems in accordance with EN 12845. These requirements only apply to locations in earthquake zones according to EN 1998-1 and for areas subject to peak ground acceleration above 9% of gravity.

In Canada, elements and components such as pipes and ducts, as well as their connections to the structure, are required to be designed to accommodate the calculated building deflections and lateral force (NBCC, 2015). NFPA 13 also provides installation requirements with respect to lateral sway bracing and horizontal seismic load to protect piping against damage.

In New Zealand, seismic design of sprinkler systems is covered in NZS 1170.5 by Section 8 Structural Design Actions – Part 5: Earthquake Actions. Fire protection systems are included in Category P.5 for high importance level (IL4) buildings, or Category P.6 for other buildings. These code clauses apply to any components which are essential for occupation of the building. There are also requirements for the seismic design of sprinkler system tanks in NZS 4541, Clause 6.6.3.

10.4 SPRINKLER EFFECTS ON FIRE SAFETY

Sprinklers have beneficial effects on fire development, property protection and life safety.

10.4.1 Effects on fire development

The majority of sprinkler systems are designed to control a fire by cooling the fire gases and the burning surfaces, and pre-wetting surrounding material to stop the fire from spreading. The design intent is to control the fire until it is finally extinguished by the fire service or staff using portable equipment. In reality, in many cases, the design intent is exceeded, and the fire is actually extinguished by the sprinkler system (CAENZ, 2008). This is probably related to the fact that the fire size at sprinkler actuation is much smaller than the design coverage area and is supported by statistics stating that 95% of all fires activate four or fewer sprinkler heads (Hall, 2010).

The effect of sprinklers on the heat release rate in fires has been studied in a number of experiments (e.g. Madrzykowski & Vettori, 1992) and in compartments with exposed CLT surfaces (Tests 4 and 5 from Zelinka et al., 2018), as shown in Figure 10.3, where the sprinklered scenarios are represented by the dashed line (Figure 10.3a) and from Tests 4 and 5 (Figure 10.3b) of which Test 5 had 20 minutes delayed manual activation. In these experiments, the heat release is reduced to almost zero after just a few minutes.

10.4.2 Property protection by sprinklers

Property protection was the initial aim of using sprinklers. Sprinklers were first used in industries of many different types, including sawmills. Now the use is extended to official and commercial buildings like assembly halls, shopping malls and other complex buildings. Testing procedures for cultural heritage applications with water mist to secure the invaluable property have also recently been developed (Arvidson, 2020). This includes several old timber churches in Norway and Sweden and their wall and ceiling paintings.

Sprinklers are especially recommended in tall timber buildings (Buchanan et al., 2014) since they create the possibility of a fire being extinguished or controlled well before the timber structure comes at risk of being involved in the fire. Building codes in many countries require all buildings to be protected by automatic sprinklers when they exceed a certain height, regardless of the type of construction (combustible or non-combustible).

10.4.3 Life safety by sprinklers

Sprinklers designed specifically to save lives were introduced in the United States in the 1970s as so-called residential sprinklers (Robertson, 2000; also see Section 10.3.2). As a result, firefighting agencies planned life and property loss-reduction strategies for handling incidents before they occurred. Residential sprinkler installation was such a strategy, but used mainly in some areas such as Vancouver, British Columbia, and in Scottsdale, Arizona (City of Scottsdale, 2022), where their use was made mandatory. Ahrens (2021) reports that fires in homes with sprinklers present resulted in a death rate of 88% lower and an injury rate of 28% lower than for fires in homes without sprinklers.

Purser (2001) found that sprinkler systems were highly effective in extinguishing fires rapidly, before conditions could threaten the occupants. Although there was significant smoke-logging, levels of heat and

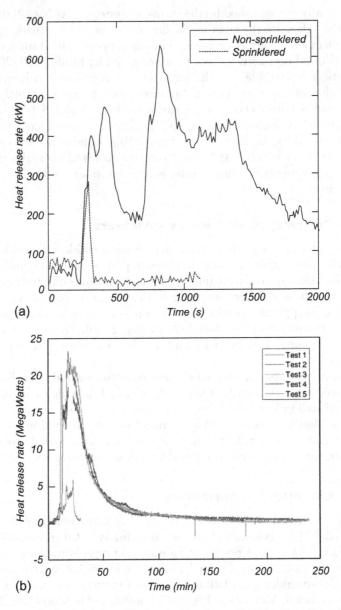

Figure 10.3 Sprinkler effect on the fire development for a sofa fuel package (Madrzykowski & Vettori, 1992) and CLT compartments (Zelinka et al., 2018).

toxic products were low, so there was ample time for occupants to escape without suffering serious injury. Further observations were that sprinklers may result in some impaired visibility during the early stages after sprinkler activation, particularly in the close vicinity to the sprinklered area. However, the use of sprinklers usually produces less loss of visibility than an equivalent unsprinklered fire in spaces contaminated with fire effluent. There was a significant benefit in terms of improved tenability resulting from a considerable decrease in heat and concentrations of irritant and asphyxiant gases.

Residential sprinklers and water mist nozzles in a residential fire scenario have recently been evaluated (Arvidson, 2017, 2022). It concludes that the performance of the water mist nozzles was comparable to or better than the residential sprinkler at approximately half of the water flow rate for the tested fire scenarios.

Sprinklers are often installed in very tall buildings in order to secure safe evacuation in cases where the fire service cannot evacuate by exterior ladders or sky lifts.

10.4.4 Cost-benefit analysis

The costs for installing sprinklers have often been used as an argument against, but the additional costs per square metre of a sprinklered residential area are usually in the order of the costs for a carpet.

Cost-benefit analyses have been used with very different outcomes, mainly depending on what is included in the analyses and how costs are calculated. In some cases, sprinklers were found to be cost-efficient in residential buildings, e.g. in Norway, where they are included in the building regulations (TEK, 2017), and also in New Zealand (Duncan et al., 2000).

BRE published an international cost-benefit study on residential sprinklers (Fraser-Mitchell and Williams, 2012). They concluded that residential sprinklers are cost-effective in homes for elderly, children and disabled persons, in blocks where costs are shared and in traditional homes with at least six bedsits per building where costs are shared.

A recent study presents a cost-benefit analysis using judgement value analysis for residential sprinklers (Hopkin et al., 2019). It provides a framework for how objective regulatory impact assessments could be undertaken in the future. It concludes that installations may offer a net benefit at least in some countries, e.g. in the United States and in Wales.

10.5 SPRINKLER RELIABILITY, PERFORMANCE AND EFFECTIVENESS

Reliability, performance and effectiveness are central parameters that need to be addressed when considering the effect of a sprinkler system (Nystedt, 2011).

10.5.1 Sprinkler reliability

Bukowski et al. (2002) define reliability as an estimate of the probability that a system or component will operate as designed over some time period.

The term unconditional reliability is an estimate of the probability that a system will operate "on demand." A conditional reliability is an estimate that two events of concern, i.e. a fire and successful operation of a fire safety system, occur at the same time. They use a term "operational reliability," i.e. a measure of the probability that a fire protection system will operate as intended when needed. The operational reliability is a measure of component or system operability and it does not take into account the possibility that system design does not match the fire hazards in the building. Therefore, there is a need to provide additional information on the likelihood that the fire development is within the design boundaries. Such measure of reliability is defined as the "performance reliability," i.e. a measure of the adequacy of the system design. A common approach to describe performance of a sprinkler system is to use terms such as Required Density Delivered (RDD) and Actual Density Delivered (ADD).

Available sources on reliability show some variability in the likelihood of successful sprinkler operation. The most likely cause of the flaws is that the collection of statistics does not recognise whether or not the fire was large enough to activate the sprinkler system or if the sprinkler system failed to operate when the fire was large. US statistics (Hall, 2010) indicates that the fire is too small to activate sprinkler heads in 65% of the fires. If this information is not considered in the collection of data, the reliability figures will be quite misleading. What is worth noting is also that a large portion of the fires either self-extinguish or is extinguished by manual intervention. Another aspect to consider when assessing the appropriate reliability figures for a specific trial design is if the system is designed in complete accordance with the standard, e.g. EN 12845, or if there are notable deviations. Reviews of sprinkler systems effectiveness (Frank et al., 2013, Ahrens, 2021) state that the reliability is in the neighbourhood of 90%.

Jensen and Haukø (2010) provide evidence on performance of sprinklers in fire by a compilation of accessible sources. The report addresses sprinklers, residential sprinklers and water mist for protection of residential, care, hospital, office, education and retail type of buildings. The information provided could be used as a knowledge base for anyone interested in sprinkler performance in various situations.

10.5.2 Sprinkler effectiveness

Fire Chief Len Garis and co-workers have published a range of detailed studies about sprinkler effectiveness, casualties, statistics on extent of damage, etc. in Canada showing that sprinklers are efficient in saving lives and property, especially in combination with alarm systems (Garis et al., 2012, 2017, 2019a and b; Maxim et al., 2013).

Reliability data on fire sprinkler systems has been collected and analysed (Fedöy and Verma, 2019). They presented an overview of studies from 1990 onwards with reliabilities over 90%. They also applied a methodology to validate scientific studies used in social sciences and concluded that there is a lack of knowledge of what an extinguishing system is, of different sprinkler systems and that they may perform differently. They suggested a systematic approach to understand the present variability in reliability data for sprinkler systems.

It should be noted that the reliability of sprinkler systems usually is higher than for many systems of passive fire protection, fire doors probably being the most obvious example with reliability levels down to 70% or less (BSI PD 7474-7).

10.5.3 Sprinkler management procedures

If a sprinkler system forms an integral part of the fire strategy for a building, it is essential to ensure that management procedures are in place to cater for those periods when the sprinkler system is not functional. Such procedures may include the following:

- Limit any planned shutdown to lower risk periods, when building occupant numbers are low.
- Isolate the area without working sprinklers from the rest of the premises by fire-resisting material.
- Avoid high-risk processes such as "hot-work."
- Train and deploy additional staff as fire patrols.
- Wherever possible, evacuation of occupants should be immediate and total with phased evacuation avoided.
- Inform the local fire and rescue service.

10.6 FIRE SAFETY DESIGN WITH SPRINKLERS: IMPLEMENTATION IN DIFFERENT COUNTRIES

The implementation of alternative fire safety design with sprinklers varies between countries. In some countries, sprinklers are required for timber buildings with more than a few storeys, as shown in Table 10.1. In other countries, sprinklers may be used for alternative fire safety design.

The data below are from a recent investigation (Östman, 2022). More information on national regulations is given in Chapter 4.

10.6.1 Countries with sprinkler requirements for taller timber buildings

The requirements may be expressed as the maximum number of storeys for a building with load-bearing timber structure or as the maximum height of a timber structure. They may be different for residential and office buildings, as shown in Table 10.1. Table 4.3 provides more comprehensive information.

Table 10.1 Countries with sprinkler requirements for buildings with load-bearing timber structure (requirements for non-sprinklered buildings in brackets)

| Country | Maximum number of storeys | | Comments |
	Residential buildings	Office buildings	
Australia	8 (3)	8 (2)	
Canada	12 (3)	12 (3)	Applies from 2020
China	5 (3)	5 (3)	
Estonia	8 (4)	8 (4)	
Finland	8 (2)	8 (2)	
Ireland	4 (3)	4 (3)	
UK	NL (3-4)	NL (10)	
USA	18 (0)	18 (5)	Applies from 2021

NL = No limit.

10.6.2 Countries with possibilities for alternative fire safety design with sprinklers

In addition to saving lives, sprinklers may allow for an alternative design of buildings. Requirements on passive fire protection to provide means of safe egress may be at least partly reduced, as shown in Table 10.2 and Figure 10.4. This will facilitate a more flexible use of alternative building products when sprinklers are installed. In some countries, wooden facade claddings may, for example, be used in sprinklered buildings, which is logical, since the risk of flames out of a window from a fully developed fire is eliminated if the sprinklers operate effectively (Nystedt, 2011).

Australia

The Australian prescriptive building regulations allow construction of all buildings up to an effective height of 25 metres from timber, as long as the timber is encapsulated with an insulating material and sprinklers are installed. This form of construction is termed "fire-protected timber." For example, external walls have traditionally been required to be completely non-combustible, over two storeys high. Fire-protected timber construction is now allowed if it meets the requirements (England, 2016).

For low- and mid-rise residential buildings, there are several prescriptive solutions in the building regulations. In most cases, there are additional reductions in fire resistance with voluntary inclusion of sprinklers.

New Zealand

The New Zealand Building Code Verification Method for calculating the required fire resistance in post-flashover fire allows a 50% reduction in

Table 10.2 Examples of reduced requirements on passive fire protection in sprinklered residential buildings in some countries

Country	Wooden facade claddings. Number of storeys	Wooden internal linings Number of storeys		Fire resistance, min	Comments
		Flats	Escape routes		
Australia	2 (2)	NL (NL)	0 (0)	Two to three storeys: load-bearing – 30 minute reduction Non-load-bearing no requirements Four to eight storeys: Non-load-bearing – 45 min	For fire-protected timber construction, the incipient spread of fire requirement may dominate the system construction, reducing the effect of fire resistance reduction
Canada	6 (3)	6 (3)	0 (0)		No limit in certain cases
Europe	–	–	–	Fire load decreased to 61% of normal fire load	EN 1991-1-2
Finland	8 (2)	8 (2)	0 (0)		Longer distance in escape routes
Sweden	8 (2)	8 (2)	0 (0)	–	Longer distance in escape routes
USA	3 (3)	NL (4)	Limits apply (4)	No reduction for sprinklers	Limits apply based on occupancy use type

Reduced requirements on passive fire protection in sprinklered buildings (requirements for non-sprinklered buildings in brackets)

NL = No limit.

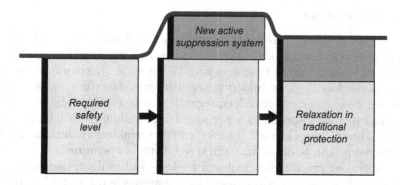

Figure 10.4 Principle for fire safety design by sprinklers. Increased fire safety by installation of sprinklers may lead to relaxations in the passive fire protection features, and still fulfil the same or higher safety level (Östman et al., 2002).

the fire load energy density if sprinklers are installed (MBIE, 2020b). The same principle is applied to the Acceptable Solution (the prescriptive code) which allows the 60-minute fire resistance rating for many buildings to be reduced to 30-minutes when an automatic sprinkler system is installed (MBIE, 2020a). The Acceptable Solution allows many other reductions in fire precautions when sprinklers are installed in buildings.

The New Zealand Building Code requirements for internal surface finishes (Clause 3.4) also include concessions for sprinklers (DBH, 2012).

Canada

The National Building Code of Canada (NBCC, 2015) prescribes the use of automatic sprinklers based on a number of factors, such as the building occupancy group, type of construction and building height (number of storeys). For example, a building of the residential occupancy group would require sprinklers if the building is made of timber construction and greater than three storeys. All buildings greater than six storeys, regardless of their type of construction, are required to be fully protected by sprinklers.

Moreover, the use of sprinklers generally allows designers to waive or reduce prescriptive fire protection requirements such as increasing the spatial separation between buildings, relaxing travel distance for safe egress, allowing interior finish materials of greater flame spread rating and eliminating the need for roofs to provide a fire resistance rating.

Given that sprinklers are already required in many prescriptive provisions of the NBCC applicable to low- and high-rise buildings, there are little opportunities to support the use of sprinklers as an alternative solution or a performance-based design intended to waive or reduce the level of fire protection requirements in Canadian buildings, although the NBCC (Clauses 3.2.2.47–3.2.2.54) allows combustible building materials and larger floor areas if sprinklers are installed.

The United States

For timber buildings, the International Building Code (IBC, 2018) has a range of different construction types, allowing non-fire-rated low-rise buildings, and buildings up to 18 storeys with 180 minutes fire resistance ratings. For low-rise buildings, sprinklers are required based on the use group, with sleeping uses requiring sprinklers, regardless of height. For offices, assembly and other uses, sprinklers are required once buildings reach four or five floors. The IBC does provide trade-offs when sprinklers are included, with additional height, building area and travel distances permitted.

As sprinkler protection is required for low- and medium-rise buildings, there is little scope for performance-based solutions to use sprinkler protection as a means by which to relax other building requirements, such as fire resistance ratings.

Europe

A reduction of the fire load if the fire cell is equipped with an automatic sprinkler system is included in the present version of Eurocode 1 (EN 1991-1-2) and also in the drafts for the next version. The fire load can then be reduced in the base case to 61% of its original value. The method is probabilistic based on the variable fire load, as described in Section 10.6.

The application of a reduction of the movable (variable) fire load can be traced back to the reliability of the sprinkler system, the fire cell size, the distribution of the fire load and the overall safety concept. Consequently, the material-independent reduction can be applied also in buildings with fire-exposed structural timber. However, the specified reduction should be applied only to the movable fire load and should not be applied to the additional fuel load–exposed timber surfaces.

Finland

In Finland, a lower fire exposure than for non-sprinklered buildings can be used when carrying out performance-based design of sprinklered buildings in which sprinklers are not the basic requirement. The height limits for timber buildings are increased when sprinklers are provided, as shown in Tables 10.1 and 10.2.

For up to two-storey timber buildings (where sprinklers are not required), the use of sprinklers enables wider use of wooden interior linings, e.g. in schools, nurseries, sport halls and office buildings.

Sweden

In Sweden, four separate reductions in fire precautions are permitted if sprinklers are installed, based on research by Nystedt (2011):

- Combustible facade cladding up to eight storeys
- Decreased requirements on surface linings in apartments in multi-storey buildings, down to class D-s1, d0 (the European reaction-to-fire class for normal wood panels)
- Decreased requirements on fire spread through windows in the same building
- Increased walking distance in escape routes

10.6.3 Examples of reduced fire precautions with sprinklers

The effect of residential sprinklers on a fire in an ordinary living room has been demonstrated in a Swedish project (Östman et al., 2002) (see Figure 10.5). It is evident that the fire damage is limited if sprinklers are

Figure 10.5 After a fire in a living room, left without sprinklers and right with sprinklers protection (Östman et al., 2002).

Figure 10.6 Exterior wood cladding permitted in a sprinklered building at the BoO1 building fair in Malmö, Sweden (Östman et al., 2002)

installed. No significant water damage has been experienced, since the water supply from sprinklers is far less than from a fire brigade arriving at a much later stage.

Modern multi-storey timber buildings with exterior wood facade cladding are examples of the implementation of fire safety design with sprinklers (see Figure 10.6). Wood facade cladding would not be permitted in such buildings with no sprinklers installed.

10.7 JUSTIFICATION FOR REDUCED FIRE PRECAUTIONS WITH SPRINKLERS

As shown in Table 10.2, and mentioned in Chapters 2 and 4, prescriptive codes in many countries allow fire resistance ratings and other fire precautions to be reduced, or fire compartment areas and travel distances to be increased when an automatic sprinkler system is installed.

It can be difficult to justify such trade-offs for the following reasons. If an automatic suppression system can be relied on with total certainty, no fire resistance or passive fire protection is necessary. However, no sprinkler system is 100% effective, so the question is what level of fire resistance should be provided, given the reliability of the sprinklers and the probability of an uncontrolled fire.

No national codes allow a total trade-off for sprinklers, but many codes allow a partial trade-off, assuming that in a sprinklered building, the probability of an uncontrolled fire is much less likely than the probability of a sprinkler-controlled fire.

As an example, if the sprinkler system fails when street water supplies are destroyed by an earthquake or an explosion, the resulting fire will have the same severity as if there had been no suppression system, so there should be no trade-off for sprinklers unless the low probability of this extremely unlikely event is taken into account.

Quantitative justification for partial trade-offs is not easy, but two possible probabilistic arguments are given:

1. If the fire resistance normally specified for burnout of a fire compartment in an unsprinklered building has an inherent safety factor of 2.0, then in the very unlikely event of such a fire due to sprinkler failure, that safety factor could be reduced to as low as 1.0, hence a 50% reduction in the fire load. Such an argument can only be justified if the method of specifying fire resistance for unsprinklered buildings is sufficiently conservative in the first instance.
2. Eurocode 1 Part 1.2 (EN 1991-1-2) specifies that for calculating fire resistance, the moveable fuel load in a sprinklered building can be taken as 61% of the design fuel load (see Section 10.6.2). This reduction results in the 80th percentile fuel load being reduced to the most likely fuel load, for design of sprinklered buildings.

If these probabilistic arguments are to be used as justification for reducing fuel load when sprinklers are installed, it is essential that automatic sprinkler systems be designed to be as reliable as possible, with enhanced reliability for tall and very tall timber buildings.

A quantitative risk assessment is a better way of justifying trade-offs resulting from active fire suppression systems. See Chapter 11 for performance-based design methods.

10.8 CONCLUSIONS

An automatic fire sprinkler system can play an important role in the fire safety design of timber buildings. Provided that they are installed correctly and operate effectively, sprinklers will control or extinguish a fire at an early stage and prevent flashover. Requirements for installing sprinklers vary considerably from country to country. An increased use of sprinklers in residential buildings would considerably decrease the number of fire victims, independent of the construction materials used in those buildings. Building designs to incorporate sprinkler systems may facilitate increased use of timber, to be used as the structural material, the internal linings or the external facade. Reliable sprinkler systems are essential in tall buildings of any material, and especially so for tall timber buildings.

REFERENCES

Ahrens, M. (2021) US Experience with sprinklers. *NFPA Research.* National Fire Protection Association. https://www.nfpa.org/-/media/files/news-and-research/fire-statistics-and-reports/suppression/ossprinklers.pdf.

Arvidson, M. (2017) *An Evaluation of Residential Sprinklers and Water Mist Nozzles in a Residential Area Fire Scenario.* RISE report 2017:40.

Arvidson, M. (2020) *Water Mist Fire Protection Systems: The Development of Testing Procedures for Marine and Heritage Applications.* Thesis, Lund University. ISBN 978-91-7895-554-1.

Arvidson, M. (2022) Active fire protection systems for residential applications in Runefors, M., Andersson, R., Delin, M. & Gell, T. (eds.) *Residential Fire Safety: An Interdisciplinary Approach.* Springer Nature, Switzerland AG.

Buchanan A., Östman B., Frangi A. (2014) *Fire Resistance of Timber Structures. Grant/Contract Reports (NISTGCR): 15-985.* https://doi.org/10.6028/NIST.GCR.15-985.

Bukowski, R. W., Budnick, E. K., Schemel, C. F. (2002) *Estimates of the operational reliability of fire protection systems, fire protection strategies for 21st century building and fire codes.* In Proceedings of the Society of Fire Protection Engineers and American Institute of Architects, Baltimore, MD.

BSI PD 7974-7 (n.d.) *Application of Fire Safety Engineering Principles to the Design of Buildings: Part 7 Probabilistic fire risk assessment,* British Standards Institution, UK.

CAENZ (2008) *New Zealand Centre for Advanced Engineering Fire Engineering Design Guide,* 3rd edition. Centre for Advanced Engineering. Christchurch, New Zealand.

City of Scottsdale (2022) *Residential Sprinkler Systems.* City of Scottsdale, Arizona. www.scottsdaleaz.gov/fire/residential-sprinkler

DBH (2012) *Schedule 1 of the Building Regulations 1992. New Zealand Building Code Clauses C1–C6 Protection from Fire.* DBH Department of Building and Housing, Wellington.

Duncan, C.R., Wade, C.A. and Saunders, N.M. (2000). *Cost Effective Domestic Fire Sprinkler Systems*. New Zealand Fire Service Commission Research Report Number 1. BRANZ, Wellington.

EN 54 (n.d.) *Fire Detection and Fire Alarm Systems, Parts 1–30. European Standard*. CEN European Committee for Standardization, Brussels.

EN 1991-1-2 (n.d.) Eurocode 1: Actions on Structures: Part 1–2: General Actions– Actions on structures exposed to fire. *European Standard*. CEN European Committee for Standardization, Brussels.

EN 1998–1 (n.d.) Eurocode 8 – Design of structures for earthquake resistance – Part 1: General rules, seismic actions and rules for buildings. *European Standard*. CEN European Committee for Standardization, Brussels.

EN 12101 (n.d.) Smoke and heat control systems and components, Parts 1–10. *European Standard*. CEN European Committee for Standardization, Brussels.

EN 12259 (n.d.) Fixed firefighting systems – Automatic sprinkler systems. Parts 1–14. *European Standard*. CEN European Committee for Standardization, Brussels.

EN 12845 (n.d.) Fixed firefighting systems – Automatic sprinkler systems – Design, installation and maintenance. *European Standard*. CEN European Committee for Standardization, Brussels.

EN 14972 (n.d.) Fixed firefighting systems. Water mist systems. Several parts. *European Standard*. CEN European Committee for Standardization, Brussels.

EN TS 17551 (n.d.) *Fixed Firefighting Systems: Automatic Sprinkler Systems: Guidance for Earthquake Bracing. Technical Specification*. CEN European Committee for Standardization, Brussels.

England, P. (2016) *WoodSolutions Technical* Design Guide 38. Fire Safety Design of Mid-rise *Timber Buildings*. Basis for the 2016 Changes to the National Construction Code, Australia.

Fedøy A., Verma A.K. (2019) *Reliability Data on Fire Sprinkler Systems: Collection, Analysis, Presentation, and Validation*. CRC Press, Boca Raton.

Frank, K., Gravestock, N., Spearpoint, M. et al (2013) A review of sprinkler system effectiveness studies. *Fire Science Reviews* 2, 6. https://doi.org/10.1186/2193 -0414-2-6.

Fraser-Mitchell J., Williams C. (2012) *Cost Benefit Analysis of Residential Sprinklers*. Final report. BRE Global.

Garis L., Clare J. (2012) *Evaluating Stakeholder Concerns with Wood Frame Buildings and Fire Risk*. University of the Fraser Valley, Canada.

Garis L., Singh A., Clare J., Hughan S., Tyakoff A. (2017) *Sprinkler Systems and Residential Structure Fires: Revisited*. University of the Fraser Valley, Canada.

Garis L., Singh A., Plecas D. (2019a) *Fire Protection System(S) Performance in the Residential Building Environment*. University of the Fraser Valley, Canada.

Garis L., Clare J., Biantoro C. (2019b) *Structure Fires in British Columbia*. University of the Fraser Valley, Canada.

Hall, J.R. (2010) *U.S. Experience with Sprinklers and other Fire Extinguishing Equipment, Fire Analysis and Research Division National Fire Protection Association*, NFPA, Quincy, Massachusetts, USA.

Hopkins, D., Spearpoint, M., Arnott, M., Van Coile, R. (2019) Cost-benefit analysis of residential sprinklers: Application of judgement value method. *Fire Safety Journal*, 106, 61–71.

IBC (2018) *International Code Council "International Building Code"*, International Code Council, Washington DC, US.

INSTA 900–1 (n.d.) *Residential Sprinkler Systems: Part 1: Design, Installation and Maintenance*. Nordic Standard INSTA InterNordicSTAndards, Standards Norway, Oslo.

ISO/DIS 20710–1 (n.d.) *Fire Safety Engineering: Active Protection Systems. Part 1: General Principles. Draft International Standard*. ISO International Organization for Standardization, Geneva.

Jensen, G., Haukø, A-M. (2010) *Sprinkler Performance Knowledge Base*, COWI Fire Research Report no. 02/2010, COWI A/S, Oslo.

Kotsovinos, P. et al. (2022). The Effectiveness of a Water Mist System in an Open-plan Compartment with an Exposed Timber Ceiling: CodeRed #03. *SPFE Extra Magazine.* https://www.sfpe.org/publications/magazine/fpeextra/fpe-extra2022/fpeextraissue74?_zs=MGA8k1&_zl=NHHD8.

Lignum (2019) *Dokumentation Brandschutz 6.1 Building Services: Installations and Bulkheads* (in German), Swiss Timber Industry, Zurich, Switzerland.

Madrzykowski, D., Vettori, R. (1992) A sprinkler fire suppression algorithm, *Journal of Fire Protection Engineering*, Vol 4, No 4, pp. 151–163. Society of Fire Protection Engineers.

Maxim, P., Plecas, D., Garis, L., Clare J. (2013) *Taller wood buildings and fire safety*. University of the Fraser Valley, Canada.

MBIE (2020a) *C/AS2. Acceptable Solution for Buildings other than Risk Group SH, for New Zealand Building Code Clauses C1-C6 Protection from Fire*. Ministry of Business Innovation and Employment, Wellington.

MBIE (2020b) *C/VM2. Verification Method: Framework for Fire Safety Design, for New Zealand Building Code Clauses C1-C6 Protection from Fire*. Ministry of Business Innovation and Employment, Wellington.

NCFPC (1973) *America Burning*, National Commission on Fire Prevention and Control, Washington DC.

NFPA 13 (n.d.) *Standard for the Installation of Sprinkler Systems*. National Fire Protection Association, Quincy, Massachusetts, USA.

NFPA 13D (n.d.) *Standard for the Installation of Sprinkler Systems in One- and Two-Family Dwellings and Manufactured Homes*. National Fire Protection Association, Quincy, Massachusetts, USA.

NFPA 13R (n.d.) *Sprinkler Systems for Residential Occupancies up to and Including Four Stories in Height*, National Fire Protection Association, Quincy, Massachusetts, USA.

NFPA 750 (n.d.) *Water Mist Fire Protection Systems*. National Fire Protection Association, Quincy Massachusetts, USA.

NBCC (2015) *National Building Code of Canada*. National Research Council Canada, Ottawa, Canada.

Nystedt, F. (2011) *Verifying Alternatives in Building with Fire Sprinkler Systems*. Department of Fire Safety Engineering and Systems Safety. Lund University, Sweden. Report 3150.

NZS 1170.5 (n.d.) *Structural Design Actions: Part 5: Earthquake Actions, Section 8. Standards New Zealand*, Wellington, New Zealand.

NZS 4541 (n.d.) *Automatic Fire Sprinkler Systems. Standards New Zealand*, Wellington, New Zealand.

Östman B. (2022) *National Fire Regulations for the Use of Wood in Buildings: Worldwide Review 2020.* Report Linnaeus University, Växjö, Sweden. ISBN 978-91-892783-39-8.

Östman, B., Arvidson, M, Nystedt, F. (2002) *Residential Sprinklers Save Lives: Experience and Fire Design with New Possibilities (in Swedish).* Trätek – Swedish Institute for Wood Technology Research. Publication 0202007, Stockholm, Sweden.

Purser, D.A. (2001) *Sprinklers and Fire Effluents: Methods for Assessing Life Threat from Sprinklered Fires.* ISO TC92/SC3 WG2. International Organization for Standardization (ISO), Geneva.

Robertson, J. (2000) *Building Codes and NFPA Standards, Session 2 Water based Fire Suppression Systems,* City of Vancouver, Canada.

TEK 17 (2017) *Building Regulations with Guidance.* Directorate for Building Quality, Norway.

Zelinka, S., Hasburgh, L., Bourne, K., Tucholski, D., Ouellette, J. (2018) *Compartment Fire Testing of a Two-story Cross Laminated Timber (CLT) Building. General Technical Report FPL-GTR-247.* U.S. Department of Agriculture, Forest Service, Forest Products Laboratory, Madison.

Chapter 11

Performance-based design and risk assessment

Paul England, David Barber, Daniel Brandon, Christian Dagenais, Gianluca De Sanctis, Michael Klippel, Dennis Pau and Colleen Wade

CONTENTS

DOI: 10.1201/9781003190318-11

SCOPE OF CHAPTER

This chapter provides an overview of the application of performance-based approaches to the fire safety design of timber buildings. Performance-based design methods are relevant for the design of tall timber buildings and other timber buildings that vary from accepted prescriptive solutions. Performance-based design approaches are commonly categorised as deterministic or probabilistic methods and should be applied in accordance with the applicable regulations, building codes and standards. This chapter provides references to detailed information that should be consulted when undertaking performance-based designs.

11.1 INTRODUCTION

11.1.1 Performance-based design

The performance-based design definition in the SFPE Engineering Guide to Performance-Based Fire Protection (SFPE, 2007a), with modifications to provide a general definition of performance-based design and to identify the need for performance criteria to be defined early in the design process, is provided below.

Performance-based design is an engineering approach based on:

- agreed or prescribed fire safety goals, objectives and performance criteria,
- deterministic and/or probabilistic analysis of fire scenarios, and
- assessment of design alternatives against the performance criteria using accepted engineering tools and methodologies.

Performance-based design is founded on the principle that a building design must meet design goals and objectives (also referred to as drivers and constraints) and the goals and objectives need to be converted into quantified performance criteria, in order to identify optimal solutions and to enable compliance of the design to be demonstrated by means of appropriate analysis methods in a transparent manner.

Methods of quantitative analysis, as identified in the above definition, are commonly classified as deterministic or probabilistic (risk-based). Whichever approach is adopted, the definition and quantification of appropriate fire scenarios is critical.

When performance-based design methods are applied to buildings, common objectives such as limiting the risk to life or property require the consideration of both the frequency of occurrence and consequences of fire scenarios and hence the performance criteria and analysis methods need to address risk. This can be achieved by either explicitly defining performance criteria in terms of risk and undertaking a quantitative risk assessment or implicitly by undertaking deterministic analysis and defining performance criteria in terms of the consequences of one or more credible worse-case fire scenarios. The selected or nominated credible worse-case scenarios infer a tolerable frequency of occurrence, although this is not normally explicitly quantified (Table 11.1).

When deterministic approaches are adopted, design fire scenarios are typically selected based on either a best estimate (also referred to as typical or credible scenarios) or more conservative estimates (referred to as credible worse- or worst-case scenarios). For life safety applications which are a major focus for building regulations and codes worse/worst-case scenarios are generally adopted. The terms worst and worse are often interchanged. In this chapter, the term credible worse case has been adopted because generally a challenging design fire scenario is derived to represent an acceptable level of risk if the performance criteria are satisfied, not the most severe

Table 11.1 Summary of design scenarios nominated in the Verification Method CV4 in the Australian National Construction Code

Ref.	Design scenario	Design scenario description
BE	Fire blocks evacuation route	A fire blocks an *evacuation route*
UT	Fire in a normally unoccupied room threatens occupants of other rooms	A fire starts in a normally unoccupied room and can potentially endanger a large number of occupants in another room
CS	Fire starts in concealed space	A fire starts in a concealed space that can facilitate fire spread and potentially endanger a large number of people in a room
SF	Smouldering fire	A fire is smouldering in close proximity to a sleeping area
HS	Horizontal fire spread	A *fully developed fire* in a building exposes the *external walls* of a neighbouring building (or potential building) and vice versa
VS	Vertical fire spread involving cladding or arrangement of openings in walls	A fire source exposes a wall and leads to significant vertical fire spread
IS	Fire spread involving internal finishes	Interior surfaces are exposed to a growing fire that potentially endangers occupants
FI	Fire brigade intervention	Facilitate fire brigade intervention to the degree necessary
UF	Unexpected catastrophic failure	A building must not unexpectedly collapse during a fire event
CF	Challenging fire	Worst credible fire in an occupied space
RC	Robustness check	The requirements of the NCC should be satisfied if failure of a critical part of the fire safety systems occurs.
SS	Structural stability and other properties	Building does not present risk to other properties in a fire event. Consider risk of structural failure

scenario that could be devised through a simultaneous failure of all components of a fire safety design at the same time as an extreme fire event occurs.

Fire safety designs for complex buildings often require multiple fire protection systems that interact with the building and occupants, and provide appropriate levels of redundancy, to address fire events where one or more fire protection systems may fail as well as addressing variations in human behaviour. It is therefore important to adopt a holistic risk-based approach to fire safety design which takes account of these interactions and probabilities of failures of fire protection systems.

The broader community tends to be more averse to hazards that lead to multiple fatalities from a single event and therefore for larger buildings with large populations, it may be necessary to consider both individual and societal risks when undertaking quantitative risk assessments to ensure that societal expectations are satisfied. The following definitions can be applied to individual and societal risk:

- Individual risk can be defined as the frequency at which an individual may be expected to sustain a given level of harm from the realisation of a specified hazard (e.g. fire) in the subject building.
- Societal risk can be defined as the frequency that a number of people may be expected to sustain or exceed a specified level of harm in the subject building(s) from the realisation of a specified hazard.

11.1.2 Early developments

Most building codes provide prescriptive building solutions which are in effect specifications that can be adopted to satisfy the applicable building legislation. These typically reference national or international product standards, testing standards and engineering design methods. They can be applied with a high level of confidence that the design will comply with the building code provided that the building is documented and constructed in accordance with the code and referenced documents. For many buildings this may be the preferred option, provided the designs are also compatible with other design drivers and constraints, and the buildings are not unusual or innovative. For large and more complex buildings or also existing buildings where many drivers and constraints and/or innovative building systems are being considered, prescriptive approaches might not be able to provide an optimal solution and an alternative pathway may be needed.

Historically, building codes and/or building regulations have included alternative pathways to allow variations and modifications to prescribed solutions for buildings on a case-specific basis. These were generally accepted on the basis of demonstrating equivalence to an accepted prescriptive solution. In its most basic form, one fire protection measure is removed, and an alternative is introduced that is considered to be at least as effective. The term "trade-off" has commonly been used to describe this approach.

The development of performance-based design and risk assessment approaches began in the early 1970s and a summary by Meacham (1996) identified key outcomes from this period:

- The observation by Fitzgerald (1985) and others in the United States that the various fire safety measures in a building combine to form a single fire safety system. This eventually led to the development of what is now known as the fire safety concepts tree (NFPA, 550, 2022)
- In the late 1970s in Australia, the application of risk assessment modelling to fire safety design of buildings was introduced by Beck (1983).

The fire safety system approach is also commonly referred to as a holistic approach.

Further developments continued throughout the 1980s and 1990s with the UK Building regulations being published in a performance-based format in 1985 with several other countries also introducing performance-based approaches in the 1990s. During this period, numerous guidance documents were developed:

- Draft national building fire safety systems code. Building Regulation Review Task Force, Microeconomic Reform: Fire Regulation, Department of Industry Technology and Commerce, Canberra, Australia (Beck, 1991)
- Fire Engineering Design Guide, First Edition. University of Canterbury, Christchurch, New Zealand (Buchanan, 1994)
- Fire Engineering Guidelines, First Edition, Sydney (FCRC, 1996)
- British Standard DD240-1(1997), Fire Safety Engineering in Buildings: Guide to the Application of Fire Safety Engineering Principles

Currently, many countries allow performance-based pathways as alternatives to prescriptive approaches, although permissible approaches and the extent to which acceptable risk levels are articulated vary.

11.1.3 Overview of the fire safety design process

A generic high-level fire safety design process for a building is shown in Figure 11.1, which is an enhanced version of a flowchart from England and Iskra (2021). It includes the following procedures which need to be undertaken in close consultation with the relevant stakeholders:

- Identification and documentation of mandatory and voluntary objectives
- Selection of compliance pathways

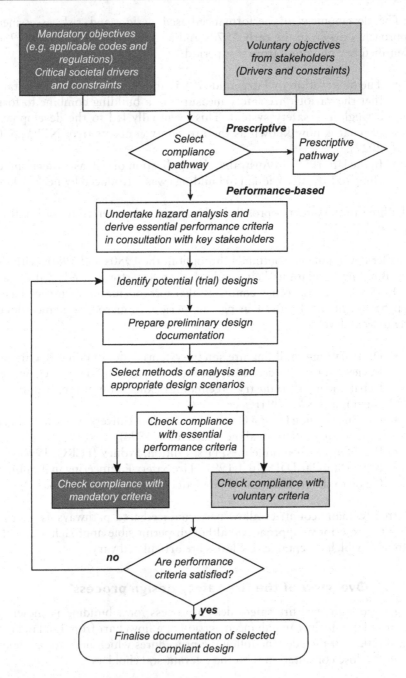

Figure 11.1 Overview of the design process.

- Hazard analysis
- Derivation of essential performance criteria
- Identification of one or more potential (trial) fire safety solutions (strategies)
- Development of preliminary design documentation of the preferred potential building solution
- Selection of methods of analysis and appropriate design scenarios
- Checking compliance against all mandatory and voluntary criteria
- Development and documentation of the compliant building solution

There are many minor variations to the above process, depending on the applicable regulations and practices. ISO 23932-1 includes another example.

Although the focus of this chapter is use of performance-based design to demonstrate compliance with applicable building codes, there are other objectives that the designer needs to account for. Some of these may be mandatory such as the application of safe design principles through workplace health and safety legislation whereby hazards during installation, maintenance and normal use of buildings need to be considered. Irrespective of the regulatory requirements, designers have a duty of care to consider these hazards even if not required by a building code.

Other objectives may be voluntary such as business continuity and enhanced levels of property protection (above minimum code requirements).

Identifying objectives (key drivers and constraints) early in the design process facilitates the adoption of a holistic approach to design so that synergies between the various drivers and constraints can be exploited and design constraints can be managed, allowing an efficient and effective design solution to be determined. For example, business continuity requirements may require a large proportion of fires to be kept small leading to the adoption of a design strategy based on early automatic suppression. Early suppression can then be incorporated in the life safety strategy if a performance-based design approach is adopted to demonstrate compliance with the relevant building code.

CIB W14 (Thomas, 1986) identified the protection of life, for both occupants and firefighters, and the protection of neighbouring property as the top-level code objectives. However, the mandatory objectives in building codes can vary between jurisdictions, particularly with respect to property protection.

There is further diversification with respect to performance criteria which may be specified in building codes as qualitative criteria, quantitative criteria or a mixture of both. To manage the uncertainty and differing interpretations of codes with qualitative performance criteria, a common approach is to establish a team of stakeholders to determine the quantitative acceptance criteria relevant to a specific project. An example of stakeholder agreement is the Fire Engineering Brief (FEB) described in the International Fire Engineering Guidelines (ABCB, 2005).

Guidelines are available describing possible processes for various jurisdictions, but there are significant variations in interpretations between
stakeholder teams. There is a growing focus on explicitly quantifying performance criteria in future building codes, e.g. Draft NCC Building Code
of Australia ABCB 2022, or implicitly quantifying the expected performance using comparative approaches, e.g. INSTA 950.

11.1.4 Pathways for demonstrating compliance

There are two pathways that can be followed to demonstrate compliance of
a building with relevant building codes, a prescriptive pathway or a performance-based pathway:

- If the prescriptive pathway is followed, demonstrating compliance of
  the overall design is relatively straightforward in that it is only necessary to select a predefined combination of provisions from the relevant
  code. Notwithstanding this, demonstrating compliance of individual
  components of the proposed design, such as the fire resistance of elements of construction and/or the reaction-to-fire performance, still
  requires significant expertise and diligence.
- The performance-based pathway provides an option for situations
  where the prescriptive pathway is too restrictive or unsuitable for
  a particular application, or there is an opportunity for innovation.
  It is expected that the application of performance and risk-based
  approaches/methods will be undertaken by appropriately qualified
  and competent engineers in the fields of fire safety and structural engineering. It is particularly important that users understand the limitations of any methodology used.

The term "Fire Safety Engineering" (FSE) is commonly applied to describe
the process of deriving and demonstrating compliance of a performance-
based design. Fire Safety Engineering comprises the application of engineering methods based on scientific and natural principles for the development
or assessment of design in the built environment by evaluating design fire
scenarios and by quantifying the risk associated with the consequences or
outcomes of these scenarios. See also ISO 13943 for a definition of Fire
Safety Engineering.

The two pathways are shown schematically in Figure 11.2.

There are two major branches in the performance-based pathway:

- Comparative, whereby the risk associated with a proposed building
  is compared against a benchmark building (usually an accepted prescriptive solution). The performance criteria generally require the fire
  risk relating to the proposed building to be at least equivalent to the
  selected benchmark building.

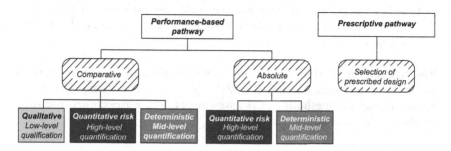

Figure 11.2 Prescriptive and performance-based pathways for demonstrating compliance with building codes and regulations.

- Absolute, whereby the performance criteria are specified either explicitly in terms of acceptable or tolerable risk or implicitly typically by means of the specification of scenarios and required deterministic outcomes for each scenario.

Methods of analysis used to determine compliance can be classified as belonging to one of the following types:

- Quantitative risk assessments
- Predominantly deterministic assessments, but estimates of frequency and probabilities may inform the selection of credible and/or worse-case fire scenarios if the fire scenarios are not fully prescribed.

A third method, qualitative assessments, is shown under the comparative branch. This method has low levels of quantification of outcomes and the frequency of occurrence. Such methods rely substantially on engineering judgements, sometimes supported by fire test data. The use is generally restricted to designs where it is possible to determine that a proposed design presents an equivalent or lower risk compared to an acceptable prescriptive solution without undertaking detailed analysis, e.g. designs with single variations and/or minor variations from an accepted benchmark prescriptive solution that can be adequately compared without the need to undertake a holistic design.

There are no clearly defined boundaries between the above types of analysis since they have many common features:

- Qualitative or semi-quantitative engineering judgements are necessary when applying all methods, particularly where directly applicable data is limited to support deterministic assessments and quantitative risk assessments.
- When undertaking a deterministic analysis, the likely frequency of occurrence of the design fire scenario infers an acceptable level of risk in conjunction with the performance criteria.

- When undertaking quantitative risk assessments, deterministic techniques may be used to determine the consequences of reference scenarios.

The frequency/probability components within each of the above analysis methods are recognised in ISO 23932-1 which broadly classifies all three of the above analysis methods as risk assessments. These aspects are discussed further in the following sections.

11.1.5 Sources of further information

The application of performance-based codes varies between jurisdictions and therefore reference should be made to the relevant codes, standards and reference material when undertaking performance based-design of a building for fire safety. Some general information and national documents are summarised below; see References for further details.

General References/International Standards

The Evolution of Performance-Based Codes and Fire Safety Design Methods (Meacham, 1996)

Code Official's Guide to Performance-Based Design Review (SFPE, 2004)

Risk Analysis in Building Fire Safety Engineering (Hasofer et al., 2007)

Fire Risk Assessment (SFPE, 2007b)

Performance-based Fire Protection (SFPE, 2007a)

SFPE Handbook of Fire Protection Engineering (SFPE, 2016)

ISO standards:

- ISO 16732-1 Fire Safety Engineering – Fire Risk Assessment – Part 1 General
- ISO 16733-1 Fire Safety Engineering – Selection of Design Fire Scenarios and Design Fires – Part 1: Selection of Design Fire Scenarios
- ISO/TS 16733-2 Fire Safety Engineering – Selection of Design Fire Scenarios and Design Fires – Part 2: Design Fires
- ISO 23932-1 Fire Safety Engineering – General Principles

Europe

Eurocode 1 Actions on Structures – Part 1–2: General Actions – Actions on Structures Exposed to Fire (EN 1991-1-2:2002)

Eurocode 5 Design of Timber Structures – Part 1–2: General – Structural Fire Design (EN 1995-1-2:2004)

Fire Safety Engineering – Comparative Method to Verify Fire Safety Design in Buildings. Inter-Nordic Technical Specification (INSTA TS 950)

Fire Safety Engineering – Guide for Probabilistic Analysis for Verifying Fire Safety Design in Buildings. Inter-Nordic Technical Specification (INSTA 951)

UK

Application of Fire Safety Engineering Principles to the Design of Buildings – Code of Practice (BS 7974)
Structural Timber Buildings – Fire Safety in Use Guidance. Structural Timber Association, UK (STA, 2020)

Australia

Draft National Construction Code 2022 (ABCB, 2021a)
Handbook – Fire Safety Verification Method (ABCB, 2019b)
Fire Safety Verification Method Data Sheets – Handbook Annex (ABCB, 2019a)
Australian Fire Engineering Guidelines (ABCB, 2021b)
WoodSolutions – Technical Design Guide 17 – Fire Safe Design of Timber Structures – Compliance with the National Construction Code (England and Iskra, 2021)

New Zealand

Verification Method C/VM2, Framework for Fire Safety Design (MBIE, 2020)
Fire Engineering Design Guide (Spearpoint, 2008)

The United States

Performance Code for Buildings and Facilities (ICC, 2021)
NFPA 5000 – Building Construction and Safety Code (NFPA 5000, 2018)
Performance-Based Fire Safety Design (Hurley and Rosenbaum, 2015)

11.2 HAZARD ANALYSIS AND FIRE SCENARIOS

11.2.1 Overview of hazard analysis process

A hazard analysis is undertaken to identify fire hazards and risk factors that have the potential to cause harm. The information obtained from the hazard analysis is then used in conjunction with a preliminary qualitative analysis, normally undertaken with key stakeholders, to derive trial fire safety strategies, to identify appropriate analysis methods and relevant design fire scenarios for evaluation.

There are various hazard identification techniques or combinations of techniques that can be applied:

- Checklists
- What If Analysis
- Hazard Identification (HAZID)
- Hazards and Operability Analysis (HAZOP)
- Failure Mode and Effects Analysis (FMEA)
- Literature review/review of historic record

11.2.2 Overview of fire scenarios

A fire scenario is defined as a qualitative description of the course of a fire with respect to time identifying key events that characterise the studied fire and differentiate it from other possible fires (ISO 16733-1).

Essentially, the nature of fires is stochastic, so there is a universe of potential fire scenarios that can apply to a specific building. It is not possible to undertake specific analyses for all potential scenarios and the number of fire scenarios needs to be rationalised for analysis.

The fire scenario definition is not restricted to the fire development and spread, but also includes the performance of various fire protection systems and building features that interact with a fire together with human response and resilience which impact on the effectiveness of evacuation or avoidance strategies and fire brigade intervention.

This highlights the importance of applying a holistic approach to fire safety by considering fire safety as a single system. Nevertheless, breaking the fire safety system down into subsystems that interact can be an effective simplification for analysis without losing the benefits of considering fire safety holistically. An example of a subsystems approach is shown in Figure 11.3, but other groupings of subsystems can also be derived.

ISO 16733-1 describes a nine-step process to identify design fire scenarios:

1. Identification of specific challenges
2. Location of fire
3. Type of fire
4. Potential complicating hazards leading to other fire scenarios
5. Systems and features impacting on fire
6. Occupant actions impacting on the fire
7. Selection of design fire scenarios
8. Modify scenario selection based on system availability and reliability
9. Final selection and documentation

To undertake a quantitative analysis, it is necessary to reduce the number of scenarios that need to be considered and quantify the related inputs for

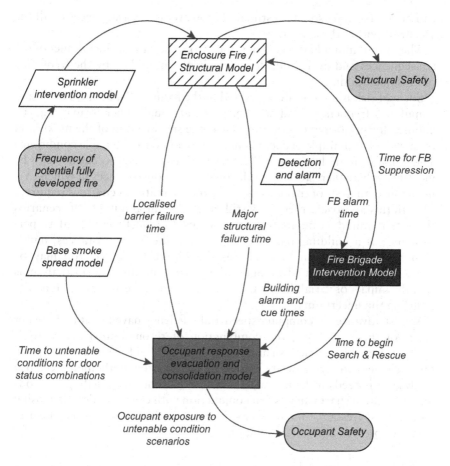

Figure 11.3 Graphic indicating subsystems adopted for a multi-scenario quantitative risk assessment of fire-protected timber construction (England, 2016).

the analyses as appropriate. The approach taken to reduce the number of scenarios depends upon the selected analysis method.

11.2.3 Rationalisation of fire scenarios for quantitative risk assessments

If a quantitative risk assessment approach is adopted, the continuum of possible fire scenarios is subdivided into clusters and for each cluster, its probability of occurrence is defined in conjunction with a representative fire scenario such that the consequences for the representative fire scenario can be used as a reasonable estimate of the average consequences of all the scenarios within the cluster.

The overall individual risk is obtained from the sum of the product of the consequences and probability for each cluster multiplied by the frequency of fires in a building.

Societal risk is commonly expressed in the form of F–N curves depicting cumulative frequencies and consequences, although other formats can be adopted for performance criteria such as the specification of the maximum frequency that multiple occupants may be exposed to untenable conditions. The amount of analysis can be further reduced if scenario clusters can be identified where there are no significant adverse consequences in which case no detailed analysis of that cluster or the representative scenario is required.

With this approach, there may still be a substantial number of scenarios that are required to be analysed to adequately model the fire safety performance of a building and its occupants. In these circumstances, multi-scenario analysis techniques such as the Monte Carlo Method may be employed. A significant advantage with the use of multi-scenario analyses is that the impact of variability of inputs is included in the analysis potentially reducing the uncertainty.

Whilst advances in computer speed and efficiency have facilitated greater usage of these methods, the combination of resource-intensive analysis methods for consequences such as computational fluid dynamic (CFD) and finite element analysis of structures in conjunction with multi-scenario analysis still needs to be managed. The use of simpler modelling methods for the multi-scenario analysis in conjunction with checks undertaken using more precise (resource-intensive) methods for critical scenarios may be necessary to provide a practical solution.

11.2.4 Rationalisation of fire scenarios for deterministic analyses

Generally, deterministic analyses are restricted to a limited number of design fire scenarios that are closely aligned to the performance criteria and are normally intended to represent worse credible scenarios. However, in applications where a building code or relevant legislation does not define, in quantifiable terms, a worse credible scenario, there is likely to be substantial variability in outcomes, especially if absolute analysis methods are employed. To some extent, this variability or uncertainty can be reduced if comparative analysis methods are undertaken.

Some common causes for variability include the treatment of the following:

- Common mode failure (e.g. failure of a detection system may lead to no automatic alarm for occupants and notification of the fire brigade, no activation of automatic smoke management measures, no release of hold open devices fitted to fire and smoke doors)

- Failures of one or more independent fire protection systems
- Treatment of distributions for inputs relating to unregulated matters such as fire loads, growth rates and ventilation
- Variability in human response, particularly in residential properties

The specification of design fires, derived from design fire scenarios, has often been recognised as a potential source of uncertainty in performing any fire safety engineering assessment. This uncertainty stems mainly from the natural variability in fuel load and its configuration, the complexity of the compartment fire dynamics (rate of heat and smoke release), the likelihood of window breakage providing ventilation, as well as the point of ignition and subsequent spread of fire to adjacent spaces.

11.2.5 Prescribed fire scenarios

Some building codes nominate high-level generic fire scenarios for analysis, but these do not exclude the adoption of other scenarios to address hazards that are not adequately addressed by the nominated scenarios. The generic scenarios may require *more* than one scenario to be evaluated. For example, there may be a large number of occupancy-specific design fire scenarios for a particular occupancy.

NFPA 5000 (2018) nominates the following general scenarios; more detailed descriptions and advice are provided in the codes:

- Typical occupancy-specific design fire scenario
- Ultra-fast developing fire in the primary means of egress
- Fire in an unoccupied room near a high-occupancy space
- Concealed space fire near a high-occupancy space
- Slow developing shielded fire near a high-occupancy space
- Most severe fire associated with the greatest fuel load
- Outside exposure fire

Verification methods referenced by the Australian and New Zealand building codes (ABCB, 2020; MBIE, 2020) also specify scenarios for consideration. The scenarios referenced in the Australian NCC verification method CV4 are summarised in Table 11.1 and are adaptations of those originally used in New Zealand and included in ISO 16733-1, but CV4 utilises a comparative approach rather than an absolute approach for the detailed analysis.

These high-level design scenarios are specified in qualitative terms with matters such as the number of fire locations, fire characteristics and frequency of the scenarios requiring further development based on the subject-building characteristics. In many cases, for each nominated design scenario, several design scenarios may need to be derived to address various locations, fire growth rates, different ventilation conditions, variations in occupant response and fire brigade intervention etc.

11.3 APPLICATION OF ANALYSIS METHODS TO TIMBER CONSTRUCTION

An overview of the general analysis methods described in the previous sections of this chapter when applied to timber construction will be provided using a multi-storey apartment building as an example.

The analysis methods should be selected early in the design process in conjunction with the stakeholders, following a hazard identification process, considering the minimum analysis methods required by building codes, regulations and approval authorities. For example, if an absolute quantitative risk assessment approach is not considered necessary for a project, a comparative approach or an absolute deterministic approach may be more suitable, depending on the circumstances.

11.3.1 Hazard identification

The following are some of the potential fire hazards associated with timber construction:

- Increase in frequency and consequences of fires during construction. This may not be a building code requirement or may only be partially addressed by building codes. Irrespective of the building code requirements, fires during construction must be fully addressed (i.e. an additional objective if not already required), as described in Chapter 13.
- Increase in fire growth rate if timber elements are exposed to a fire source compared to non-combustible elements. In some occupancies such as multi-residential buildings, combustible timber linings may be commonly used within apartments provided that the reaction-to-fire performance can be shown to be acceptable. Higher levels of performance may be required in exit routes. Protection can be fire-retardant treatments, non-combustible facings or gypsum plasterboard.
- Potential increase in the frequency of fires starting either within wall and floor cavities or on the surface of walls and ceilings where the surface of the timber is exposed. The following are the options to address this hazard:
  - Detailing of timber to avoid the creation of cavities with exposed timber elements as far as practical
  - Where cavities exist, using cavity barriers and non-combustible materials for sound and thermal insulation
  - Detailing of service penetrations to minimise the need for hot works within cavities
  - Restricting the use of exposed timber to areas where the likelihood of ignition is low
  - Addition of automatic sprinkler protection

- If a fully developed fire occurs and the timber elements are involved in the fire:
  - Increase of external flaming at the facade due to additional fuel from the construction materials inside the compartment may facilitate external fire spread to the floors above, horizontal external fire spread to other buildings and/or fire spread across any exposed external timber elements.
  - The fire duration may be increased due to the additional fire load causing structural failure and/or failure of the fire compartmentation.
  - The severity and duration of the fire may present an increased risk to occupants and firefighters and may also reduce the probability of successful suppression by the fire service.
- Potential mitigation measures to reduce the risk include the following:
  - Encapsulation or partial encapsulation of all timber surfaces
  - Encapsulation of most timber surfaces except for specific nominated elements, e.g. a feature wall or ceiling, or beams and columns
  - Addition of automatic sprinkler protection
  - Fire brigade intervention

Additional hazards that need to be considered, associated with the potential mitigation measures, include gross defects associated with the encapsulation systems, the cavity barriers or the automatic sprinkler system and no effective fire brigade intervention.

11.3.2 Preliminary qualitative and quantitative analysis

A preliminary *qualitative analysis* is usually the first step following hazard identification, to derive potential fire safety designs for further analysis, and also to determine and confirm the methods of analysis and the scenarios to be considered.

For relatively minor variations from the prescriptive requirements, a comparative qualitative analysis may be sufficient to demonstrate compliance with the building code, if that type of solution is permitted.

However, a *quantitative analysis* would typically be required to evaluate the risks associated with potential fully developed fires for a building of predominantly timber construction if the relevant jurisdiction mandates non-combustible construction for structural elements. Some quantitative analysis of the pre-flashover stages of the fire may still be necessary to determine fire and smoke spread prior to flashover, to estimate fire detection and alarm times, to identify compromised evacuation paths and to account for occupant response, fire brigade intervention and access for firefighting purposes, etc. This information is used to estimate the number and location of occupants at risk after flashover and the probability of successful fire brigade intervention, amongst other things.

11.3.3 Fire scenarios for quantitative risk assessment

If a quantitative risk assessment is to be undertaken, or in order to inform the selection of a worse credible scenario for a deterministic analysis, it is useful to determine the frequency of flashover fires.

Figure 11.4 shows a generic graph of enclosure temperatures versus time for a small fire enclosure, together with potential fire behaviour and interventions that may modify the fire. The top (solid) line in this graph shows a fire scenario where the fire reaches flashover and passes through a decay phase. The dotted line shows a cooling phase assuming no re-growth, secondary flashover or extended periods of smouldering combustion occurring. The dashed lines show various interventions that can reduce the severity of a fire.

For many structural and fire-separating elements, the structure and/or barriers will only be challenged if they are subjected to a fully developed fire.

If a quantitative risk assessment is being undertaken, a simple event tree can be constructed, as shown in Figure 11.5. With an estimate of the frequency of ignition (F_{ig}), an analysis of fire data, modelling of fire brigade intervention and further analysis as appropriate, the probabilities of

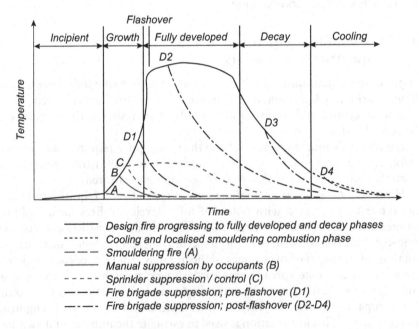

Figure 11.4 Typical design fire for a small enclosure fire.

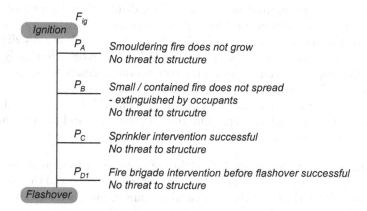

F_{ig}

Ignition

P_A Smouldering fire does not grow
No threat to structure

P_B Small / contained fire does not spread
- extinguished by occupants
No threat to strucutre

P_C Sprinkler intervention successful
No threat to structure

P_{D1} Fire brigade intervention before flashover successful
No threat to structure

Flashover

Figure 11.5 Event tree for occurrence of a flashover fire threatening a structure.

outcomes P_A to P_{D1} can be estimated so that the frequency of a flashover fire occurring (F_{fo}) can be calculated as follows:

$$F_{fo} = F_{ig}\left(1 - P_A\right)\left(1 - P_B\right)\left(1 - P_C\right)\left(1 - P_{D1}\right)$$

In this particular scenario, there is no need to evaluate the event tree branches that do not lead to flashover any further, because the fire is not expected to challenge the structure and/or barriers, simplifying the analysis considerably.

If a deterministic approach is being adopted, the frequency of flashover fires is a useful input when determining an appropriate worse credible scenario.

With a sprinkler system design that is fit for purpose and adequately maintained, the probability of successful sprinkler intervention is relatively high compared to other interventions. From a risk management perspective, it is better to prevent a large fire from occurring rather than deal with the consequences, and therefore consideration should be given to improving the reliability of sprinkler systems through measures such as monitored valves, enhanced water supplies and duplication of water supplies in addition to regular maintenance and inspection.

11.3.4 Quantitative risk assessment of structure and barrier performance

The fire dynamics of fully developed fires is covered in detail in Chapter 3. This section will focus on the identification of fire scenarios and factors to be considered when deriving probabilities and distributions.

The fire enclosure will be assumed to comprise encapsulated timber except for one internal wall with an exposed massive timber surface.

If a multi-scenario analysis is undertaken, distributions can be adopted for ventilation conditions and moveable fire loads with an additional contribution estimated from the exposed timber. This will allow a distribution of fully developed design fires to be generated that can be individually analysed or grouped into clusters yielding representative design fires, each with an associated probability of occurrence.

The following fire brigade interventions should be considered, see Figure 11.4 for a graphical representation of the timing and stage of the fire when the interventions occur:

- D1 – prior to flashover: Probability of occurrence used as an input to determine probability of flashover not being reached and therefore no significant threat to the structural adequacy, integrity and insulation performance of fire-resistant structural members or barriers. Refer to Section 11.3.3.
- D2 – fully developed phase: Enclosure temperatures are very high at this stage, hence a low probability of successful suppression by the fire brigade would be assumed because of the relatively short time available prior to flashover. Priority will be given to search and rescue activities. The risk to fire brigade personnel from direct exposure to the fire will be significant. Firefighting from the doorway may be viable for small enclosures. Similarly, in low-rise buildings, external firefighting may be viable and therefore a small probability of success may be justified in some cases.
- D3 – mid decay phase: As the fire passes through the decay phase, the fire will tend towards fuel control with a lower likelihood of large plumes projecting from doorways, improving the probability of successful suppression and firefighter safety. If the fire transitions from flaming combustion to predominately smouldering or char oxidation, enclosure temperatures will fall further and access for firefighters within the enclosure for short periods will increase the likelihood of successful suppression. If fire brigade intervention does not occur, depending upon the enclosure configuration and properties of the timber elements, the fire may redevelop, potentially generating a secondary flashover. It is more likely that smouldering combustion will be substantially reduced and temperatures will drop. Some smouldering may continue in localised areas (e.g. at corners, joints and connections where localised radiative feedback between timber surfaces may occur), which is represented by D4.
- D4 – end of decay phase: At this stage, the firefighters are able to gain safer access to the enclosure and there will be a very high probability of successful suppression.

Close liaison with the relevant fire brigades should be undertaken when estimating these probabilities using appropriate models and data sets if available, and scenarios relating to the times to detection and alarm from the pre-flashover phase. Further details relating to fire brigade intervention are included in Chapter 14.

When quantifying the performance of encapsulation systems, the potential for gross defects such as the substitution with non-fire-resistant protection systems should be considered. Whilst this may be considered a very low-probability event, recent experience with external cladding systems indicates that a scenario evaluating the hazard resulting from gross defects should be included in the analysis. A simple flowchart/event tree can be used to assign probabilities of occurrence, as shown in Figure 11.6.

For estimating performance of a properly installed encapsulation system, it is common to assume a uniform distribution about a mean value at the level of specified performance. However, in addition, the probability of a gross defect needs to be incorporated. For example, it could be assumed that standard plasterboard is used instead of a fire protective plasterboard, e.g. Type X.

For each scenario, the times and probabilities of successful fire brigade intervention should be determined, and the timing and probabilities of the possible failure of structural elements and barriers predicted.

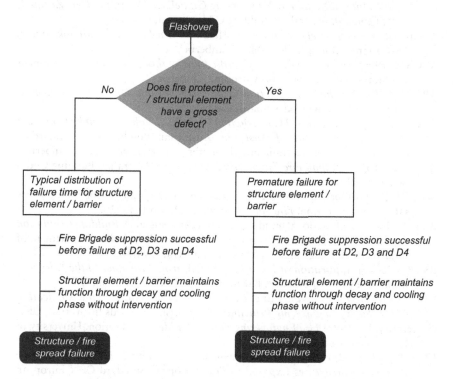

Figure 11.6 Flowchart for probability estimates of failure of structural elements or barriers.

These times should be used in conjunction with the holistic analysis of the building to predict the consequences of the failures. For timber encapsulation systems, the additional contribution to the fuel load from the areas of timber exposed by failure of the encapsulation system should be considered.

This general approach can be applied to absolute or comparative analysis risk-based approaches. Provided that effective compartmentation is maintained, well-designed mass timber tends to perform well compared to other protected structural elements, potentially increasing the time available for evacuation even though the fire severity may be increased due to additional fuel load.

If a deterministic approach is adopted due to the complex interactions between components of the building fire safety system and low-probability high-consequence events, the comparative approach has significant advantages over the absolute approach because of the difficulty of defining credible worst-case scenarios.

REFERENCES

ABCB (2005) *International Fire Engineering Guidelines (Australia, New Zealand, USA, Canada)*. Australian Building Codes Board, Canberra.

ABCB (2019a) *Fire Safety Verification Method Data Sheets: Handbook annex*. Australian Building Codes Board, Canberra.

ABCB (2019b) *Handbook: Fire Safety Verification Method*. Australian Government and States and Territories of Australia, Canberra.

ABCB (2020) *National Construction Code 2019 Volume One Ammendment 1*. Australian Building Codes Board, Canberra.

ABCB (2021a) *Draft National Construction Code 2022 Volume 1 Public Comment Draft. Building Code of Australia*. Australian Building Codes Board © Commonwealth of Australia and the States and Territories 2022, Canberra.

ABCB (2021b) *Australian Fire Engineering Guidelines*. Australian Building Codes Board, Canberra.

Beck, V. (1983) Outline of a stochastic decision-making model for building fire safety and protection. *Fire Safety Journal*, 6(2), pp.105–120.

Beck, V. (1991) Draft national building fire safety systems code. *Building Regulation Review Task Force, Microeconomic Reform Fire Regulation*. Department of Industry Technology and Commerce, Canberra, Australia.

BS 7974 (2019) *Application of Fire Safety Engineering Principles to the Design of Buildings: Code of Practice*. BSI British Standards Institution, UK.

BS DD240-1 (1997) *Fire Safety Engineering in Buildings Guide to the Application of Fire Safety Engineering Principles*. BSI British Standards Institution, UK.

Canterbury, U. (1994) *Fire Engineering Design Guide*. First Edition. University of Canterbury, Christchurch.

EN 1991-1-2 (2002) *Eurocode 1 Actions on Structures: Part 1–2: General Actions: Actions on Structures Exposed to Fire*. European Standard CEN European Committee for Standardization, Brussels.

EN 1995-1-2 (2004) *Eurocode 5 Design of Timber Structures: Part 1–2: General: Structural Fire Design*. CEN European Standard CEN European Committee for Standardization, Brussels.

England, P. (2016) *WoodSolutions Technical Design Guide 38 Fire Safety Design of Mid-rise Timber Buildings. Basis for the 2016 Changes to the National Construction Code*. Forest and Wood Products, Melbourne, Australia.

England, P. and B. Iskra (2021) *WoodSolutions Technical Design Guide 17 Fire Safe Design of Timber Structures-Compliance with the National Construction Code*. WoodSolutions, Melbourne.

Buchanan, A.H. (1994) *Fire Engineering Design Guide*. First Edition. CAENZ New Zealand Centre for Advanced Engineering, University of Canterbury, Christchurch.

FCRC (1996) *Fire Engineering Guidelines*. First Edition. Fire Code Reform Centre, Sydney, NSW.

Fitzgerald, R. W. (1985) Risk analysis using the engineering method for building fire safety. In *Proceedings of the 1st International Symposium of Fire Safety Science*. Hemisphere, Washington, DC.

Hasofer, A. M., V. R. Beck and I. D. Bennetts (2007). *Risk Analysis in Building Fire Safety Engineering*. GB, Elsevier, UK.

Hurley, M. J. and E. R. Rosenbaum (2015) *Performance-based Fire Safety Design*. CRC Press Taylor & Francis Group, LLC, Boca Raton, US.

ICC (2021) *Performance Code for Buildings and Facilities*. ICC International Building Code Council.

INSTA 951 (n.d.) *Fire Safety Engineering: Guide to Probabilistic Analysis for Verifying Fire Safety Design in Buildings*. Nordic INSTA (InterNordicSTAndards) Standard, Norway.

INSTA TS 950 (n.d.) *Fire Safety Engineering: Comparative Method to Verify Fire Safety Design in Buildings*. Nordic INSTA (InterNordicSTAndards) Technical Specification.

ISO 13943 (n.d.) *Fire Safety: Vocabulary*. ISO International Organization for Standardization, Geneva.

ISO 16732-1 (n.d.) *Fire Safety Engineering: Fire Risk Assessment: Part 1 General*. ISO International Standardization Organization, Geneva.

ISO 16733-1 (n.d.) *Fire Safety Engineering-Selection of Design Fire Scenarios and Design Fires Part 1 Selection of Design Fire Scenarios*. ISO International Standardization Organization, Geneva.

ISO 23932-1 (n.d.) *Fire Safety Engineering: General Principles Part 1 General*. ISO International Standardization Organization, Geneva.

ISO/TS 16733–2 (n.d.) *Fire Safety Engineering: Selection of Design Fire Scenarios and Design Fires: Part 2: Design Fires*. SO International Standardization Organization, Geneva.

MBIE (2020) C/VM2 verification method. In *Framework for Fire Safety Design for New Zealand Building Code Clauses C1–C6 Protection from Fire (Including Amendment 6)*. Compliance Document. The Ministry of Business, Innovation and Employment, Wellington.

Meacham, B. J. (1996) *The Evolution of Performance-based Codes and Fire Safety Design Methods*. National Institute of Standards and Technology, Gaithersburg.

NFPA 5000 (2018) *Building construction and safety code.* NFPA: National Fire Protection Association, US.

NFPA 550 (2022) *Guide to the Fire Safety Concepts Tree.* National Fire Protection Association, US.

SFPE (2004) *Code Official's Guide to Performance-Based Design Review.* SFPE Society for Fire Protection Engineering.

SFPE (2007a) *Engineering Guide to Performance-Based Fire Protection.* SFPE Society for Fire Protection Engineering.

SFPE (2007b) *Fire Risk Assessment.* SFPE Society for Fire Protection Engineering.

SFPE (2016) *Handbook of fire protection engineering.* SFPE Society for Fire Protection Engineering.

Spearpoint, M (2008) *Fire Engineering Design Guide, Third Edition.* CAENZ New Zealand Centre for Advanced Engineering, University of Canterbury, Christchurch. https://ir.canterbury.ac.nz/handle/10092/15062

STA (2020) *Structural Timber Buildings: Fire Safety in Use Guidance.* Structural Timber Association, UK.

Thomas, P. (1986). Design guide: Structure fire safety CIB W14 Workshop report. *Fire Safety Journal* 10(2), pp.77–137.

Chapter 12

Robustness in fire

Michael Klippel, Andrea Frangi,
Robert Jockwer, Joachim Schmid,
Konstantinos Voulpiotis and Colleen Wade

CONTENTS

SCOPE OF CHAPTER

With the increasing number of complex and tall timber buildings with a significant area of unprotected timber surfaces, questions arise about the robustness of these buildings in extreme fire scenarios. In recent building projects, measures for robustness have been implemented on an ad hoc basis in agreement between the designers and the authorities. This chapter discusses general approaches to achieve structural robustness with regard to fire design and evaluates them to give guidance for robust fire design of timber structures.

DOI: 10.1201/9781003190318-12

12.1 BASICS OF STRUCTURAL ROBUSTNESS

Regardless of the building material, structural robustness is an important characteristic of a structure that prevents damage that is disproportionate to the original cause (Agarwal et al., 2012). Structural robustness tends to be less important in areas subject to severe earthquakes because buildings designed for high seismic loads tend to have a high degree of inherent structural robustness for other load cases, including severe fires.

The mathematical definition for robustness can be described with the equation from Starossek and Haberland (2010) to describe disproportionate collapse $P(C)$ as in Equation 12.1:

$$P(C) = P(E) \times P(D|E) \times P(C|D) \qquad \text{12.1}$$

where
    $P(C)$ is the probability of disproportionate collapse
    $P(E)$ is the probability of accident occurring or "Exposure"
    $P(D|E)$ is the probability that damage occurs given this accident or "Vulnerability"
    $P(C|D)$ is the probability that collapse occurs given the occurrence of damage

However, for the expression to be complete, the consequences of the said damage or collapse need to be evaluated. Without knowing the magnitude of the consequences, an insignificant progressive damage which is disproportionately larger than another insignificant initial damage can appear to be very serious according to Equation 12.1. The updated expression to include the direct consequences (C_{Dir}, caused by the initial damage) and indirect consequences (C_{Ind}, caused by the progressive damage) is given by the expectation of total consequences E[C] in Equation 12.2, as derived from Baker et al. (2008):

$$\mathbf{E}[C] = \left(P(E) \times P(D|E)\right) \times C_{Dir}$$
$$+ \left(P(E) \times P(D|E) \times P(C|D)\right) \times C_{Ind} \qquad \text{12.2}$$

We recommend the reader to refer to Voulpiotis et al. (2019) for further information on the state-of-the-art quantification of robustness.

For tall buildings, *collapse* is generally not acceptable, thus tall buildings must be designed with an extremely low probability of structural collapse, even in extreme loading. In case of fire, this may lead to the concept of *design to withstand burnout* (assuming failure of all other safety measures). The term "burnout" is discussed in Chapter 3. Robustness extends beyond only structural behaviour in fire, but also includes other performance objectives. For example, the extreme case of the fire in the Grenfell tower clearly shows that, even if collapse did not occur, many people died due to the uncontrolled fire spread in the building.

12.2 BASICS OF ROBUSTNESS AND FIRE SAFETY ENGINEERING

Fire safety engineering (FSE) incorporates many different aspects related to the performance of a structure in a fire situation, and the resulting safety for people and society. In European countries, the load-carrying capacity of timber members in a fire situation is mainly regulated in Eurocode 5 (EN 1995-1-2, 2004), while the fire loads are regulated in Eurocode 1 (EN 1991-1-2, 2002). Interestingly, in contrast to prior versions (ENV 1995-1-2, 1994), the latest version of Eurocode 5 does not contain a "system effect" in the fire situation, taking into account the robustness of a structural system. General guidance about FSE is available in handbooks such as the SFPE *Handbook of Fire Protection Engineering* (Hurley et al., 2015) with its section about timber design currently under revision.

In most countries, rules concerning the spread of fire in a structure (i.e. cladding, insulation materials, cavities, sprinklers, etc.), as well as serviceability considerations (i.e. the non-structural aspects related to the evacuation of a structure) are mostly regulated on a national level. Required fire safety concepts contain structural, organisational and active fire protection measures that must be designed in parallel. All of these measures are subject to uncertainty, which is an important aspect to consider in the planning of a robust fire safety concept, which aims at reaching a fire safety goal. This fire safety goal can be defined on a project basis, including all relevant stakeholders. Consequently, fire safety goals and accepted or tolerable risks are the basis for the definition of required safety measures. A holistic design approach towards the robustness of buildings in a fire situation can be achieved by tailoring the fire safety concept for the particular building and including fire-related risks and the fire safety goals.

12.3 NORMATIVE FRAMEWORK AND ROBUSTNESS

There are three approaches to quantify robustness and design for it, in increasing complexity. They are listed in the following and discussed in more detail in Adam et al. (2018):

1) Deterministic methods, such as Alternative Load Path Analysis (ALPA) and minimum tie forces. They aim to satisfy assumed damage scenarios such as failure of a single column.

2) Reliability approaches, which compare the failure probabilities of damaged and undamaged states.
3) Risk approaches, which compare the direct and indirect consequences and their probabilities.

Today, most standards for robustness in the built environment are written on a prescriptive rule basis, which is implementing deterministic methods only. There is, however, room for performance-based design in cases where the choice of verification is agreed between parties, e.g. building owner, designer, insurance and authority representatives. The help that building codes provide to the designer of robustness is known to be notoriously vague, and sometimes no explicit requirement for robustness design exists (Huber et al., 2019). A survey carried out by Bita et al. (2019) studied the experience of practising structural engineers in the field of robustness and came to the conclusion that although prescriptive design is the current primary approach to implementing robustness design in building codes, a performance-based design approach would be preferred.

The Eurocode EN 1991-1-7 (2010) defines robustness as "the ability of a structure to withstand events such as fire, explosion, impact or the consequences of human error so as not to cause damage that is disproportionate to the cause of the damage." The corresponding ISO standard (ISO 2394) gives further possibilities to increase robustness:

(i)   By avoiding critical events
(ii) By dimensioning of individual components
(iii) By enabling alternative load paths
(iv) By reduction of the consequences

In general, it can be stated that in order to increase the robustness of a structural system, the methods or strategies selected must either

(1) reduce the probability of failure or
(2) limit the consequence of a failure

For the former, the designer can increase the size of certain structural members or reduce the probability that the damage occurs in the first place. For the latter, which is not well addressed in literature, a sound conceptual design is required, for example a design concept that uses compartmentation effectively to keep the fire within a room without further spread.

12.4 EXPOSURE TYPES

Considering the different types of exposure is paramount for an effective robustness design. Exposure may be from structural actions such as accidental loading or material weakness, or from unplanned events such as explosions or unexpected fires. While distinctions such as whether the exposure is static or dynamic, cyclic or monotonic, short term or long term are useful when considering the conceptual design of a building, the primary distinction that needs to be taken into account for robustness design is whether the exposure is localised or systematic.

A localised exposure is one that affects only one, or a very small number of elements. A systematic exposure is one that affects many or all elements (e.g. columns located in one fire compartment). It is very important to realise that the vast majority of known and acceptable robustness measures are only addressing localised exposures. Worse, if a systematic exposure occurs in the structure, those robustness measures may actually worsen, rather than improve, the performance of the structure. A well-known example of the described systematic exposure in timber engineering is the background of the collapse of the Bad Reichenhall Arena, where a so-called progressive collapse behaviour appeared, i.e. the failure of an individual main support led to the chain-reaction collapse of the entire arena roof structure (see Winter and Kreuzinger, 2008).

Fire exposure is more complex than other accidental loads. Both local and systematic exposures need careful consideration before robustness measures are implemented. In general, a fire load starts as a localised exposure in the initial stages (localised exposure), but can spread quickly to become a large event (systematic exposure). Its extent is typically determined by taking into account multiple variables, e.g. the type of fire load (with randomly distributed total value and a heat release rate within certain bounds), the failure modes of the glazing, the availability of combustible surfaces arranged vertically and/or horizontally, the size and efficiency of the fire compartment and countermeasures. In general, it must be made clear that fire design is actually addressing a calculated accidental load case where certain aspects have been investigated in the past, as opposed to generic robustness for "unforeseen events." Therefore, different strategies may be needed to address robustness in localised and systematic cases, and some of these strategies may be contradictory.

12.5 CONSEQUENCES RESULTING FROM A FIRE EVENT

In the event of a fire, active fire protection and organisational measures might be sufficient to fight the fire before any flashover occurs, in which case only a limited and localised fire exposure will occur. If a fully developed fire or a travelling fire happens, the fire will affect several structural and separating members, which means that the structural fire safety measures and the firefighting strategy are of key importance. Firefighter access may be compromised in some cases, after a major earthquake for example.

Smoke, elevated temperatures and fire spread are among other relevant consequences that must be addressed. As the strategies to fight the fire in the fire growth phase and the fully developed fire phase are significantly different, it is apparent that both cases must be considered separately, i.e. very different goals are followed, e.g. during evacuation and after flashover in the same fire compartment.

Selected different direct and indirect consequences related to a fire event can be structured, as presented in Figure 12.1 and given as examples in Table 12.1 (please note that this list is not intended to be exhaustive and can be extended).

12.6 EVALUATION OF IMPROVEMENTS PERFORMED IN PRACTICE

In the following, the direct application of a robustness measure taken from normal-temperature design is presented. This is followed by a general view on the robustness and finally its application for the fire situation for tall timber buildings. Additional robustness is not necessary for low-rise timber buildings because the consequences of failure are less severe.

12.6.1 Prevention of progressive collapse for the fire situation

Recently, several tall timber buildings have been finished, among others being Mjøstårnet in Norway (see Figure 12.2) and HoHo in Austria.

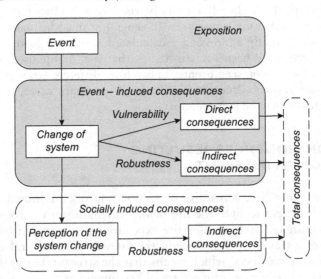

Figure 12.1 Overview of consequences (modified from JCSS 2001 and Schubert 2009).

Table 12.1 Examples of potential fire events, direct and indirect consequences and related robustness measures

Event	"Failure" and direct consequence	Indirect consequence	Robustness measures
Fire	Combustion	Creation of toxic gas and smoke, fire spread	Reduction of combustible surface material, compartmentation, sprinkler system, fire detection system
Failure of manual suppression	Increased fire	Increased risk for tenants	Fire protection measures, training, fire detection system
Failure of secondary elements	Localised failure	Alternative load paths might be activated	Include alternative load-paths, over-dimensioning of structure
Smouldering combustion	Continuation of fire after burnout	Reignition and later collapse	Fill gaps/voids with non-combustible materials and create airtight joints
Failure of integrity	Spread of fire (vertically and/or horizontally)	More systematic exposure	Increased resistance to fire spread. Extra inspection of workmanship
Fire detector failure	Delayed evacuation, delayed filling of dry sprinkler pipes and delayed alert of fire brigade	Personal injury and property damage, limited effect of sprinkler	Regular integrated tests of detector systems, improvement of automated fire detection
Sprinkler failure	Fully developed fire	Damage of complete compartment and possible further spread	Regular integrated tests of sprinkler systems, no reduction of fire resistance of structural system due to sprinklers
Spread of fire	Larger effected area and more members	Change from localised to systematic exposure	Change robustness concept
Exposure to elevated temperature	Reduction of strength and stiffness	Failure of member and connection	Encapsulation of members and connections
Early failure of encapsulation	Charring of timber, reduction of anticipated failure time	Failure of connection and/or member	Design structure with additional fire resistance. Inspection of encapsulation
Fire exposure	Charring of timber, additional fuel load	Failure of member, increase of temperature and extension of fire	Increase of load-bearing capacity/fire resistance R of all or selected members, limit the share of visible combustible surfaces
Combustion volatiles (gas/smoke)	Leakage of smoke, visibility	Increased number of fatalities	Improved compartmentation, improved design for reaction to fire
Collapse or partial collapse of timber building	Singular monetary loss	End of success story of tall timber buildings	Perform adequate fire safety engineering for all timber buildings

Figure 12.2 (a) Mjøstårnet, Norway. At 84.5 metres, it is currently one of the world's tallest timber buildings (photo Peter Lang, Rothoblaas). (b) HoHo in Austria (photo proHolz Austria / Bruno Klomfar).

In projects such as the HoHo, the robustness in fire has been addressed by activating alternative load paths, where it was shown that any single column could fail without leading to a progressive collapse. Other projects are applying this approach by verifying design where one or more members can be assumed to have failed. This deterministic approach gives information of the behaviour of the structure with regard to well-defined exposures, but little or no information on the effectiveness of certain measures to increase robustness.

It should be stated that timber structures often use columns with pinned supports and single span beams; thus, they are sensitive to disproportionate overall structural failure in the event of a single element failure. Such a structural design concept should not be applied to tall timber buildings.

Evaluating the actual structural boundary conditions, it should be highlighted that structural elements such as simply supported beams and columns are fixed in their position not only by frictional forces, but also by engineered connections. Thus, these connections allow the transfer of additional tensile and shear forces which could be activated in case of the failure of one member, see schematic illustration in Figure 12.3. In that example, the stabilising effect is created by the additional resistance of the diagonals and their connections, generating forces not considered in the normal design.

Figure 12.3 shows that a failure of an element does not lead immediately to a total collapse of the system when the actual connection design activates alternative load paths (ALPA). However, in the fire accidental case, the scenario that only one member is exposed is quite unrealistic since travelling

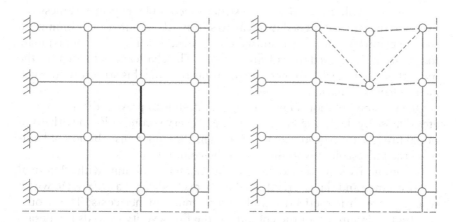

Figure 12.3 Static system with pinned supports of beams and columns before failure of the column in the fire compartment (left) and after column failure with additional diagonal elements (right).

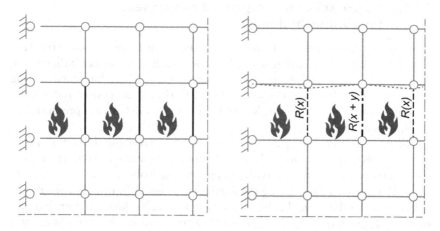

Figure 12.4 Static system with pinned supports in fire (a) and a possible situation at an advanced stage (b) where a member with increased fire resistance $R(x + y)$ is arranged. Note that all exposed columns are in the same fire compartment.

fires or full compartment fires are quite likely when a significant share of combustible structure is present. Figure 12.4a illustrates a more realistic scenario with a significant area of a compartment in fire.

Consider the structure shown in Figure 12.4a where the fire severity exceeds that considered in the prescriptive design. If the fire duration exceeds the column's fire resistance, e.g. $R(x)$ (providing x min load-bearing resistance), failure of the all exposed columns can be expected. In consequence, collapse of the entire structure could be expected, unless the overall structure is designed in a way to survive the failure of all columns in the fire compartment. The resulting consequences will be disproportionate to the marginal exceedance of fire severity. The challenge is to identify the most effective measure to increase the robustness of this structure for such an accidental scenario.

One possible solution shown in Figure 12.4b is to prevent disproportionate collapse by designing key elements, e.g. every second column with additional fire resistance (indicated as $R(x + y)$), allowing redistribution of loads from the failed columns to the remaining columns.

An alternative solution would be to design all the columns with additional fire resistance and thus as reinforced elements (indicated as $R(x + y)$), which would be much less effective and lead to much higher costs. This would still lead to failure of all the columns if the fire severity exceeds the design scenario; however, the probability of this exceedance is greatly reduced. Further, it should be noted that this can be achieved as well by combining active fire protection with enhanced structural fire engineering provisions.

12.6.2 Approaches for improved robustness for timber buildings

The key element approach is addressing the vulnerability of the structure as per Equation 12.1. Because of this, and along with other challenges of key elements (e.g. architectural, element protection), it has been a long debate whether "the key element approach" should be considered a valid "robustness measure" or not. More details on this debate are presented by Voulpiotis et al. (2021).

The focus for improving robustness must be on reducing the overall probability of disproportionate collapse of the building without having to debate the categorisation of the individual approaches. Robustness improvements should be assessed considering the particular project's boundary conditions.

Contrary to the typical robustness designs which address the probability of collapse given damage $P(C|D)$ for the fire situation, the safety measures may also need to address

(1) the reduction of the probability of the occurrence of events (exposure), $P(E)$, and/or
(2) the reduction of the structure's vulnerability $P(D|E)$.

An optimised combination of measures can be most beneficial. In particular, the following measures can be considered:

(i) Reduction of the probability of critical fire events by sprinklers, including further measures such as independent water supplies, redundant piping and pumps

(ii) Selected load-bearing timber elements are designed to meet a prescriptive $R(x + y)$ requirement, while other elements are designed for $R(x)$

(iii) All load-bearing timber elements are designed to meet a prescriptive $R(x + y)$ requirement, although this may imply over-dimensioning

(iv) Use of fire-resistant detailing and creation of additional or redundant load path (as mentioned above): e.g. beams and slabs supported by direct bearing rather than supported by connections which are vulnerable to fire attack

(v) The dimensioning of the supporting system for the case of fire is carried out with more realistic fires instead of the simplified standard fire exposure

(vi) The structural fire protection and active fire protection are carried out in parallel holistically, e.g. without reducing the structural fire design requirements because of the introduction of sprinklers.

12.6.3 Improvement of the robustness for structural timber buildings

In the following, the potential improvements for the robustness of buildings with major elements made from timber are presented. While most of the approaches diverge only very little from non-combustible buildings, some ideas address the combustibility of wood.

For the following consideration, structural collapse is defined as a failure mode, which should be avoided. This implies that a fire with limited consequences is still an acceptable event. For the evaluation and analysis of the risk that structural collapse is reached, an event tree may be created, as shown in Figure 12.5.

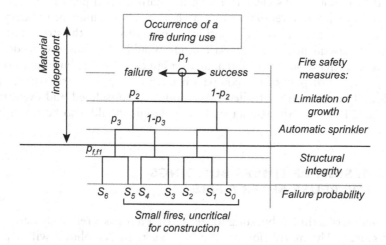

Figure 12.5 Event tree analysis of a fire event, with the probability of occurrence P_i and the individual event failure probability S_i.

For simplicity reasons, Figure 12.5 focuses on the fully developed fire (in contrast to the growth phase where evacuation can be ensured) and does not show details such as early failure of the passive fire protection of structural members (e.g. gypsum plasterboards). From Figure 12.5, it can be seen that many elements above the horizontal heavy line are material-independent. Looking at the event tree, it becomes further apparent that many elements of the event tree are important before reaching the demand of the structural fire design for the fully developed fire (e.g. fire occurrence or starting fire, failure of occupants and fire brigade in stopping the fire, failure of sprinkler in stopping the fire). This is case S_6 in Figure 12.5, whereas $P_{f,fi}$ is the probability of a failure in a fully developed fire. Consequently, any measure as presented in Section 12.6.2, may only be favourable for this case. As long as this measure is not replacing other measures (e.g. redundant sprinkler feed), it may be an improvement of the overall robustness.

In the following, the focus is the probability of collapse (case S_6), when the structural fire design must provide for structural survival in the fully developed fire after failure of all other fire safety measures. The main difference between structural timber and other building materials is the combustibility of wood.

Structural timber contributes to the structural fuel corresponding to the fire development, i.e. the more severe the fire, the higher the additional fire load from exposed wood surfaces. Design must ensure that the additional fuel load from the consumption of timber members does not lead to structural collapse. Further, the consumption of the combustible structural elements might increase the fire duration and, consequently, challenge other measures (e.g. service penetrations, compartmentation walls, fire curtains and also the protection of load-bearing steel elements in the same compartment). It is not always clear if a long fire with limited peak temperatures (e.g. due to limited ventilation) or a short fire with high peak temperatures (e.g. due to increased ventilation) is more severe for the structural elements. Consequently, a parametric study would be essentially needed to answer this question for the particular design case (i.e. the building, storey, compartment or part of the compartment). In the parametric study, both fuel- and ventilation-controlled fires should be considered and eventually assessed. To do so, the particular facade design should also be taken into account.

12.7 DESIGN OF TIMBER BUILDINGS FOR REUSE AFTER A FIRE

The design of a timber building for reuse after a fire is a relatively new topic in research. The motivation for reuse rather than demolition will depend on the extent and severity of any fire. Limited information is available, see e.g., Matzinger (2019), but in-depth information or planning guidelines are missing. In general, the consideration to reuse a timber structure must consider load-bearing and compartmentation aspects as well as smoke damage. Furthermore, possible damage induced by extinguishing water might be relevant. In the design process, the reuse after a fire scenario might be considered with additional covering of the timber members, and design to replace some load-bearing and non-load-bearing elements after the fire. Charring and possible water damage must be taken into consideration and the structural integrity reassessed. Furthermore, indoor air quality and the functionality of any fire safety measures must be checked. In general, it seems that a well-thought-through design-for-deconstruction approach that allows the de-installation of structural members seems to be beneficial in this context.

12.8 DISCUSSION AND CONCLUSION

The need for robustness in fire design increases as the height and complexity of the building increase.

On one hand, robustness in fire design can be increased by reducing the likelihood of a fire event. This appears to be independent of the building material. However, considering the severity of a fire event, the combustibility of a building material may have an influence, e.g. when looking at the fire spread on combustible, vertical exterior surfaces (e.g. timber facades). This hazard may be addressed by reducing the time to flashover or by considering the increased possibility for a full compartment fire rather than a travelling fire.

On the other hand, robustness of a structure can be increased by increasing the redundancy of structural elements. In this case, a simplified event tree analysis may help to find out redundant and complementary elements in the case of a fully developed fire.

Looking at the design of structural members made from timber, increased robustness can come from an extended parametric study in the course of a performance-based fire design which looks more closely at effects of single parameters affecting the fire severity, e.g. ventilation conditions.

In general, it can be stated that comprehensive fire design and the creation of a robust fire safety concept can be used to increase the robustness of a structure. The concept must be able to guarantee the safety of the occupants and fire brigades for all considered fire scenarios. It must be recognised that measures to achieve robustness against fire can be very different from measures to achieve robustness against some other localised accidents. This is because fire is a systematic exposure, which potentially affects a large number of elements simultaneously. Therefore, the conceptual design of the structure and the structural detailing are keys to providing a robust structure. This is valid independent of the structural material.

REFERENCES

Adam, J.M., Parisi, F., Sagaseta, J. and Lu, X. (2018) Research and Practice on Progressive Collapse and Robustness of Building Structures in the 21st Century. *Engineering Structures*, 173, pp.122–149.

Agarwal, J., Haberland, M., Holický, M., Sykora, M. and Thelandersson, S. (2012) Robustness of Structures: Lessons from Failures. *Structural Engineering International*, 22(1), pp.105–111.

Baker, J.W., Schubert, M. and Faber, M.H. (2008) On the Assessment of Robustness. *Structural Safety*, 30(3), pp.253–267.

Bita, H.M., Huber, J.A., Voulpiotis, K. and Tannert, T. (2019) Survey of Contemporary Practices for Disproportionate Collapse Prevention. *Engineering Structures*, 199, p.109578.

EN 1991-1-2 (2002) Actions on Structures. General Actions. Actions on Structures Exposed to Fire. In *European Standard CEN*. European Committee for Standardization.

EN 1991-1-7 (2010) Actions on Structures Part 1–7 General Actions: Accidental Actions. In *European Standard CEN*. European Committee for Standardization.

EN 1995-1-2 (2004) Design of Timber Structures Part 1–2 General: Structural Fire Design. In *European Standard CEN*. European Committee for Standardization.

ENV 1995-1-2 (1994) Design of Timber Structures: Part 1–2 General Rules: Structural Fire Design. In *European PreStandard CEN*. European Committee for Standardization.

Huber, J.A., Ekevad, M., Girhammar, U.A. and Berg, S. (2019) Structural Robustness and Timber Buildings: A Review. *Wood Material Science & Engineering*, 14(2), pp.107–128.

Hurley, M.J., Gottuk, D.T., Hall Jr, J.R., Harada, K., Kuligowski, E.D., Puchovsky, M., Watts Jr, J.M. and Wieczorek, C.J. eds. (2015) *SFPE Handbook of Fire Protection Engineering*. Springer.

ISO 2394 (n.d.) *General Principles on Reliability of Structures. International Standard*. International Organization for Standardization.

JCSS (2001) *Probabilistic Model Code*. Joint Committee on Structural Safety.

Matzinger, I. (2019) *Brand- und Wasserschadensanierung, Presentation at Beiratssitzung des Österreichischen Ingenieurholzbauverbandes (IHBV)*. Ingenieurholzbauver band

Schubert, M. (2009) *Konzepte zur informierten Entscheidungsfindung im Bauwesen* (Vol. 322). vdf Hochschulverlag AG an der ETH Zürich.

Starossek, U. and Haberland, M. (2010) Disproportionate Collapse: Terminology and Procedures. *Journal of Performance of Constructed Facilities*, 24(6), pp.519–528.

Voulpiotis, K., Köhler, J., Jockwer, R. and Frangi, A. (2021) A holistic framework for designing for structural robustness in tall timber buildings, Engineering Structures. DOI: https://doi.org/10.1016/j.engstruct.2020.111432.

Winter, S. and Kreuzinger, H. (2008) The Bad Reichenhall Ice-Arena Collapse and The Necessary Consequences for Wide Span Timber Structures. In Proceedings WCTE 2008 Conference, Miyazaki, Japan.

Chapter 13

Building execution and control

Andrew Dunn, Ed Claridge, Esko Mikkola,
Martin Milner and Birgit Östman

CONTENTS

DOI: 10.1201/9781003190318-13

SCOPE OF CHAPTER

This chapter covers control of workmanship, fire safety during construction, responsibilities and enforcement, fire detection and suppression and emergency procedures. Quality and inspection of workmanship are vital for high-quality buildings, whether of timber or other construction materials. Timber buildings require certain precautions due to the risk for greater exposure of combustible materials. Furthermore, not all fire safety measures for the final building will be in place throughout construction; consequently, adequate processes are required to maintain the fire safety of building sites. All construction sites require formalised fire safety management systems, including auditing of contractors and subcontractors.

13.1 INTRODUCTION

The main objectives of this chapter are to ensure that fire protection measures are installed correctly and to avoid any danger of fires during construction. Several guidelines on building execution and control have been published around the world (AWC, 2014; Canadian Wood Council, 2016; Campbell, 2017; NCC, 2019; STA, 2017a; WoodSolutions, 2014) and one standard (INSTA 952).

Guidelines on prevention of fire spread within a building are given in Chapters 6 and 9. Active fire suppression systems are described in Chapter 10.

13.2 CONTROL OF WORKMANSHIP

Timber buildings consist of a large number of components made from different wood-based materials, all of which are designed and installed to fulfil multiple performance functions such as fire safety, acoustic performance, weather protection, energy and thermal efficiency and so on. The methods used for assembling and erecting these multiple layers are vital to ensure adequate performance.

Building practices vary throughout the world. In some countries, most of the installation is undertaken in the factory, sometimes including windows, doors and service installations, and only final assembly is undertaken at the site. In other countries, only the structural components are manufactured off-site, with the majority of lining and installation work undertaken on-site. In any case, both delivery systems will generally have the same fire safety systems installed, just at different times.

Fire safety during the construction process requires as much attention as fire safety in the final completed building. The fire safety strategy during construction will necessarily be different due to a different consequence of risks in the partially completed building.

13.2.1 Installation of fire protection measures

Where fire protection is required, success of the fire safety strategy relies on the correct installation of all the fire-protective components. As an example, insulation such as mineral wool must be mounted carefully in direct contact with wooden structural elements to provide the designed protection. Voids around wood elements can lead to premature exposure in the event of a fire and can lead to earlier charring, loss of strength and therefore decreased fire resistance. Another example is fasteners (nails or screws) used for securing fire-resistant cladding (or lining). If the fasteners are too short, the cladding/lining may be prone to premature loss of fixity (fall-off), and structural elements will be exposed to fire at an earlier stage, leading to increased charring and loss of strength which will reduce fire resistance of the element.

13.2.2 Installation of fire stops and cavity barriers

Fire stops enhance the fire-resisting properties of fire-protective elements. Cavity barriers provide compartmentation or closure at ends of a space between elements, such as in an external wall facade (external cladding) or an internal load-bearing wall element.

The correct installation of fire stops and cavity barriers within the building, as well as in facade gaps or voids, the erection and connectivity of penetrations and building services systems at the construction site are essential to ensure the fire performance of a timber structure. The installation of cavity barriers and fire-stop details can only be checked during the construction period, prior to being hidden behind linings. The quality of workmanship of such details must be monitored closely when they are accessible.

Common modes of failures are as follows:

- Fire stopping missing around penetrations and building services
- Cavity barriers missing around voids through which a service penetration passes
- Lack of an effective cavity barrier to close off voids around an external wall opening or internal compartment wall
- Fire stopping missing at compartment envelope junctions where the internal dry lining (internal cladding) is absent
- Incorrect or change of specification of fire stopping or cavity barrier

13.3 INSPECTION DURING CONSTRUCTION

Fire safety systems in buildings include a combination of both active and passive systems. Active systems include smoke hazard management and automatic or manual fire suppression systems. Passive systems include the construction of fire-rated walls and ceilings, fire-stop collars separating floors and provision for safe evacuation from the building. All of these components require inspection, and these inspections should be at a time when they are accessible and able to be clearly viewed.

13.3.1 Inspection of passive fire protection measures

In high-rise or multi-level buildings, there will likely be significant numbers of service penetrations through fire-rated construction. As the construction programme continues, these penetrations will be progressively covered, making visual inspection difficult, so the inspection schedule needs to complement the building programme. This inspection schedule may require site visits at regular intervals to ensure that passive systems at each floor level are appropriately inspected.

It is essential to inspect items such as walls, ceilings and service penetrations for appropriate fire resistance as construction proceeds. It may be convenient to inspect the construction of those same building elements that include acoustic construction.

The use of digital photo records with location and date labels provides evidence of installation for an audit of a building at a hand-over stage.

13.3.2 Inspection of active fire protection systems

Active fire safety systems such as fire sprinkler systems, fire detection and alarm systems are generally inspected later in the construction programme to coincide with testing and commissioning. Depending on the type and complexity of the actual system installed, the inspection may occur near the completion of the building work or at various stages during construction. Identifying non-compliant fire safety requirements early in the construction process permits corrective or remedial action to be taken without causing further delays or additional costs to the completed project.

13.3.3 Coordination of interacting trades

The responsibilities of interacting trades must be clearly stated, and overarching project management must be communicated and enforced at the beginning of every building project. Checks carried out by third parties, such as building inspectors, are only random checks. An inspection schedule should be established that considers the complexity of all fire safety requirements.

Hand-over check sheets between trades provide a good means of communicating compliance at each stage of the building process. The *Structural Timber Association Fire Safety in Use: Vol. 2 Cavity Barriers* (STA, 2020) has a set of check lists for trades to adopt as part of the hand-over process between trades.

Fire engineering design involves architects, designers and fire engineering consultants, specialist designers, contractors and suppliers of services systems, and producers of hardware and equipment. It is important that each trade's responsibilities are well defined and that inspections or any changes are recorded and the records retained. A person with overall responsibility for fire design and safety should be appointed for each major construction project. In most countries, the main contractor is legally the primary contact. Errors often occur in the interface between different trades, and focus on these areas is recommended.

13.3.4 Documentation

Good quality fire protection documentation should always be produced, normally by the principal designer with support from the fire engineer for the building. This documentation is mandatory in some countries. Inspection plans and checklists should be produced for design and execution and be communicated to all parties, specifying in detail the inspection areas and responsibilities. Critical areas need special attention, such as interfaces between various trades and control functions.

Most of these inspections needs apply to all buildings, but additional care should be taken in timber buildings. Additional inspection by a third party might be considered essential in tall or complex timber buildings.

13.4 FIRE SAFETY DURING CONSTRUCTION

The construction phase of any building provides a unique fire risk for a relatively short period of time in the lifespan of the building. The risks are very different from those in the completed building because they occur during a time when many of the fire protection measures in the completed building are not yet installed or activated.

An analysis of fires in buildings in England from 2009 to 2012 (TRADA, 2012) showed that fires in timber buildings (dwellings and non-residential) under construction had, on average, larger areas of damage compared to other construction materials. The cause of these fires was surveyed by the Cost Action 1404 project (Martin and Klippel, 2018), who found that of the fires with an identifiable source of ignition, over 54% were due to arson, 31% due to hot works and the remaining 15% due to either a propane heater, smoking or an electrical fault.

In the UK, changes to the approval for permission to build are being introduced with "fire safety gateways." These gateways include fire service reviews so that firefighting views are addressed and included in the building fire safety strategy. The current objective of these new UK requirements is to promote fire safety at the planning stages for buildings.

1. *Gateway One – Planning*
    Applicants are required to submit a planning document demonstrating fire safety.
2. *Gateway Two – Technical Design and Construction*
    A Building Control application and this gateway act as a "hard stop" for complex and high-rise buildings. Construction cannot begin until final approvals from the Building Safety Regulator (including the fire service) have been received for the project.
3. *Gateway Three – The Final Certificate*
    In this stage, the Building Safety Regulator will conduct a final inspection and then issue a completion certificate. Here prescribed documents and information on the as-built building will be required, and information must be handed over to the person(s) responsible for the occupied building.

Different types of timber buildings present different risk profiles. Small section timbers in light timber framing provide a much higher risk of rapid fire spread than post and beam structures with significant distances between members or mass timber buildings which have large flat surfaces more difficult to ignite.

The UK Structural Timber Association, working with other safety bodies, has produced a comprehensive set of design guides for the building industry to create fire-safe timber structures. The guidance is divided into two sections:

1) Preventing a fire from occurring is covered in the security and method procedures in *16 Steps to Fire Safety* (STA, 2017a).
2) In the event of a fire, providing safe egress for persons on the building site, and providing a building design that will not present a fire risk

to adjacent buildings or people. This advice is in *Separating Distance Guidelines* (STA, 2017b).

Fire services should observe local timber buildings under construction to understand how they are built. Their understanding of the layers of fire-resistant construction, cavity barriers and detailing for robustness may assist fire services in future fire rescue and fire suppression activities (STA, 2020, 2021).

13.4.1 Recommended fire precautions during construction

To provide the best fire safety during the construction of a building, the following measures must be addressed:

1. Appoint a person responsible for managing construction fire safety, both at the design stage and the construction phase
2. Take preventive measures to prevent any fire from starting
3. As far as possible, provide adequate fire detection and suppression if a fire starts
4. Provision of a comprehensive fire safety plan for the construction site
5. Implement a risk assessment and resultant actions to stop any fire spread to neighbouring buildings

Figure 13.1 illustrates the three main areas to be addressed, with recommended activity under each measure. These fire precautions are then described in more detail below (Martin and Klippel, 2018).

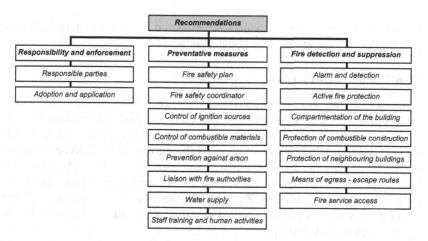

Figure 13.1 Measures required to prevent construction fires. Based on Cost Action 1404 project (Martin and Klippel, 2018).

13.5 RESPONSIBILITY AND ENFORCEMENT

13.5.1 Responsible parties

All parties involved in a construction project are responsible for preventing construction fires. The main contractor is generally responsible for implementing security and safety on a construction site. The main contractor should also provide a safety coordinator for the site. However, the project's owner and consultants should consider the fire safety measures on a construction site before the main contractor is chosen. Some measures need to be taken into account in the early stages of the design, such as by the engineers and architects, as many hazards can be addressed by good design and planning before they become an issue. Examples include compartmentation, either permanent or temporary and construction sequencing to allow for fire service access during construction.

The building designers deciding on the location and material specification of the building have a duty to ensure that in the event of a fire, there is no risk to neighbouring buildings or people (STA, 2017b).

13.5.2 Adoption and application

The recommendations aim to improve the fire safety of buildings during the construction phase, irrespective of building materials. The information applies to the design and planning stages and the actual construction phase. It does not apply to the completed structure. Minimum building standards for fire safety in completed buildings are addressed in Chapter 4 of this guide.

13.6 PREVENTIVE MEASURES

The following preventive measures are essential activities to avoid the possibility of a fire occurring during construction.

13.6.1 Fire safety plan

A fire safety plan should be developed for each project and updated as the project progresses. It is generally developed by the main contractor. The fire safety plan organises the site activities and nominates who is responsible for carrying out all activities. It clarifies the duties for each decision-maker and involved parties. All the measures and activities for fire safety are contained in the plan, which becomes the reference document for any questions regarding fire safety on-site. As each project and site is unique, the fire safety plan is specific to that site and remains under constant review during the entire project duration.

All persons involved in the construction process or visiting the site should be made aware of the importance of fire prevention and the contents of the fire safety plan. This should include what to do in the event of a fire, emergency procedures, location of assembly points and any housekeeping practices, i.e. smoking.

The following are the critical steps in the creation of the fire safety plan that is developed in a risk assessment strategy:

1. Analyse the risks and factors arising from the construction, operations, implementation schedule and different phases of work. This analysis primarily consists of identifying
   a. the potential causes of fire
   b. the factors contributing to increasing its effects
   c. the site-specific sources of ignition, fuel and oxygen
2. Develop the policies, procedures and systems to prevent and control the risks.
3. Analyse the available resources on and outside the site boundary (off-site), including allocation of key staff regarding fire and emergency duties, consultation with the emergency services to obtain their feedback and addressing any parties' concerns.
4. Develop a protocol of emergency procedures for various individuals with roles and responsibilities in a fire emergency, such as sounding alarms, calling rescue services and shutting down hazardous operations.

The fire safety plan should include the following points, as a minimum:

- The organisation of, and responsibilities for, the fire safety on-site
- Risk assessments and fire prevention reports requiring specific fire safety measures
- Fire safety training/instruction given to site personnel and visitors, including required actions in case of fire
- Procedures for reporting emergencies to the fire service
- Fire service access to the site and the buildings at all times
- Fire protection provisions: portable fire extinguishers, standpipes, hydrants, hose reels and water supplies (and during the final stage of construction: automatic fire sprinklers, automatic fire detection and alarm systems, temporary emergency lighting)
- Evacuation plan and procedure for emergency notification and evacuation (escape route inside the building, including corridors and stairwells) of all persons in the building under construction
- Fire prevention measures, including security requirements and control of ignition sources
- Procedures for hot work permit operations, cutting and welding
- Electrical supplies and equipment (maintenance of temporary electrical installations)

- Compliance with "no smoking" policies (locations of designated smoking areas when these are provided)
- Plant equipment and vehicles
- Prohibition of open fires
- Materials storage and waste control (control/reduction of combustible materials; control of flammable liquids and gases, proper storage and disposal of waste materials)
- Separation from adjacent buildings and other hazards
- Security measures to minimise the risk of arson

13.6.2 Fire safety coordinator

For all projects, the main contractor must identify a Fire Safety Coordinator who will be responsible for all the fire safety issues on-site for the entire construction period.

The Fire Safety Coordinator has several duties:

- Develop the fire safety plan in coordination with the local fire service
- Control the application and enforcement of the measures provided in the fire safety plan, e.g. supervising hot work permit programme
- Inspect and check escape routes, emergency lighting, fire detection and alarm devices, firefighting equipment and site access
- Ensure correct housekeeping and storage
- Check for potential fire sources, e.g. electrical equipment and rubbish
- Ensure instruction and conduct periodic training of the workers on-site in fire protection equipment use, e.g. extinguishers and hose reels
- Ensure liaison and contact with the local fire services
- Manage the work of the security personnel
- Maintain written records of every measure and monitor fire protection system impairments.

13.6.3 Control of ignition sources

For a fire to start, three things are needed: a source of ignition, fuel and oxygen. A fire cannot start if any one of these elements is missing. Taking measures to avoid the three elements coming together may reduce the chances of a fire occurring.

There are many ignition sources on a building site. Other than arson, the lead ignition source is hot works from cutting, grinding, welding, brazing and soldering, and heat-applied materials.

Hot works

Hot works include any activities that generate heat, open flames or sparks that could initiate fires or explosions:

- Cutting and grinding
- Welding, brazing and soldering (Figure 13.2)
- Thermal spraying
- Use of oxyacetylene torch or blow torch
- Installation of heat-applied materials

As far as reasonably practicable, all activities involving hot works should be avoided and replaced with alternative methods. Where hot works cannot be avoided, they can be controlled by requirements of written permission (a permit) to be obtained prior to commencement of the hot works. Appropriate activity may involve an inspection of the hot works area before work begins to ensure that

- Combustibles have been moved or are adequately protected
- Appropriate fire extinguishers are on hand, fully charged and operable
- Evacuation paths are available
- Any equipment that operates at surface temperatures exceeding 75°C must not come into contact with combustible materials

A suitably trained and equipped person must be assigned to watch all hot works activities, including for an appropriate period after works have been completed. Hot works areas should be inspected at the end of the day's work.

Electrical equipment

The use of electrical equipment and supply systems can be another source of ignition during construction. All electrical supply installations, both permanent and temporary, must be installed and maintained in accordance with relevant standards by a registered electrician.

All portable electrical devices and extension cords should be regularly inspected and tagged. Faulty or damaged equipment must be removed from

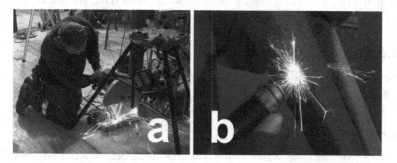

Figure 13.2. Examples of hot works on a construction site (photo Per Rohlén, Brandskyddslaget).

use immediately and be labelled accordingly. Electric cabling should be protected against possible damage from construction site activities.

Prepare fragile components, such as temporary lights, with guards to prevent accidental damage where they are exposed to the risk of an impact. Low-voltage equipment should be used where practicable. Mobile heat-producing equipment, such as air heaters, bitumen heating tanks or steam cleaners, must be placed outside the structure. Provide a safe, fire-resistant and fire-alarmed area for overnight battery charging. The charging points should be regularly inspected.

Smoking

Smoking materials are a significant ignition source of fires on construction sites. A non-smoking policy should be established throughout the site because hazardous materials, such as flammable gases, may be used in open as well as enclosed areas. Smoking restriction zones must be clearly identified, signposted and strictly enforced.

The risk of smoking materials being discarded around the perimeter of the site should also be considered, and, if the risk is significant, precautions should be implemented. These may include providing hoardings constructed from fire-preventative coverings.

Other ignition sources and fuel sources

All possible further ignition sources should be managed. Equipment powered by internal combustion engines (compressors and generators) should be positioned in the open air or a well-ventilated non-combustible enclosure and as far as possible from combustible materials.

Procedures on the fuel temporary storage and refuelling must be a part of the fire safety plan. Fuel tanks must not be filled whilst engines are running or still hot. Vehicles should only be fuelled in designated areas. Fuel storage is not allowed within the structure under construction.

Open fires, including the burning of waste materials, should be prohibited on the construction site.

13.6.4 Control of combustible materials

The following detailed actions are required to reduce the storage of combustible waste material or combustible building materials.

Stored and waste materials

Remove combustible waste materials, dust and debris from the building and its immediate vicinity at the end of each shift or as soon as practicable. Store materials susceptible to spontaneous ignition, such as oily rags, in clearly labelled non-combustible containers and remove them from the site as soon as practicable. Unless specific items of vegetation are planned to be retained, all dry vegetation should be removed from the sites for a distance of 20 m from buildings under construction and work areas.

Storage of combustible building materials

As far as possible, plan the delivery of combustible materials to minimise the time they are stored on-site. Where significant volumes of combustible building materials are to be stored on-site, they should be kept in a secure area at least 10 m away from buildings and any location where hot works are undertaken.

Where there are no reasonably practicable alternatives and combustible building materials have to be stored within or close to the building under construction, the area used for storage should:

- Have controlled access
- Be remote from possible ignition sources such as hot works
- Have firefighting equipment close by
- Be protected by preventative fire covers (e.g. fire-resistant, or non-combustible sheeting)

Exposed combustible materials during construction

During the construction process, combustible materials may be temporarily exposed in locations such as the facade or as parts of wall or ceiling linings. The following are the typical examples:

- Shade cloths, tarps and other covering around scaffolding separating work areas and around the site perimeter
- Combustible facade materials
- Timber structural components

For buildings of four or more storeys, where the exposed facade is combustible, or the construction is predominantly of combustible construction, one or more of the following additional controls may be specified:

- If an automatic fire sprinkler system is to be provided, the sprinkler system should be progressively commissioned so that the number of unprotected storeys with significant exposed combustible materials is limited to two below the current construction level.
- Early installation of permanent or temporary fire compartments can limit fire spread in the event of an uncontrolled fire. Protection of door openings, windows, shafts and service penetrations need to be addressed.
- A temporary alarm system may need to be provided and evacuation procedures modified to address the expected rate of fire spread.

Protection of light timber frame construction

Light timber framing can be vulnerable to fire safety before the fire-protective linings are attached. The fire risk can be reduced in the following ways:

- Protecting the structure with temporary non-combustible lining materials
- Applying the permanent fire-protective linings as soon as practical
- Applying the fire-protective linings off-site onto prefabricated panels
- Where possible, design, plan and install fire protection as the building progresses

13.6.5 Prevention against arson

Arson is the largest threat to combustible construction, and therefore specific activities to address this threat should be taken:

- Provide fencing to ensure that the site is secured against unauthorised entry
- Provide good site illumination or motion-activated lightening
- Install (temporary) windows and doors after the construction of the first floor to make entry more difficult
- Install video surveillance systems and clear signs of surveillance
- Avoid storage of combustible waste near the building or in open containers
- Provide security guards to patrol out of normal working hours
- Collect data of arson and vandalism history in the local area to update safety measures
- For significant projects, seek support from the public to assist in preventing arson and vandalism

13.6.6 Liaison with fire authorities

During the design phase, the designer should contact the local fire service to identify their requirements, such as emergency access on the site. At the commencement of the construction, the Fire Safety Coordinator should make contact with the local fire services and invite them to review and recommend adjustments to the fire safety plan. Provisions for water supplies should be agreed upon at this time.

Regular liaison with the fire service is necessary, mainly where there are changes to the site conditions or safety plan details. An initial site plan should be prepared, and a process for updated details to be available in a fire emergency should be agreed upon.

The site plan should include:

- Fire service access points to the site
- Fire service grab pack (site information such as hydrants and fire safety plan)
- Any special provisions for firefighting activities
- Emergency escape routes and stairs
- Positions of hydrants and hose reels that are operative
- Location of booster connections
- Any other operative fire safety systems that have been provided
- Locations of assembly points and registers of persons currently on the site
- Details of temporary accommodation and storage areas, including a location for storage of hazardous items such as flammable liquids and gas cylinders

13.6.7 Water supplies

The building's construction programme should be planned, as far as reasonably practicable, to always maintain adequate firefighting water supplies throughout the site. A regular update should be provided to the fire service on the hydrants and hose reels that are operational and of any potential or actual interruptions to the water supplies.

If the firefighting water supplies are interrupted, the Fire Safety Coordinator should take appropriate action such as prohibiting hot works, notifying site workers and take any other additional actions considered necessary.

13.6.8 Staff training and human activities

Fire safety awareness

All persons working on or visiting the site should be made aware of the importance of fire prevention and the content of the fire safety plan, including what to do in the event of a fire. A "fire-safe" working environment must be promoted, ensuring that all processes and precautions are applied and maintained in partnership with the main contractor and subcontractors.

Training and fire drills

Construction personnel and security staff must be able to use the portable firefighting equipment provided on-site. Therefore, training should be undertaken at regular intervals for the following tasks:

- Use of portable firefighting equipment
- Safety precautions for those undertaking hazardous operations

- Site-specific emergency procedures, e.g. evacuation, meeting point locations, contact of responsible persons and emergency services
- Regular fire drills

Fire checks

At the end of each working day or shift, a fire check must be undertaken, particularly in areas where hot work has been undertaken. Where 24-hour security is provided, fire checks should be undertaken throughout the night, during holiday periods and at weekends.

A checklist for fire-safe construction sites has been developed in a Swedish project (Bengtson et al., 2012) and published in a Nordic guideline (Östman et al., 2012) (see Table 13.1).

13.7 FIRE DETECTION AND SUPPRESSION

Fire detection and suppression are activities or infrastructure implemented in the case of a fire occurring, and there is a need to suppress a fire. Strategies are discussed in the following sections.

13.7.1 Alarm and detection

The permanent automatic detection and alarm system should be progressively installed so that the number of storeys with significant exposed combustible materials is limited to two storeys below the current construction level. The detection system should be linked to an alarm-receiving centre unless there is a 24-hour security presence on-site.

Where activities are likely to activate the detectors, e.g. hot works, the detection system may need to be voluntarily deactivated, as addressed in the fire safety plan.

Where it is impractical to commission the permanent automatic detection and alarm systems during construction, an alternative means of warning of fire and other emergencies must be established to allow staff to raise the alarm across the site and alert the fire service. Manual devices may be utilised, provided that the following criteria are fulfilled:

- They are distinctive and clearly audible above background noises in all areas
- All staff and visitors are trained/instructed so that they can recognise the fire/emergency alarm and understand what action to take
- The devices are distributed throughout the site, and staff are trained in their use

Table 13.1 Proposed checklist for a fire-safe construction site

Measures	Checkpoints
Organisation	• Organisation plan • Protective round • Random checks • Follow-up on any remarks
Education	• Extinguishing equipment, location, handling, etc. • Escape routes • How to alert the rescue service • Restrictions on flammable hot work • Handling of flammable goods • Fire alarm systems
Order • Smoking • Waste management	• General smoking ban • Storage in suitable containers, safety distance to hot work, buildings, etc. • Emptying continuously, but no later than at the end of the working day • The risk of self-igniting materials
Gas cylinders	• Storage outdoors at the end of the working day, in a special gas container or other, approved place • The storage area must be fenced and provided with warning signs
Flammable hot work	• Permits for flammable hot work must always be available • May only be performed by personnel with special training • Protective distance of at least 2 m to combustible material or equivalent protection, e.g. screens, fire protection fabric, etc. • Extinguishing equipment must always be available
Hand fire extinguisher	• Must be within 25 m
Material flows	• Combustible building materials, packaging, garbage, etc. that are not used during the working day are stored outdoors or in a fire-technically separated place, e.g. garage, garbage room • Within the work area indoors, a maximum of approx. 1,000 kg of combustible material may be stored outside EI 30 / E 30-separated storage
Fire technical separation	• Fire technical class of temporary building construction • Penetrations are sealed with approved sealing methods • Fire doors must not be left open with wedges or similar. Doors for evacuation must not be locked
Evacuation	• Walking distance to the nearest escape route must not exceed 45 m • "Dead ends" are multiplied by 1.5 • Evacuation alarm
Warning marking	• Extinguishing equipment • Gas container • Flammable liquid • Evacuation • No smoking

(Continued)

Table 13.1 (Continued) Proposed checklist for a fire-safe construction site

Measures	Checkpoints
Opportunities for action for the rescue service	• Current action plans • Rescue routes • External fire hydrants
Electrical installations	• Protected from mechanical damage • Moving lighting devices, protective glass/grille, stable location • No halogen lamps • Design as a so-called 5-conductor system • Earth fault circuit breaker • Fire seal when crossing a fire cell boundary

(Bengtson et al., 2012)

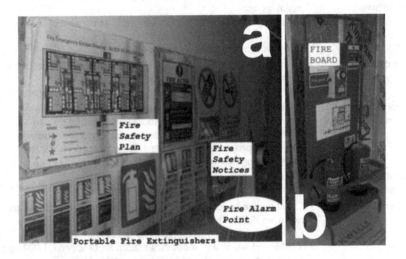

Figure 13.3 Fire point boards (STA, 2017c).

All staff should be given emergency services contact details and instructions to call emergency services on their personal telephone if a fire occurs (Figures 13.3–13.5).

13.7.2 Active fire protection

The following active fire protection systems should be installed or be available.

Portable fire extinguishers

Portable fire extinguishers must be provided at fire points on each floor, covering an area no greater than 500 m² per extinguisher. At least one fire extinguisher suiting all fire risks and electrical fires must always be

Figure 13.4 Tidy and untidy policy (STA, 2017c).

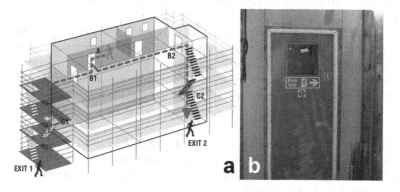

Figure 13.5 Escape routes (STA, 2017c).

provided at each fire point, exit way and stairway. Their exact locations must be indicated in a plan as part of the fire safety plan.

In addition, portable fire extinguishers should be provided for the fireguard, while hot works are being undertaken and at any other locations determined as a result of risk assessments. All fire extinguishers should be maintained and regularly inspected, and all staff should be trained in the use of manual firefighting equipment (Figure 13.6).

Automatic sprinkler system

When automatic fire sprinkler systems are provided in the completed building, the project should be planned to achieve their installation and operation as soon as reasonably practicable. In some jurisdictions, there are regulations that prescribe when automatic sprinklers must be installed. There are

Figure 13.6 Fire safety board and fire alarm point on a UK housing site (Courtesy of M. Milner, STA).

significant advantages in progressively bringing the sprinkler system into service on each floor level as it is constructed.

This approach is particularly effective in buildings where the design strategy relies on a sprinkler system to supplement fire separations or control fire spread when combustible materials are exposed during construction. As a minimum, automatic sprinkler systems should be progressively installed so that the number of unprotected storeys with significant exposed combustible materials is limited, for example, to two storeys below the current construction level.

Another strategy is to install temporary sprinkler systems during construction. It is recommended to at least install temporary sprinklers in stairways and fire hazardous areas, such as areas with combustible materials or areas with a high risk of ignition.

Hydrants

All fire hydrants and booster connections required by the national building regulations for the completed building must be fully operational as soon as reasonably practicable. The hydrants should be progressively brought into service on each floor level as it is completed. Risers should be progressively installed as construction is undertaken. Hydrants and hose reels required by the completed building must be progressively commissioned as soon as possible.

13.7.3 Compartmentation of the building

Fire compartments are often the primary strategy for reducing the fire spread within a completed building. This strategy could also be applied during construction and can include the following:

- Early installation of permanent or temporary fire compartments
- Early installation of fire doors when their frames are made fire-resistant
- Use of temporary doors and windows
- Installation of vertical fire breaks
- Installation of temporary fire-resistant barriers such as reusable fire-resistant textiles

13.7.4 Protection of combustible construction

Structural framing is often protected by means of fire-protective linings, and the same strategy can be employed during the building's construction. Therefore, some measures can be applied to protect the frame:

- Protect the structure with non-combustible lining materials
- Where possible, design, plan and install fire protection as the building progresses (Figure 13.7).

13.7.5 Protection of neighbouring buildings

Where there are timber frames that are yet to have their fire protection installed, a fire that is involved in the frames may have a higher intensity than when protected. The consequence of this is that the fire may cause

Figure 13.7 Use of non-combustible sheathing on a timber frame in the UK (Courtesy of M. Milner, STA).

immense heat that could reach neighbouring buildings. Therefore, measures should be taken to counter this hazard, such as the installation of the fire-protective lining progressively throughout the building's construction, particularly on the bounding walls within a fire compartment (STA, 2017b).

13.7.6 Means of egress – Escape routes

Construction programmes should be planned to ensure that construction personnel always have more than one path of travel to exit the building. The travel paths should take into account the number of people, activities being undertaken and occupant capabilities. Paths of travel should be clearly marked and illuminated, and the temporary storage of construction materials must not occur.

13.7.7 Fire service access

Fire service access to the construction site must always be clear and unobstructed. If this is not possible, the fire service should be immediately notified of any changes or restrictions to the access points. If practicable, significant changes to the access to the site should be discussed with the fire service before being implemented.

13.8 EMERGENCY PROCEDURES

Written emergency procedures must be displayed in prominent locations and be given to all employees and visitors on-site. Typically, they should include the following:

- Emergency contact details for key personnel, for example, fire wardens, floor wardens and first-aid officers
- Contact details for local emergency services, for example, police, fire service and poison information centre
- Description of the mechanisms for alerting people at the workplace to an emergency, for example, siren or bell alarm
- Evacuation procedures, including arrangements for assisting any hearing, vision or mobility-impaired people
- A map of the workplace illustrating the location of fire protection equipment, emergency exits and assembly points
- Triggers and processes for advising neighbouring businesses about emergencies
- Procedures for testing the emergency plan, including the frequency of testing

- Instructions to nominated personnel, such as the security guards, to open gates or barriers and provide ready access to the site for the fire service in the event of an emergency

Clear signs must be provided and maintained in prominent positions indicating the locations of fire service access routes, escape routes, positions of dry riser inlets and the fire extinguishers provided for use by trained staff. Signs should be reviewed regularly and replaced or repositioned as necessary.

Regular checks should be undertaken to ensure that travel paths are maintained clear of obstructions and provided with clear signage. Inspections should be undertaken daily or weekly, depending on the risks associated with the site. The frequency should be increased if significant hazards such as blocked exits are observed.

REFERENCES

AWC (2014) *Fire Department's role in Prevention and Suppression of Fires During Construction of Large Buildings.* American Wood Council, Leesburg, VA, USA.

Bengtson, S., Dittmer, T., Rohlén, P., Östman, B. (2012) *Fire Safety at Construction sites: Guidance (in Swedish).* SP Report 2012:11. Technical Research Institute of Sweden, Stockholm.

Campbell, R. (2017) *Fires in Structures Under Construction, Undergoing Major Renovation, or Being Demolished.* NFPA: National Fire Protection Association, USA.

Canadian Wood Council (2016) *Fire Safety During Construction for Five and Six Storey Wood Buildings in Ontario: A Best Practice Guideline.*Canada

INSTA 952 (n.d.) Fire Safety Engineering: Review and Control in the Building Process. Nordic INSTA (InterNordicSTAndards) Standard.

Martin, Y., Klippel, M. (2018) Fire safety of (timber) buildings during construction dissemination document N216–07: COST action FP 1404, https://ethz .ch/content/dam/ethz/special-interest/baug/ibk/costfp1404-dam/dissemination-documents/N216-07-Guidelines%20fire%20safety%20of%20(timber) %20building%20during%20construction.pdf

NCC (2019) *National Construction Code Building Code of Australia*, Volume 1.Australian Building Codes Board, Australia.

Östman, B., Karlsson, B., Mikkola, E., Stenstad, V., Just, A., König, J., Schmid, J., Jensen, G., Buksans, E. (2012) *Fire Safe Timber Buildings, Version 3: Nordic-Baltic Knowledge Review and Guideline (in Swedish).* SP Report 2012:18. Technical Research Institute of Sweden, Stockholm.

STA (2017a) *16 Steps to Fire Safety.* Structural Timber Association, Clackmannanshire, UK.

STA (2017b) *Design Guide to Separating Distances during Construction.* Structural Timber Association, Clackmannanshire, UK.

STA (2017c) *Operative Fire Safety Induction Pack. Available to STA Members on the Web Site*. Structural Timber Association, Clackmannanshire, UK.

STA (2020) *Structural Timber Buildings Fire Safety in Use Guidance. Volume 1: Pattern Book Systems. Volume 2: Cavity Barriers and Fire Stopping*. Structural Timber Association, Clackmannanshire, UK.

STA (2021) *Compliance Routes for Fire Resistance for Mass Timber: Vol 6*. Structural Timber Association, Clackmannanshire, UK.

TRADA (2012) *Fire Safety on Timber Frame Construction Sites*. Trada Technology, High Wyccombe, UK.

WoodSolutions (2014) *WoodSolutions Guide No 20. Fire Precautions during Construction of Large Buildings*. WoodSolutions, Melbourne, Australia.

Chapter 14

Firefighting considerations for timber buildings

Ed Claridge, Christian Dagenais, Andrew Dunn,
Claudius Hammann, Kamila Kempna,
Martin Milner and Jan Smolka

CONTENTS

DOI: 10.1201/9781003190318-14

SCOPE OF CHAPTER

This chapter discusses fire service considerations relevant to timber buildings. Firefighting practices may be different in timber buildings compared with non-combustible construction. Internationally, fire services have raised concerns regarding the increased use of wood within buildings and specifically the use of timber structural elements of tall buildings. These concerns often stem from lack of knowledge of timber performance in fire, and firefighter experience from fires in non-combustible steel and concrete construction and traditional low-rise timber buildings. This chapter discusses relevant concerns of firefighters regarding large and tall timber buildings.

14.1 INTRODUCTION

All types of construction require careful design, good quality construction and ongoing maintenance to perform adequately in the event of a fire. Tall timber buildings are a special concern because they are a relatively new concept with designs and configurations that have not been considered previously.

Low-rise buildings of any materials are often not designed to withstand burnout and may have very little resistance to severe fires. Such buildings may collapse early in a fire, even prior to arrival of the fire service. Firefighting operations can be conducted from the exterior of domestic-scale low-rise buildings, so decisions about making entry to the building are much easier than for tall and large multi-occupancy low-rise buildings. Multiple options will typically be available, externally through windows and other openings at ground level or higher up the building using ladders and aerial appliances.

Buildings that are too tall for external firefighting, or those with a very large low-level footprint, require completely different firefighting approaches. Firefighting operations undertaken internally increase the risk considerably, particularly if the firefighters need to travel significant distances from the entry point. Tall buildings will require internal access using lifts and stairs, not only to the locations of the fire, but into other areas of the building, including levels above the fire. Very tall buildings have limited access and egress options, reduced facilities for firefighting and the water supply for firefighting will be restricted to that provided by the building infrastructure (inbuilt firefighting risers).

A wealth of experience has been gained by firefighters in masonry and concrete construction, including likely fire performance and possible failure mechanisms. Such experience is not yet available for tall timber buildings. There is a concern in the firefighting community about the behaviour of fires in large or tall timber buildings, and, for example, how such fires will behave in high winds and the suitability of the compartmentation.

As it is likely that the initial approach to firefighting in large timber buildings will be similar to other buildings, firefighters need to understand the fire behaviour of structures with combustible building materials where the dangers of fires in voids and cavities, and performance of protection materials such as gypsum plasterboard encapsulation, and whether the building's fire compartments have been designed to withstand burnout (the ability for flaming and smouldering combustion to stop without intervention).

Understanding the scope of structural fire resistance of timber buildings during the later stages of the fire duration is especially important given that the design concepts of burnout in traditional buildings may not apply to timber construction. Timber that is not fully protected against charring with dry linings (fully encapsulated) will begin to char later in the fire, which may result in smouldering combustion continuing until eventual collapse, unless there is intervention by firefighters. Partial protection is the common approach in low-rise domestic-scale buildings as the life safety function of the fire strategy in the building structure will have been met before the failure of the dry lining/cladding.

When making decisions to commit to internal firefighting, there must be confidence that the firefighters will not be put at risk from unforeseen fire spread or structural collapse. It is important for the emergency responders to know how the building has been constructed. Confusion can arise in an incident where there is a lack of information and experience readily available on how the fire and the building may behave. Exposed timber structures can be particularly problematic as there is a perception that they do not have much fire resistance, whereas the opposite may be true. Fire services need to know before entering large timber buildings that these buildings will perform in a predictable and manageable way during the fire. In some countries, the regulations for buildings over a certain height ask for a grab pack with information to the fire service on how the building is constructed and may perform in the case of fire. This is a major step in supporting the fire service to tackle events in tall buildings and will support timber building techniques.

14.2 TRADITIONAL FIRE KNOWLEDGE

The traditional knowledge of fire services about fires in timber buildings is often based on older timber buildings that are not designed for current fire-resistance standards, such as historical light timber frame houses or post and beam timber buildings. These historical buildings may have little or no fire resistance and their fire performance is completely different from modern mass timber structures which have large panelised timber elements. The many different types of timber construction are described in Chapter 1.

Firefighting in low-rise buildings is predominantly an external firefighting strategy focused on limiting fire spread to neighbouring properties. An enhanced firefighting strategy is required for taller buildings irrespective of their construction. For tall timber buildings, especially those with vulnerable occupants, the fire engineering design will be based on robust compartmentation that is designed to stop the fire from spreading so that the fire can burn itself out and maintain the stability of the building. Many codes will require the provision of active fire suppression, typically sprinklers, which will control the size of the fire, so the building structure is not likely to be challenged. Even with sprinkler protection and the very good history of their success in reducing fire loss, no building codes allow for complete reliance on sprinklers to provide fire safety in tall buildings. An appropriate risk-based approach is required for the design of complex buildings that fall outside of the common building types accommodated in building codes.

The severity of a fire in a timber building may depend on the amount of intentional or eventual exposed structural timber which can add to the fuel load and influence the fire behaviour. Small-sized exposed timber elements such as isolated beams or posts, which have been features of buildings for centuries, provide less potential fuel than exposed mass timber walls or ceilings. When all the timber surfaces are encapsulated by non-combustible fire-resisting linings/claddings, a fire is expected to behave in the same way as in a building of non-combustible materials, so reduced concerns should be anticipated from fire services.

There are added complications for firefighters to appreciate when comparing historical buildings with modern timber structures. Different considerations may need to be given when considering the appropriate firefighting techniques and tactics, depending on the age of the building and its type of construction. Table 14.1 shows typical differences between historic and modern buildings.

14.3 FIRE SERVICE CONCERNS RELATED TO MASS TIMBER BUILDINGS

In recent years, fire services around the world have expressed concerns about fire safety in mass timber buildings, especially tall timber buildings.

Table 14.1 Different firefighting issues between historical and modern buildings (Smolka et al., 2018)

Historic buildings	Modern buildings

• Unknown materials
• Unknown impact of fire on the structure
• Premature collapse
• Unknown plan of building
• Hidden fire spreading in cavities
• Retrofitting and changes of structure unknown to fire-fighters
• Possible absence or malfunction of fire protection
• Void spaces and attics
• Buildings in wildland interface
• Large open stairs
• Limited compartmentation
	• Increasing synthetic materials in the fire compartment
	• Smart Technologies
	  • Energy-saving technologies
	  • Photovoltaic technologies
	  • Electric vehicles
	  • Energy storage
	• Larger building footprint
	• Complicated and sophisticated building footprint
	• Void spaces and attics
	• Flammable external cladding
	• An increasing number of storeys
	• Open floor plan

There are different approaches to building codes in different countries. In the UK, there are newly imposed restrictions on allowable combustible materials in external walls for residential-type buildings, whereas building codes in some other countries are changing to allow more use of combustible materials. In 2018 and 2019, US and Canadian firefighters expressed concerns about proposed changes in building codes to allow taller mass timber buildings (Havel, 2018; O'Brocki, 2019).

Other concerns relate to the limited field experience of firefighters in dealing with tall timber buildings, poor understanding of the associated fire behaviour, increased demands on firefighting resources and the need for more investigation on the necessary firefighting response (Grimwood, 2017, 2019; Smolka et al., 2018). Research is currently underway, such as the TIMpulse (2022) project in Germany, which is considering the fire environment and its potential impact on firefighting as well as updates to the building regulations for increased use of timber construction in high-rise buildings (Engel et al., 2021) (Figure 14.1).

Specific concerns expressed by firefighters include the following:

- Faster fire growth and greater total heat release rates
- Earlier flashover, including the possibility of multiple flashovers
- Increase in fuel load producing longer duration fires
- Increased fire fighting water demands
- Greater requirements for resources inside the building, including access above the fire floor
- Hidden fire spread in voids and ongoing combustion behind encapsulation
- Increased severity of external flaming from windows and openings

Figure 14.1 Firefighter team during the timber fire test scenario at the Technical University of Munich, 2021.

- Increased chance of fire spread to adjacent buildings
- Greater reliance on fixed fire protection systems
- Increased production of carbon monoxide due to ongoing smouldering combustion
- Increased influence of wind-driven fires

A number of these concerns apply to any type of construction, especially fuel loads and other issues outside the normal management of the building. However, when comparing otherwise similar buildings, the use of combustible construction when compared to non-combustible construction will inherently present a greater fuel load and additional challenges to fire safety design and potential firefighting.

As a response to the concerns raised in North America when considering taller timber buildings, the proposed building code changes included increased redundancy and more robust fire protection features (O'Brocki, 2019).

14.4 LIGHT TIMBER FRAME CONSTRUCTION

Chapter 1 provides details of various types of timber structures and wood products. Concerns about firefighting in timber structures often do not distinguish between the specific types of timber construction. Several categories are identified in this discussion; light timber frames with solid timber members, light timber frames with engineered wood products and mass timber structures.

14.4.1 Light timber frame structures with solid timber members

Light timber frame construction is primarily used for low- to mid-rise residential buildings. If unprotected, the small section sizes and connection methods of light timber framing make these buildings vulnerable. Fires in unprotected light timber frame buildings (generally houses) account for over half of all fire fatalities in the United States (Fire Protection Association, 2011).

The fire performance of light timber frame construction has been of concern in the firefighting community since the 1970s. There have been warnings about possible collapse without warning of roof and floor truss systems (NIOSH, 2005). This is a particular danger if firefighters are working on the roof of the building. Failures often result from fire damage to the connections rather than the timber framing, especially failure of steel hangers and punched metal truss plates which have poor fire resistance when unprotected. Some studies show failure of unprotected light timber structures within 6–13 minutes of fire exposure (Harman and Lawson, 2007; Björkman, 2012).

14.4.2 Light timber frame structures with engineered wood products

The use of Engineered Wood Products (EWP) to replace solid timber construction has been increasing for a number of years, raising concerns in the firefighting community. The biggest concern is with timber I-joists, which are vulnerable because of their small cross-sectional dimensions.

In the United States, recent experiments to provide comparable fire performance between traditional light timber construction and EWPs including I-joists were carried out by Underwriters Laboratories (2008). The results are summarised in Figure 14.2. The main findings were that an unprotected timber floor assembly with no ceiling, representing typical "legacy" construction with solid wood joists, had less fire resistance than the same assembly with a gypsum lath and plaster ceiling. The fire resistance was much less for timber I-joists or unprotected timber trusses with metal truss plates.

In Europe, the fire resistance of I-joist, metal web and solid timber joist floors must all meet tested criteria of fire resistance times that will match the statutory regulation times, mostly requiring partial protection with gypsum plasterboard. The UK Structural Timber Association is expecting to release new information on fire resistance of generic EWP floor assemblies. EWP floors and solid timber floors can also be designed to Eurocodes for appropriate fire resistance, as described in Chapter 7.

14.4.3 Charring in protected light timber frames

When a light timber framed assembly passes a fire resistance test, the acceptance criteria are the containment ability of the assembly and the

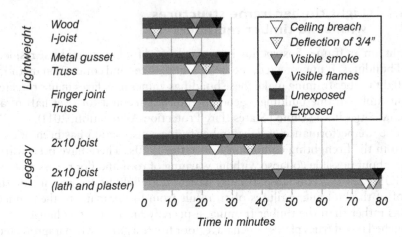

Figure 14.2 Failure times in US light timber frame assemblies, comparing modern lightweight EWPs with "legacy" solid timber systems (Underwriters Laboratories, 2021).

load-bearing capacity, as described in Chapter 2. When an assembly has passed a fire resistance test, significant charring may have taken place under the protective linings. Firefighters need to be aware that such charring can occur behind protective linings in real fires in light timber-framed buildings. This may require the removal of large areas of protective layers of gypsum plasterboard before leaving a building that has experienced a significant fire.

14.5 MASS TIMBER STRUCTURES

Mass timber or heavy timber structures have historically been associated with industrial buildings, including those with large open compartments and no concealed spaces. Although large fires have occurred in these types of structures, the performance of the heavy timber structural components is typically well understood. Figure 14.3 shows how severely fire-damaged wooden beams retained sufficient strength to support steel beams following the San Francisco earthquake and fire in 1906.

14.5.1 Strategies for protection of mass timber structures

Most mass timber buildings will be protected by an automatic sprinkler system. The following discussion refers to the low-probability event of a fire which is not controlled by the sprinkler system.

Encapsulation is a fire protection strategy intended to protect the timber from charring throughout the duration of the fire. In timber buildings where

Figure 14.3 Severely damaged large wooden beams retain sufficient strength to support failed steel beams.

all the timber is fully encapsulated to withstand burnout, fire safety will be no different from buildings of traditional non-combustible materials.

Partial encapsulation uses protective linings to prevent charring of the timber structure during the early stages of a fire. An encapsulation lining that is designed for 60 minutes standard fire exposure will prevent charring of the timber for that condition, but some charring may then occur. Most countries have prescriptive guidance for compliance with building regulations that require full or partial encapsulation using protective linings.

The encapsulation concept is seen in the objectives of the code change in Australia where the objective of encapsulation of timber was changed

> to prevent or delay ignition of the timber structural member so that the response to an enclosure fire will be similar to that for a building constructed of non-combustible elements such as masonry or concrete during the growth period. The fire-protected timber element is still required to achieve the Deemed-to-Satisfy Fire Resistance.
>
> (Forest and Wood Products Australia, 2016)

This encapsulation concept is demonstrated in a comparative fire test conducted on a timber-framed compartment and a similar steel-framed compartment, with the framing in both cases protected by two layers of 13-mm fire-rated gypsum plasterboard on the walls and two layers of 16-mm fire-rated gypsum plasterboard on the ceiling. Identical compartment sizes and fuel loads were used. Figure 14.4 shows that the time–temperature curves of both compartment fires were identical (Exova Warringtonfire, 2011).

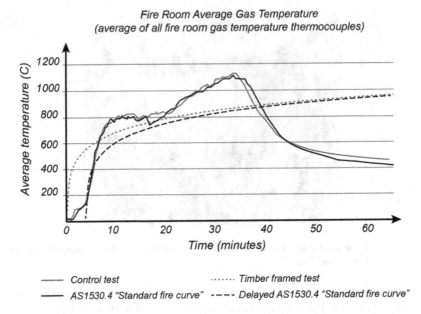

Figure 14.4 Time–temperature curves of fires in combustible and non-combustible framed compartments with protective lining (Exova Warringtonfire, 2011).

Similar findings were reported by the National Research Council Canada through a series of four large-scale apartment fire tests (Su and Lougheed, 2014). Two apartments were of light timber frame construction, one was of cross-laminated timber (CLT) and the fourth was of cold-formed steel construction – used as the benchmark for fire performance in the National Building Code of Canada.

This lack of contribution of the encapsulated timber structure to the fuel load was also seen in the research work undertaken by the Southwest Research Institute (Janssens, 2018; Zelinka et al., 2018; Brandon et al., 2021; Su et al., 2018), where six test rooms were made and set on fire. Each room was made from CLT covered with varying layers of fire-protective plasterboard or left exposed. The structure that had three layers of 16-mm fire-protective grade gypsum plasterboard to both the ceiling and walls resulted in no fire damage to the CLT, as shown in Figure 14.5.

14.5.2 Exposed timber structures

Exposed combustible material inside a building may speed up the growth of the fire and reduce the time to flashover. The reaction to fire-class performance and consequently the surface spread of flame performance of the wall and ceiling are key characteristics for the growth period of the fire. These effects on the fire growth and time to flashover can be attributed

Figure 14.5 Undamaged CLT walls and ceiling after removal of the fire-damaged gypsum plasterboard, following a compartment fire test (Su et al., 2018).

to increased fuel load from burning of the compartment's structure in addition to the compartment's contents. This increase in fuel load can increase the production of volatiles and smoke, and the severity of flaming, both inside and outside the compartment. The increase in fuel load may increase the total heat release rate and cause a longer burning duration (see Chapter 3).

The use of combustible construction materials has the potential to introduce several issues not traditionally accepted within building codes, including combustible materials in fire-isolated egress paths, within concealed spaces and on external balcony areas.

Despite the additional fuel load, exposed timber structures can be designed to meet or exceed relevant regulations and guidance by using local building codes or test evidence and contemporary fire engineering methods taking into account the most recent research results (STA, 2022).

14.5.3 Combustible linings

Historically, combustible surfaces and wall linings have always featured in buildings. Many building codes allow combustible linings to be used within a fire compartment, regardless of the combustibility of the structural materials. They typically limit the fire hazard by prescribing surface finish limitations as described in Chapter 5, to control early fire growth and increase the time available for occupants to escape.

Fire engineers need to distinguish between the moveable fuel load contained within a compartment and that which is provided by the fixed structure. In the case of combustible linings, consideration may need to be given to whether they provide additional fuel load above what is already assumed in the design of the building and how the lining may further interact with the timber structure.

For many mass timber buildings, showcasing the timber and exposing it is desirable and one of the main architectural drivers. However, where there are other requirements such as acoustic performance requirements and specific uses of buildings that require the installation of equipment and other services, the timber may be concealed behind the finished wall, ceiling and floor surfaces. Figure 14.6 shows a raised floor being installed and the framing of the walls prior to the final linings of a CLT building during construction. In this case, the framing and floor also included timber, increasing the fuel load further and creating additional voids.

14.5.4 CLT structures

Modern mass timber elements such as CLT panels can also provide high levels of fire resistance. This is due to the inherent nature of thick timber members to char slowly allowing mass wood structures to maintain significant structural capacity when exposed to fire. Many standard fire resistance tests have been undertaken on various CLT panels to confirm the various fire resistance ratings available. Compartment fire tests are also increasingly indicating the ability for CLT panels to withstand real fire scenarios under various conditions (Zelinka et al., 2018). If glueline failure results in falling off of charred layers of CLT, this raises the possibility of additional fuel causing secondary flashover and continued burning. For this reason, CLT manufacturers internationally are moving towards the use of fire-resistant adhesives (STA, 2022).

Figure 14.6 Timber flooring on a raised timber frame and wall framing prior to lining within a tall CLT building (photo E. Claridge).

14.6 TALL TIMBER BUILDINGS

14.6.1 Sprinkler systems

Tall timber buildings will be protected by active fire protection such as an automatic sprinkler system. Most of this chapter refers to the effects of a severe fire, in the unlikely event of a fire not being controlled by the sprinkler system in its early stages. This rare event might be caused by loss of water due to an explosion, an earthquake or maintenance difficulties. As with non-combustible construction, there is a case to be made for very tall timber buildings to have extra redundancy built into active fire systems, such as a dual water supply. See Chapter 10 for more on active fire protection.

14.6.2 Firefighter access

All high-rise buildings present challenges to fire services because they generally contain a greater number of occupants, take longer for occupants to evacuate and place greater reliance on fire service resources, such as taller ladders and specialist aerial appliances. Once the building height exceeds that of the available fire service resources, all firefighting operations will necessarily need to be undertaken internally, increasing the complexity and difficulty of search and rescue, fire containment and extinguishment operations. The available options will depend on the location of the building and local fire service resources. Major metropolitan fire services will typically include a range of specialist aerial appliances providing access to various building heights, including in some cases up to eight storeys, as shown in Figure 14.7.

14.6.3 Burnout

Firefighting in a tall timber building can be more complex, hazardous and present additional complications than in a tall non-combustible structure. The main difference between combustible and non-combustible structures is design for burnout (see Chapters 2, 3 and 11 for more information on burnout).

The main concern regarding structural stability after burnout includes greater risks of fire spreading through cavities and continued smouldering after the moveable fuel has been consumed. Failure to identify these risks and fully extinguish the fire and any hot spots could lead to potential instability of the structure and eventual collapse, long after the fire has apparently gone out.

Current research projects are being undertaken to explore the mechanisms that may lead to extinguishment of charring or smouldering timber without fire service intervention (Future Timber Hub, 2022). In view of this uncertainty, it is essential for firefighters to ensure that all charring and smouldering has ceased before they leave the building after a severe fire.

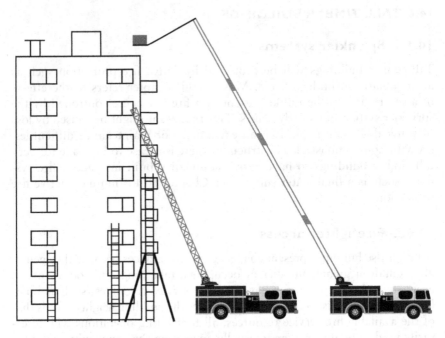

Figure 14.7 Firefighting appliances for buildings up to about eight storeys.

14.7 FIREFIGHTING CONSIDERATIONS

Firefighters will normally undertake a 360-degree survey or "size up" of a building fire to assess the risk and establish the tactical response prior to entry or external attack. Where design approaches to mass timber buildings include encapsulation, there is no reason to anticipate that the fire environment inside a timber building would be any different to a non-combustible building. In this sense, firefighters should not hesitate to enter a timber building or assume that it would perform any worse than traditional construction. However, understanding the environmental cues and how timber buildings may perform differently to traditional construction in a severe fire should be recognised, so that different command decision-making and tactics can be deployed, if necessary.

14.7.1 Firefighting water supplies

Firefighting water supplies have been typically based on historical events and an understanding of the fire environment that go back many decades. Many fire codes have established firefighting water supplies based on calculations to support the flow rates required for riser mains/standpipes and traditional firefighting practices. In recent years, there has been a call in the fire design community to base firefighting water flow requirements on more modern performance-based practices allowing for the specific design of buildings, including such factors as fuel loading and available ventilation rates through windows. More recently, means to calculate building-specific firefighting water requirements on the needs of responding firefighters has been the subject of research and discussion (Grimwood and Sanderson, 2015).

For timber construction, it may be useful to consider increased heat release rates and therefore larger water demands. Increasing the firefighting water supplies to timber buildings may be relevant if issues such as multiple hose streams might be necessary above and beyond conventional system design requirements, especially during the construction period.

14.7.2 External fire exposure to surrounding buildings

Increased external flaming should be anticipated from timber-lined compartments. This places a greater level of fire exposure via received radiation on to the building's cladding and adjacent properties. Anticipating such exposure with mass timber buildings should inform the firefighting response, so that more resources can be provided earlier, to manage external fire spread and fire exposure to adjacent buildings.

Additional debris and fire brands, especially under wind-driven conditions may also be experienced in timber buildings that do not feature encapsulation or non-combustible protective linings.

14.7.3 Combustible cores and vertical enclosures

Many recent tall timber buildings are hybrid structures with the main structural cores including egress stairs, lift shafts, corridors and firefighting intervention shafts, made from non-combustible materials, typically concrete.

The recently completed Brock Commons building in Canada, shown in Figure 14.8, and the taller Mjøstårnet building in Norway have concrete stairwells and lift shafts. This is required by building codes in many countries. Such requirements support the importance of these parts of the buildings for firefighting intervention and provide greater confidence to firefighters in the robustness and resilience of this type of construction. However, many countries may allow combustible timber construction for vertical shafts for services and access, even though the fire dynamics of such complex geometry may not be fully understood.

An understanding of the construction components, particularly those relevant to fire service intervention, may be important in providing confidence to firefighters (Figure 14.9).

Figure 14.8 Brock Commons building featuring concrete cores (photo Naturally Wood).

Figure 14.9 Hydrant riser system being progressively installed during construction (photo E Claridge).

14.7.4 Void spaces and cavities

Fires in cavities and void spaces often represent hazards for firefighters, because they can allow a hidden spread of fire in structures where observation is challenging, and extinguishment can be very difficult. In addition, spread of fire in voids and cavities can be affected by positive stack-effect, negative stack-effect, external wind and pressurised ventilation. This applies to all buildings, regardless of materials.

Light timber frame construction systems often have voids or concealed spaces, including attics and roof spaces, and void spaces in floors and walls, both within the fire-rated system or adjoining fire-rated systems such as suspended ceilings, multiple ceilings or cavities in facades. In addition, there may be shafts for vertical services, vents or garbage disposal.

Building regulations in many countries prohibit or limit the size of void spaces or require sprinklers to be installed in them. Often cavity barriers are installed to reduce the void length, preventing spread within the building. Figure 14.10a and b shows details from national guides.

A strategy often employed is to limit the length of the void to the size of the neighbouring fire compartment. For example, an apartment is a single fire compartment; therefore, cavity barriers should be installed within cavities of the compartment at the corners of the compartment or apartment. Figure 14.11a and b shows common locations for cavity barriers from national guidance. More information on cavity barriers is available in Chapter 9.

There have been many incidents where cavity fires have caused significant challenges to the fire service in timber buildings and in some cases ongoing combustion resulted in significant damage and collapse of a timber structure following extinguishment of the main fire. In 1995, the British Government undertook a large-scale project to investigate the performance of larger timber frame buildings. The Timber Frame 2000 project included full-scale fire tests in a six-storey light timber frame test building to assist development of design guidance for medium-rise timber frame structures (Enjily, 1996). The final BRE project report (BRE, 2003) addressed the project objectives regarding compartmentation and stairwells. The report did not address a fire that occurred several hours after a room fire test was completed, initiated by embers that had entered an external wall which was not inspected during the fire test procedure. A follow-up report by Chiltern Fire into the causes of the cavity fire made it clear that a high standard of workmanship is required in timber frame buildings, including adequate cavity stops to prevent the unseen spread of fire in cavities. They noted that firefighters are not always familiar with detecting fires of this type which start inside cavities (Lavender et al., 2002).

14.7.5 Identifying voids and fires within voids

Extinguishing fires in voids is difficult, as often the source is not easily recognisable. To increase the ability to identify a fire inside a void, infrared or thermal imaging cameras can be employed (see Figure 14.12). For tall buildings, identifying fires within external cavities or in voids behind wall cladding may be especially difficult without external access. The infrared or thermal imaging camera can also be used to assist in identifying the makeup of the structure, as frame elements often have higher thermal mass and can show up in infrared or thermal images, as colder colours. However,

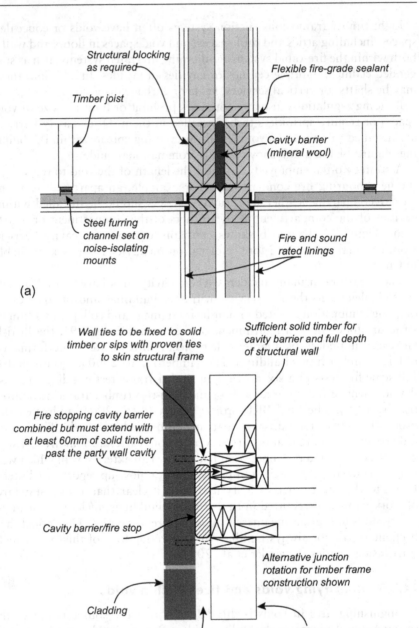

Figure 14.10 (a) Cavity barrier in a floor wall junction of a light timber-framed fire-rated structure (Forest and Wood Products Australia, 2016). (b) Cavity barrier in structural timber buildings – plan view at junction of compartment wall to external wall cladding (STA, 2020).

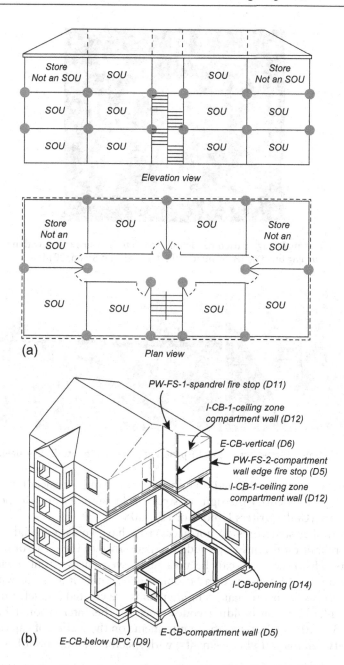

Figure 14.11 (a) Locations where cavity barriers should be installed (SOU = sole occupancy unit) (Forest and Wood Products Australia, 2016). (b) Cavity barrier in structural timber building locations (STA, 2020).

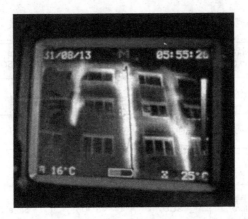

Figure 14.12 Thermal imaging used to identify the extent of fire travel within voids in a burning building (Björkman, 2013; Östman and Stehn, 2014).

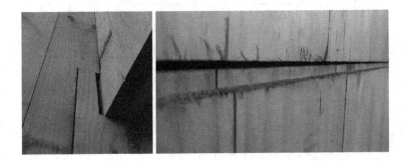

Figure 14.13 Gaps created between CLT panels at junctions (photo E. Claridge).

care must be given as thermal imaging cameras can be misleading if being used to try and identify the signs of structural instability and the potential for collapse (Underwriters Laboratories, 2008).

Mass timber structures do not necessarily have voids, especially if they are not reliant on linings to provide fire resistance to the structural timber elements. However, gaps can be present between heavy timber elements and the quality of construction becomes increasingly important with timber structures to ensure that no hidden voids are created and left untreated (Figure 14.13). Some building codes, such as the International Building Code (IBC, 2018), place specific requirements on the sealing of gaps at junctions between mass timber elements, with specific sealants or adhesives.

14.7.6 Extinguishing fires within a void

Fires in void spaces are mainly ventilation controlled, which can present challenges to fire suppression due to their unpredictable fire behaviour.

The most important aspect of extinguishing work is to avoid opening the void and introducing oxygen to the fire before the fire is under control. Extinguishing media are best applied through small openings, such as piercing nozzles and cutting extinguishers.

Brick clad buildings with cavities and combustible materials within the cavity such as wood-sheathed timber framing may allow the fire to spread rapidly from one floor to another if the cavity barriers are not correctly installed or missing. Cavity barriers slow the spread down, but do not necessarily stop fires, so a delay in attending to a fire in a cavity may lead to increased fire spread in a combustible cavity.

Tests have shown (Sæter Bøe and Hox, 2017) that, among water-based extinguishing media, penetration nozzles or cutting-type extinguishers are the most efficient for fires in a cavity as well as utilising the least amount of water. Other possible extinguishing agents are nitrogen or carbon dioxide, but techniques and tactics when using these extinguishing media are relatively new, and further research is required. The main drawback of these media is the limited cooling capabilities of the surfaces and the volume of gas required.

Training of fire services is recommended in methods on how to extinguish fires in voids and cavities. The first action when there may be a hidden fire in a cavity is to identify the structure and materials of the building and the location of cavity barriers.

Where a fire in a void has been suppressed, it is necessary to remove any linings to ensure that the fire has been fully extinguished. Where cavity barriers have not been installed, this may involve removal of linings in neighbouring fire compartments and may significantly increase the damage to the building during overhaul operations. To reduce the likelihood of smouldering fires, reigniting it may be necessary to place a fire watch within the building for several hours after a cavity fire event.

More information on prevention of fire spread in voids and cavities is given in Chapter 9.

14.7.7 Extinguishing fires in wood-based materials

As the construction and use of new materials develop, compartment firefighting techniques and procedures are continually being developed. This section is a brief review of firefighting in buildings with timber building materials.

14.7.8 Extinguishing agents

Water is still the most common extinguishing agent used in fire suppression. Water has a much higher heat capacity than other extinguishing agents and is generally readily accessible. Some fire services use a Compressed Air Foam System (CAFS) for better wettability, cooling effect and minimal water damages.

The most significant property of water is its heat capacity. The heat required to raise the temperature of 1 kg of water at 1°C is approximately 4,200 Joules per kilogram per degree Celsius (J/kg°C). This property is important for extinguishing and especially for cooling hot gases during firefighting. An additional important factor of water is the expansion ratio. The expansion ratio of water at 100°C is about 1:1,700, which means that 1 litre of liquid water expands to make 1,700 litres of steam (Smolka et al., 2018).

During firefighting, the cooling effect of water in the ceiling layer of hot gases depends on the efficiency of its evaporation. Smaller drops of water can be better evaporated and better absorbed by the heat in the hot gas layer. During firefighting operations, not all water is evaporated, and only part of this water reaches the lining of the compartment.

Equipment which allows the application of water under higher pressure and with small droplets is a more efficient extinguishing method than standard nozzles. One item of such special equipment is a piercing nozzle which produces a pressurised water mist (Smolka et al., 2018).

Fires in combustible structures often require a higher volume of water than that needed to extinguish fires in non-combustible structures. This is due to the combustible structure providing additional fuel beyond the building's contents and linings. Firefighters need to consider methods of delivering this additional water.

14.7.9 Non-direct attack

A safe way of extinguishing enclosure fires is with a non-direct attack. This can be done by special equipment such as penetrating or piercing nozzles and the application of pressurised water mist. The most common items of equipment for this application are piercing nozzles and cutting extinguishers (Cobra Cold Cut, 2022), see Figure 14.14.

Figure 14.14 Cobra cold cut system (photo Cobra Cold Cut).

Application of piercing nozzles allows firefighting to operate without introducing additional ventilation that may cause backdraft and other dangerous fire phenomena. Piercing nozzles allow firefighters to deliver the extinguishing agent into the compartment or void without entering or opening the cavity. The water piercing nozzle is applied through a small hole in the wall or the structure, and water mist is delivered under high pressure. The application of the piercing nozzle should be combined with the application of a thermal vision camera to find the origin of the fire.

14.7.10 Comparison of extinguishing equipment

Sæter Bøe and Hox (2017) compared specific firefighting equipment used for extinguishing fires in cavities and attics. In these experiments, three extinguishing methods were compared: standard nozzle, piercing nozzle and cutting extinguisher (Cobra Cold Cut System) techniques.

Experiments were undertaken in a full-scale model compartment with cavities and an attic, as shown in Figure 14.15. Two different fire scenarios were compared: scenario A is a fire behind the outer wood cladding (area 1), and scenario B is a fire inside a cavity behind a timber wall connected to a cavity above a false ceiling (area 2).

Table 14.2 compares the water consumption and the required time for extinguishing the fires. Scenario A shows rapid extinguishment with low water consumption. Scenario B presents a more complicated scenario for finding the origin of the fire and extinguishing it. A large difference in water consumption can be seen, with the standard nozzle requiring the most water,

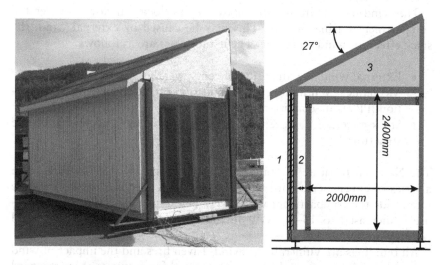

Figure 14.15 Experimental compartments used to compare extinguishment methods in a low-rise construction (Sæter Bøe and Hox, 2017).

Table 14.2 Comparison of water consumption and time to extinguishment of different nozzles (Sæter Bøe and Hox, 2017)

	Scenario A		Scenario B	
	Extinguishing time (mm:ss)	Water used (L)	Extinguishing time (mm:ss)	Water used (L)
Cutting extinguisher	02:32	≈135	03:06	≈150
Chain saw and normal nozzle	02:32	≈220	18:14	>1,200
			04:49	≈450
Piercing nozzle	–	–	02:18	≈400

with the piercing nozzle and the cutting extinguisher (Cobra Cold Cut System) having much lower water consumption and faster extinguishment.

14.8 WIND-DRIVEN FIRES

The impact of wind on fire development within compartments has been the subject of research for some time. Research has studied the effects of wind on the fire heat release rate within tunnels, on facade fires and external fire spread to neighbouring buildings. More recently, specific research has been undertaken as to the implications of wind on firefighting tactics (Kerber and Madrzykowski, 2009). There is limited research on the effect of wind on fires in buildings with internally exposed timber linings, although wind is known to have an influence (Sjöström et al., 2021) and is a concern to firefighters regarding fires in tall buildings (Grimwood, 2019).

The wind-driven fire usually causes high-velocity airflow to enter the compartment during the fire. This may be caused by external wind, the stack effect, or window failure. The main causes are as follows:

- External wind, e.g. direction of vents, weather conditions or area
- Mechanical ventilation, e.g. application of Positive Pressure Ventilation (PPV), air-conditioning systems
- Stack-effect, e.g. elevator shafts, tall buildings, geometry of the structure

The National Institute of Standards and Technology (NIST) observed in a series of experiments with wind velocity 9–11 m/s that the heat release rate inside the compartment was 1 MW before window failure, after failure it increased to 15–20 MW in the post-flashover stage (Kerber and Madrzykowski, 2009).

All buildings are vulnerable to wind-driven fires and the impact of wind on combustible structures has been recognised for centuries as experienced in many great fires such as the Great Chicago Fire in 1871 and the Great

Fire of London in 1666. Wildland–urban interface events that include fires spreading between vegetation and structures are also recognised events heavily dictated by wind conditions. The impact of wind-driven fire conditions on firefighter safety and the more severe fires that can occur with combustibles structures is of most interest for tall timber buildings. Figure 14.16 shows the relationship between wind speed and building height and location.

Figure 14.17 describes some parameters affecting wind-driven fires, based on recent research by Smolka and Kempna (2021), complementary to fire behaviour indicators used during compartment fire behaviour training of firefighters.

14.8.1 Influence of wind on fire intensity

Limited testing has quantified the impacts of external wind velocity on the severity of fires inside enclosures. Typically, more research has focused on

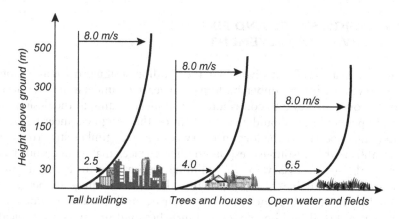

Figure 14.16 Wind speed changes with height and building location (RLS Energy).

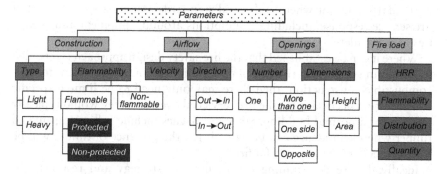

Figure 14.17 Fire compartment parameters affecting compartment wind-driven fire vulnerability (Smolka and Kempna, 2021).

the impact of wind on external fire plumes and on small compartments with limited openings and without internal flow paths to other parts of buildings. Notwithstanding these limitations, studies have shown that fire growth rates can increase by a factor of 10 and that for timber fuels, the heat release rates can increase by up to 70% (Brandon and Anderson, 2018) when compared to a structure that is completely non-combustible. However, care is required as most modern structures are often lined with combustible linings, reducing this difference.

Research on single small-compartment fires does not consider the impact on adjacent compartments. The effects of wind on facade fires have been presented by Sjöström et al. (2021). Of most interest for tall timber buildings and future research will be the impact of wind-driven fire and forced ventilation conditions on the cessation of flaming and smouldering combustion, especially if this is a strategy that is to be relied upon for design purposes.

14.9 DESIGN STAGE AND FIRE SERVICE INVOLVEMENT

Fire services are typically involved in the building design process and should be involved with tall timber buildings, at least to understand the design process, outcomes and to contribute their understanding to the design process. In particular, it should be incumbent on the design engineer to understand any concerns of the local fire services towards timber structures and to establish what resources responding fire services may or may not have. It may also be necessary for the building design team to ascertain what resources are available locally, especially where tall buildings are being considered in an area that does not have extensive tall building resource and experience. Local building codes and jurisdictional practices are typically based on a historic expectation of building construction and may not take into account any local resourcing challenges that may be present. This is particularly relevant where codes and standards assume a specific level of fire service response, including assumptions around equipment and capacities of responders.

Where it can be established that the fire safety design approach is intended to design in such a way that the timber building can be equivalent to non-combustible construction, then there may only need to be limited specific consideration necessary for the fire service above what they would typically provide for a normal building. Such approaches include fully encapsulated timber where confidence can be gained that the presence of timber will be fully protected in the event of a fire.

Ideally, before determining the firefighting strategy and response for a building on fire, firefighters would need to understand the fire design strategy for the building and what level of structural fire resistance can be expected.

The UK Structural Timber Association have issued guidance on fire safety strategy for timber buildings which supports the early engagement of the fire service for complex and tall buildings during the design process (STA, 2021).

14.9.1 Liaison with fire authorities

Depending on the local jurisdiction, some codes require the involvement of the fire service as part of the building design process. This may involve liaison to consider fire service requirements from an operational response perspective, but also to involve them directly with design decisions that may be influenced by fire service experience and resources.

During the design phase, the designer should contact the fire service to identify any specific requirements they may need to consider. Identifying that the building is to be of timber construction is important to ascertain if the fire service has any specific concerns or requirements for the finished building or during the construction phase. Information about how firefighters typically interact with building features and fire protection systems during fires can be found within various sources, including guidance provided by OSHA (2015) and NFPA 1700.

14.10 PRE-INCIDENT PLANNING

14.10.1 Tall timber buildings

Adequate pre-incident planning is essential for all buildings that present specific and significant hazards. It is particularly important for fire services that may not have experience of dealing with tall timber buildings. Pre-incident planning should include a full understanding of the concepts that the building designer used to provide an acceptable level of safety and in particular assure its structural performance in fire.

The NFPA 1620 Standard on Pre-Incident Planning (2020) calls for detailed pre-incident plans for all structures and fire service personnel should familiarise themselves with all the buildings in their first-alarm district, especially high-rise buildings, complex structures and any target hazards that they identify.

Fire service personnel must have an understanding of the fire protection features and systems within buildings. In most cases, they may be the same features that would be present for any building. However, with limited experience in dealing with tall timber buildings and with varying levels of knowledge or understanding of timber performance in fire, pre-incident planning is essential for building confidence and ensuring that incorrect assumptions regarding timber performance are not made.

14.10.2 Fire service involvement during construction and demolition

The construction and demolition stages of any building increase the fire risk to the building and likely consequences should a fire occur. The risks are different from a completed building as the fire protection features expected to be available in the completed building may not yet be installed or activated. Challenges are presented to responding fire services, particularly access around the construction site as well as into the building with changes to site conditions occurring frequently. Many building codes require specific precautions to be adopted during the construction stages of a building, typically ensuring fire service access to the site and water supplies be maintained. The International Fire Code, Chapter 33, includes requirements during the construction and demolition stages and includes specific requirements for combustible construction and incorporated changes specific to mass timber buildings (ICC, 2020). IFC 3303.5 includes requirements for standpipes, water supplies and the need to protect exposed combustible surfaces as the building height increases.

Specific risk mitigation strategies for timber structures are often required to prevent fire spread to adjacent buildings and persons in and around the construction site. Chapter 13 provides greater details on construction fires.

At the commencement of the construction of the building, the principal contractor, or the Fire Safety Coordinator as the representative for fire safety, should make contact with local emergency services and invite them to undertake adjustments to the fire safety plan. Provisions for access to the site and water supplies should be agreed at this time.

Regular liaison with the fire service is necessary, mainly where there are changes to the site conditions or safety plan details. An initial site plan should be prepared, and a process for updated details to be available in a fire emergency should be agreed. Furthermore, updated site plans should be available at the firefighter access points to the site, as described in Chapter 13.

The fire services should also be made aware of any specific features or building design/performance solutions that could affect firefighting operations.

14.10.3 Fire system impairment

It will be inevitable during the lifetime of a building that features provided to support its fire design will require maintenance and eventual replacement. Active fire safety systems typically require ongoing regular maintenance and inspection. During these activities, the systems may not be operational, which is a major threat to fire safety. Many systems standards require that the local fire service be notified of impairments to active fire systems, especially where facilities such as standpipes or hydrant systems may not be available. Such impairments are to be anticipated, so that fire services can increase their response and the number of firefighters that may respond to the building in the event of an alarm.

For tall timber buildings with exposed wood surfaces, the impairment of a fire safety system may leave the building more vulnerable than a building of non-combustible construction. Understanding that the building is of timber construction and its design basis may be relevant to the fire service response. Where a timber building is undergoing refit or alterations such that multiple systems or building features may not be available, specific consideration may be necessary for temporary compensating features to support an effective fire service, see Chapter 13.

14.10.4 Fire service site training and familiarisation

As buildings become taller and more complex, so does the fire service response to such buildings. Specific and complex features provided that may require fire service interaction such as smoke control systems and evacuation management systems require training and familiarisation if there is any expectation that the responding firefighters be able to use them as expected and as required.

Emergency exercises form a crucial part of the commissioning and ongoing operation of complex buildings and should include local fire services so that they can become familiar with how the building operates during an event, such as an evacuation exercise, and familiarise themselves with all the features of the building.

14.11 POST-EARTHQUAKE FIRES AND FIRE SERVICE RESPONSE

Major fires following earthquakes have occurred over the centuries causing widespread loss, exacerbating the impact of the initial earthquake. Major fires such as those experienced following the San Francisco, USA, earthquake in 1906, and more recently Kobe, Japan, in 1995 indicate the challenges to the built environment where building systems are damaged and fire services are unable to respond. For this reason, historical urban conflagrations have often resulted in restrictions on combustible building materials. For modern timber buildings, especially those in active seismic areas, greater resiliency in fire protection concepts may be warranted compared with the requirements for traditional non-combustible construction.

14.12 FUTURE NEEDS

There is a significant lack of knowledge and practical experience with firefighting in tall timber buildings. The fire environment associated with exposed timber surfaces is different to historic construction and may present new hazards and risks for occupants as well as for firefighters. There is a need for collection of data, knowledge and case studies from firefighting events, in order to develop a better understanding and confirm strategies and approaches for firefighting in tall timber buildings.

As combustible structures become larger, taller and more complex, the robustness and resilience of the buildings and their fire safety features also need to increase. This will require appropriate changes to national and international building codes.

There is a critical need for more understanding of the smouldering combustion of large timber elements after fire exposure. As structural concepts progress and move away from the traditional assumption of burnout, a greater reliance will inevitably be placed on firefighters to extinguish fires and ensure that continued smouldering of the timber structure does not occur.

Lastly, it is essential that all emergency responders be knowledgeable and have an understanding of how combustible structures and tall timber buildings perform in the event of severe fires. Without sufficient fire-ground experience of mass timber buildings, the emergency response must be informed by education and research that considers the needs of the responders. As buildings evolve, so must the firefighter response, with new strategies to ensure the most favourable outcomes for all stakeholders.

REFERENCES

Björkman, C. (2013) *Brand i flerbostadshus i Luleå Klintvägen. Fördjupad olycksundersökning* (in Swedish). Umeå kommun Brandförsvar & Säkerhet, Luleå, Sweden.

Björkman, J. (2012) *Timber Joints with Punched Metal Plate Fasteners Subjected to Fire*. Master Thesis. Luleå University of Technology. Department of Civil. Environmental and Natural Resources Engineering.

Brandon, D., Anderson, J. (2018) *Wind Effect on Compartment Fires*. RISE Report 2018:72. Research Institutes of Sweden.

Brandon, D., Sjöström, J., Temple, A., Hallberg, E., & Kahl, F. (2021) *Fire Safe Implementation of Visible Mass Timber in Tall Buildings: Compartment Fire Testing*. RISE Report 2021:40. Research Institutes of Sweden.

BRE (2003) *Multi Storey Timber Frame Buildings: A Design Guide*. BRE Report 454. Building Research Establishment, Watford, UK.

Cobra Cold Cut (2022) https://www.coldcutsystems.com/

Engel, T., Brunkhorst, S., Steeger, F., Butscher, D., Kurzer, C., Werther, N., Winter, S., Zehfuß, J., Kampmeier, B., Neske, M. (2021) *Schlussbericht zum Verbundvorhaben TIMpuls „Brandschutztechnische Grundlagenuntersuchung zur Fortschreibung bauaufsichtlicher Regelungen im Hinblick auf eine*

erweiterte Anwendung des Holzbaus". Fachagentur Nachwachsende Rohst-
offe, Gülzow-Prüzen.

Enjily, V. (1996) *Fire Test on a 5 Storey Timber Construction. Centre for Timber Technology and Construction*. Building Research Establishment, Watford, UK.

Exova Warringtonfire (2011) *Extension of the Concession which Allows Timber Framed Construction in Class 2 Buildings to Include Class 3 Buildings, 2404400-RPT12-1*. Warringtonfire, Australia.

Fire Protection Association. (2011) Building, Design and Management. Fire in timber frame buildings. A review of fire statistics from the UK and the USA. BDM14. RISCAuthority. Version 01. 2011

Forest and Wood Products Australia (2016) *Fire Safety Design of Mid-Rise Timber Buildings: Basis for the 2016 changes to the National Construction Code*, Melbourne, Australia. https://www.fwpa.com.au/forwood-newsletters/1078 -national-construction-code-ncc-mid-rise-timber-buildings.html

Future Timber Hub (2022) *Exploring the Self-extinguishment Mechanism of Engineered Timber in Full-scale Compartment Fires*. Australian Research Council. https://futuretimberhub.org/projects/exploring-self-extinguishment -mechanism-engineered-timber-full-scale-compartment-fires.

Grimwood, P (2017) *Euro Firefighter 2. Firefighting Tactics and Fire Engineer's Handbook*. D&M Heritage Press. Huddersfield, UK

Grimwood, P (2019) WARNING! Danger of Fire & Collapse. https://www.linke-din.com/pulse/warning-danger-fire-collapse-paul-grimwood/

Grimwood, P & Sanderson, I. (2015) A performance based approach to defining and calculating adequate firefighting water using s.8.5 of the design guide BS PD 7974:5:2014 (fire service intervention). *Fire Safety Journal 78*. 155–167. 10.1016/j.firesaf.2015.08.007.

Harman, K. & Lawson, J. (2007) *A Study of Metal Truss Plate Connectors When Exposed to Fire, NIST Interagency/Internal Report (NISTIR), NISTIR 7393*. National Institute of Standards and Technology, Gaithersburg, MD.

Havel, G. (2018) Construction Concerns: IBC 2021 Heavy Timber Proposal. *Fire Engineering*, December 2018.

IBC (2018) *International Building Code*. International Code Council, Washington, DC.

ICC (2020). 2021 International Fire Code. *Section 3303 Owner's responsibility for fire protection*. International Code Council, Inc. Illinois. USA.

Janssens, M. (2018) *Development of a Fire Performance Test Method for Evaluating CLT Adhesives*. Southwest Research Institute, San Antonio, USA.

Kerber, S. and Madrzykowski, D. (2009) *Firefighting Tactics Under Wind Driven Fire Conditions: 7-Story Building Experiments. NIST Technical Note 1629*. U.S. Department of Commerce, Gaithersburg.

Lavender J, Bullock M and Lennon T. (2002) *Understanding fire risks in combustible cavities*. Report ref. Chilt/E01006TR, PII ref. no. O-CIF004. High Wycombe. UK.

NFPA 1620 (2019) *Standard for Pre-Incident Planning*. 2020 Edition. National Fire Protection Association, Quincy, Massachusetts. USA.

NFPA 1700 (2020) *Guide for Structural Firefighting*. 2021 Edition. National Fire Protection Association, Quincy, Massachusetts. USA.

NIOSH (2005) *Preventing Injuries and Deaths of Firefighters Due to Truss System Failures*. National Institute for Occupational Safety and Health. DHHS (NIOSH) Publication Number 2005–132. 2005. Cincinnati, Ohio. USA.

O'Brocki, R. (2019) Tall Mass Timber Buildings and Fire Service Concerns. *Fire Engineering*. www.fireengineering.com/articles/2019/05/tall-mass-timber -buildings-and-fire-service-concerns.html.

OSHA (2015) Fire Service Features of Buildings and Fire Protection Systems. *Occupational Safety and Health Administration OSHA 3256–09R 2015*. U.S. Department of Labor, Washington, D.C. USA.

Östman, B., Stehn, L. (2014) *Fire in Residential Multi-storey Timber Buildings: Analyses, Recommendations and R&D Needs* (in Swedish). SP Report 2014:07. SP Technical Research Institute of Sweden.

RLS Energy [Online] *Wind Speed Data* [cit.2022-01-29] http://rlsenergy.com/prod-ucts/wind-power/wind-speed-data/.

Sæter Bøe, A., Hox. K. (2017) *Full Scale Fire Tests with Comparison of Different Extinguishing Techniques to Fires Developing Inside Cavities*. Report RISE Fire Research, Norway.

Sjöström, J., Brandon, D., Temple, A., Hallberg, E., and Kahl, F. (2021) *Exposure from Mass Timber Compartment Fires to Facades*. RISE Report 2021:39. Research Institutes of Sweden.

Smolka, J., Kempna, K. et al. (2018) *Guidance on Firefighting and Bio-Based Materials, COST Action FP1404*. https://ethz.ch/content/dam/ethz/special -interest/baug/ibk/costfp1404-dam/dissemination-documents/N225-07-Fire -fighting%20Guideline.pdf

Smolka, J., Kempna, K. et al. (2021) *Wind Driven (Forced Draught) Impact on Enclosure Fires*. Application of Structural Fire Engineering, Ljubljana, Slovenia.

STA (2020) *Structural Timber Buildings Fire Safety In Use Guidance, Volume 2 – Cavity Barriers and Fire Stopping*. Structural Timber Association, Clackmannanshire, UK.

STA (2021) *Advice Note 7.3: Robustness against Fire: Fire Safety Strategy for Structural Timber Buildings*. Structural Timber Association, Clackmannanshire, UK.

STA (2022) Web site: https://www.structuraltimber.co.uk/sectors/clt-special-inter-est-group/. Structural Timber Association, Clackmannanshire, UK.

Su, J. Z. & Lougheed, G. D. (2014) *Report to Research Consortium for Wood and Wood-Hybrid Mid-Rise Buildings: Fire Safety Summary – Fire Research Conducted for the Project on Mid-Rise Wood Construction (Report No. A1-004377.1)*. National Research Council Canada, Ontario, Canada.

Su, J., Lafrance, P. S., Hoehler, M., & Bundy, M. (2018) *Fire Safety Challenges of Tall Wood Buildings – Phase 2: Tasks 2 & 3 – Cross-Laminated Timber Compartment Fire Tests*. NFPA Fire Protection Research Foundation, Quincy, Massachusetts. USA.

TIMpulse (2022) *Research Project TIMpuls. Faculty of Civil, Geo and Environmental Engineering*, Technical University of Munich. https://www .bgu.tum.de/timpuls/startseite/

Underwriters Laboratories Inc. (2008) *Report on Structural Stability of Engineered Lumber in Fire Conditions. NC9140*. Underwriters Laboratories Inc. Northbrook, IL.

Underwriters Laboratories Inc. (2021) *Structural Stability of Engineered Lumber in Fire Conditions. Fire Safety Academy Training Package.* UL Fire Safety Research Institute. Northbrook, IL.

Zelinka, S. L., Hasburgh, L. E., Bourne, K. J., Tucholski, D. R., & Ouellette, J. P. (2018) *Compartment Fire Testing of a Two-Story Mass Timber Building. General Technical Report FPL-GTR-247.* U.S. Department of Agriculture, Forest Service, Forest Products Laboratory, Madison, WI.

Index

Printed in the United States
by Baker & Taylor Publisher Services